The IMA Volumes in Mathematics and its Applications

Volume 165

Tadele Mengesha • Abner J. Salgado
Editors

A^3N^2M: Approximation, Applications, and Analysis of Nonlocal, Nonlinear Models

Proceedings of the 50th John H. Barrett Memorial Lectures

Springer

Editors
Tadele Mengesha
Department of Mathematics
University of Tennessee
Knoxville, TN, USA

Abner J. Salgado
Department of Mathematics
University of Tennessee
Knoxville, TN, USA

ISSN 0940-6573 ISSN 2198-3224 (electronic)
The IMA Volumes in Mathematics and its Applications
ISBN 978-3-031-34088-8 ISBN 978-3-031-34089-5 (eBook)
https://doi.org/10.1007/978-3-031-34089-5

Mathematics Subject Classification: 26A33, 65N12, 65N30, 65N50, 65N55, 65F10, 65M15, 65M60, 35J70, 65R10, 65J10

This Springer imprint is published by the registered company Springer Nature Switzerland AG
The registered company address is: Gewerbestrasse 11, 6330 Cham, Switzerland

Preface

The John Barrett Memorial Lectures is one of the very few long running lecture series in mathematics in the southeastern United States. This lecture series was inaugurated in 1970 as a tribute to Dr. John H. Barrett, an influential Mathematics Department Head at the University of Tennessee Knoxville (UTK). These lectures have been held annually, with very few exceptions, since 1970. Initially, the series centered on Ordinary Differential Equations, one of the research areas of John Barrett. Since the mid-1980s, however, the topics of the lectures have rotated yearly and have been chosen to reflect research interests within the Department of Mathematics at UTK. The lecture themes have traversed the mathematical landscape, focusing on mathematics education, computational and applied mathematics, discrete mathematics, stochastics, general relativity, nonlinear partial differential equations, topological data analysis, and topological quantum field theory. A testament to the breadth and depth of topics of previous lectures is the many proceedings that have been produced over the years [2–4, 9–11].

The year 2020 was set to mark the 50th anniversary of the Barrett Lectures. To celebrate this occasion, the topic of *Approximation, Applications, and Analysis of Nonlocal, Nonlinear Models* was chosen as the theme of the lectures. Incidentally, the mathematical study of fractional derivatives and fractional differential equations, a class of nonlocal equations, was the subject of John Barrett's doctoral dissertation [6] and early research [5]. To this day, these topics continue to be of great interest for the mathematical community with many applications in engineering and the sciences.

The grand goal of this installment of the lectures was to bring together experts from the computational, scientific, engineering, and mathematical communities who work with nonlocal models and provide a platform for their interaction and exchange of ideas. The lectures were to survey the state of the art in computational practices, mathematical analysis, and applications of nonlocal models, while exploring new application domains and promoting new collaborations between participants. The Lectures were to bring together specialists working with nonlocal models who might otherwise not interact directly, owing to their diverse fields of study and specialization. The expectation was that the mixture of computational scientists, mathematicians, and application specialists would lead to new, productive collaborations.

We had gathered a stellar set of invited speakers. We procured internal, NSF, and IMA financial support. While everything was set for the lectures to take place in May 2020, on March 11, 2020, the World Health Organization declared COVID-19 a global pandemic [1]. That forced us to postpone the Barrett Lectures with the intention of having them at a future date, hopefully during the Fall of the same year. As the Fall approached, it became clear that there were no signs of the situation significantly improving for this meeting to safely take place. For this reason, it was decided that the lectures were to take place online May 17–19, 2021. The lectures were dedicated to the memory of Lida K. Barrett, who had passed away earlier that year.

Despite the virtual format, the lectures were a success with over 150 registered participants from all over the world. The lectures consisted of a series of plenary lectures by three experts, who presented a survey of the field:

- Vassili Kolokoltsov (Warwick University, UK): *Unifying approach to various classes of generalized fractional integrals and derivatives and to the solutions of corresponding fractional PDES.*
- Luis Silvestre (University of Chicago, USA): *Conditional regularity estimates for the Boltzmann equation.*
- Ricardo H. Nochetto (University of Maryland, USA): *Approximation of linear and nonlinear fractional diffusion.*

In addition to the plenary lecturers, there were research level talks by experts (in alphabetical order):

- Mark Allen (Brigham Young University, USA): *The fractional unstable obstacle problem.*
- Mark Ainsworth (Brown University, USA): *Fractional order modeling of crystalline structures.*
- Olena Burkovska (Oak Ridge National Laboratories, USA): *Nonlocal phase-field models permitting sharp interfaces.*
- Marta D'Elia (Sandia National Laboratories, USA): *Data-driven learning of nonlocal models: From high-fidelity simulations to constitutive laws.*
- Qiang Du (Columbia University, USA): *Nonlocal conservation laws and traffic flow modeling.*
- Max Gunzburger (Florida State University, USA): *A multifidelity method for a nonlocal diffusion model.*
- Xingjie Li (University of North Carolina, at Charlotte, USA): *Study of dispersion relations for coupling nonlocal and local elasticities.*
- Robert Lipton (Louisiana State University, USA): *Nonlocal elastodynamics and fracture.*
- Robin Neumayer (Carnegie Mellon University, USA): *Quantitative stability for minimizing Yamabe metrics.*

- Petronela Radu (University of Nebraska, USA): *Nonlocal operators and solutions: Properties, decompositions, and convergence.*
- Pablo Seleson (Oak Ridge National Laboratories, USA): *Analysis of the overall equilibrium in local-to-nonlocal coupling.*
- Stewart A. Silling (Sandia National Laboratories, USA): *Origins and uses of nonlocality in mechanics.*
- Mariana Smit Vega Garcia (Western Washington University, USA): *Regularity of almost minimizers with free boundary.*
- Pablo R. Stinga (Iowa State University, USA): *Harnack inequality for fractional nondivergence form elliptic equations.*
- Xiaochuan Tian (University of California, San Diego, USA): *Trace theorems for nonlocal function spaces.*

We thank all speakers as well as other participants of the lectures who have made it a success.

This volume contains contributions from all three plenary lecturers and five research-level articles from speakers and participants on topics related to the theme of the lectures. The nature and timeliness of the subject should make this book a useful reference for researchers in all aspects of approximation, analysis, and applications of nonlinear nonlocal models at all levels. We thank all contributing authors for this volume.

The 50th Barrett Lectures, in their original inception, were partially funded by a grant from the National Science Foundation (DMS-2001695), by the Institute for Mathematics and its Applications (IMA) at the University of Minnesota, through its Participating Institution (PI) Program, and the Barrett Memorial Fund at the Department of Mathematics, UTK. The organizers are grateful for the support.

The organizers are especially indebted to all the Mathematics Department staff members that made all the organizational details run smoothly, and perfectly. In alphabetical order: Shameca Gandy-Woods, Juvy Melton, Angie Prine, Ben Walker, Angela Woofter, and Amanda Worsham are the true heroes behind the scenes. Without them the lectures would not have been as successful.

Finally, we thank the IMA and Springer for their enthusiasm and help during the preparation of this book.

Knoxville, TN, USA
2022

Tadele Mengesha
Abner J. Salgado

John H. Barrett (1922–1969)

John H. Barrett was born in Randall, Kansas on December 20, 1922. He received his M.S. and Ph.D. from the University of Texas in 1946 and 1951, respectively. As mentioned above, his Ph.D. thesis was on differential equations of non-integer order [5,6], and was directed by Hyman Joseph Ettlinger. He held faculty positions at the University of Delaware (1951–1956), Yale University (1955–1956), and the University of Utah (1956–1961) before coming to the University of Tennessee (1961–1969). In January 1969, John Barrett died from complications following a kidney transplant, in the prime of his mathematical and administrative career.

John Barrett was an expert on oscillation and disconjugacy theory of linear differential equations, a subject that bears his permanent mark. One of his widely known contributions was the extension of the classical Prüfer transformation for 2nd order scalar differential equations to systems of differential equations [7]. He found a transformation analogous to the well-known Prüfer polar-coordinate transformation for a real self-adjoint linear homogeneous differential equation of the second order.

John Barrett came to the University of Tennessee in the fall of 1961 as a Professor of Mathematics. Upon his arrival, he immediately initiated a seminar in ordinary differential equations which attracted a number of graduate students. Through this seminar, his classes, and his personal contact, he was a major force in the Graduate Program in the Department of Mathematics in the early 1960s.

Professor Barrett became Head of the Mathematics Department in 1964. Through his efforts and influence, several active research groups were established at the University of Tennessee and by 1967, the number of faculty in the department had more than doubled. The Barrett Lecture series is a fitting reminder of his many contributions to the University and research community.

Lida K. Barrett (1927–2021)

Lida K. Barrett had a varied and remarkable career. She was the first female head of the Department of Mathematics at the University of Tennessee, Knoxville and the second female president of the Mathematical Association of America. She began her study of higher mathematics at the University of Texas, Austin under the tutelage of R. L. Moore. She met and married John Barrett, another mathematics graduate student at Texas. John graduated before Lida and accepted a position at the University of Delaware. Lida commuted from Newark to Philadelphia where she was continuing her graduate work in topology. She completed her Ph.D. at the University of Pennsylvania [8]. Her dissertation supervisors were John Kline and R. D. Anderson.

Lida Baker Kittrell was born on May 21, 1927 in Houston, Texas. Lida received her bachelor's degree from Rice University in 1946 and immediately took a job as a mathematician at the Schlumberger Well Surveying Corporation and then taught at the Texas College for Women during the 1947-48 academic year. After receiving her Ph.D. Lida taught one year at the University of Connecticut while John was a postdoc at Yale. Then the Barretts moved to Salt Lake City where both held faculty positions at the University of Utah. They moved from Utah to Tennessee where John became Head of the Mathematics Department at the University of Tennessee. Because of nepotism rules, Lida could teach only part time and began consulting with mathematics book publishers and in the ceramics group at the Oak Ridge National Laboratory.

After John's death related to a kidney transplant, Lida became a full-time professor at Tennessee and became department head in 1973. She moved on to become the Associate Provost at the University of Northern Illinois, after which she became Dean of the College of Arts and Sciences at Mississippi State University. After retiring from Mississippi State she served as Senior Advisor to the Assistant Director for Education and Human Resources at the National Science Foundation. She ended her academic career after a three year appointment as Professor of Mathematics at the U.S. Military Academy at West Point. Lida received the Yueh-Gin Gung and Dr. Charles Y. Hu Award for Service to Mathematics from the Mathematical Association of America in 2008.

References

1. WHO Director-General's opening remarks at the media briefing on COVID-19—11 March 2020. https://www.who.int/director-general/speeches/detail/who-director-general-s-opening-remarks-at-the-media-briefing-on-covid-19---11-march-2020. Accessed: December 7, 2022.
2. *Special issue on Barrett Lectures in control theory*. World Scientific Publishing Co. Pte. Ltd., Singapore, 1998. Papers from the 1997 Barrett Lectures held at the University of Tennessee, Knoxville, TN, March 20–22, 1997, Math. Models Methods Appl. Sci. **8** (1998), no. 7.

3. David F. Anderson and David E. Dobbs, editors. *Zero-dimensional commutative rings*, volume 171 of *Lecture Notes in Pure and Applied Mathematics*. Marcel Dekker, Inc., New York, 1995.
4. Garth Baker and Alexandre Freire, editors. *Nonlinear partial differential equations in geometry and physics*, volume 29 of *Progress in Nonlinear Differential Equations and their Applications*. Birkhäuser Verlag, Basel, 1997. The 1995 Barrett Lectures.
5. J. H. Barrett. Differential equations of non-integer order. *Canad. J. Math.*, 6:529–541, 1954.
6. John H. Barrett. *ORDINARY DIFFERENTIAL EQUATIONS OF NON-INTEGRAL ORDER*. ProQuest LLC, Ann Arbor, MI, 1951. Thesis (Ph.D.)–The University of Texas at Austin.
7. John H. Barrett. A Prüfer transformation for matrix differential equations. *Proc. Amer. Math. Soc.*, 8:510–518, 1957.
8. Lida Kittrell Barrett. *REGULAR CURVES AND REGULAR POINTS OF FINITE ORDER*. ProQuest LLC, Ann Arbor, MI, 1954. Thesis (Ph.D.)–University of Pennsylvania.
9. Theodora Bourni and Mat Langford, editors. *Mean Curvature Flow*. De Gruyter, Berlin, Boston, 2021. Proceedings of the John H. Barrett Memorial Lectures held at the University of Tennessee, Knoxville, May 29–June 1, 2018.
10. Xiaobing Feng, Ohannes Karakashian, and Yulong Xing, editors. *Recent developments in discontinuous Galerkin finite element methods for partial differential equations*, volume 157 of *The IMA Volumes in Mathematics and its Applications*. Springer, Cham, 2014. 2012 John H. Barrett Memorial Lectures, Selected papers from the workshop held at the University of Tennessee, Knoxville, TN, May 9–11, 2012.
11. Xiaobing Feng and Tim P. Schulze, editors. *Recent advances in numerical methods for partial differential equations and applications*, volume 306 of *Contemporary Mathematics*. American Mathematical Society, Providence, RI, 2002. Selected papers from the John H. Barrett Memorial Lectures held at the University of Tennessee, Knoxville, TN, May 10–12, 2001.

Organization

General Chair:	Tadele Mengesha University of Tennessee
General Chair:	Abner J. Salgado University of Tennessee

Organizers

Tadele Mengesha	University of Tennessee
Abner J. Salgado	University of Tennessee

Contents

Part I
Plenary Lectures

CTRW Approximations for Fractional Equations with Variable Order

V. N. Kolokoltsov

Abstract The standard diffusion processes are known to be obtained as the limits of appropriate random walks. These prelimiting random walks can be quite different, however. The diffusion coefficient can be made responsible for the size of jumps or for the intensity of jumps. The diffusion limit does not feel the difference. The situation changes if we model jump-type approximations via CTRW with non-exponential waiting times. If we make the diffusion coefficient responsible for the size of jumps and take waiting times from the domain of attraction of an α-stable law with a constant intensity α, then the standard scaling would lead in the limit of small jumps and large intensities to the most standard fractional diffusion equation. However, if we choose the CTRW approximations with fixed jump sizes and use the diffusion coefficient to distinguish intensities at different points, then we obtain in the limit the equations with variable position-dependent fractional derivatives. In this chapter, we build rigorously these approximations and prove their convergence to the corresponding fractional equations for the cases of multidimensional diffusions and more general Feller processes.

Keywords Variable order fractional equations · Continuous time random walks (CTRWs) · Subordinated Markov processes

V. N. Kolokoltsov (✉)
Department of Statistics, University of Warwick, Coventry, UK
National Research University Higher School of Economics, Moscow, Russia
Moscow State University, Moscow, Russia
e-mail: v.kolokoltsov@warwick.ac.uk,

T. Mengesha, A. J. Salgado (eds.), *A^3N^2M: Approximation, Applications, and Analysis of Nonlocal, Nonlinear Models*, The IMA Volumes in Mathematics and its Applications 165, https://doi.org/10.1007/978-3-031-34089-5_1

1 Introduction

The standard diffusion equation

$$\frac{\partial u}{\partial t}(t,x) = \frac{1}{2}a(x)\frac{\partial^2 u}{\partial x^2}(t,x), \tag{1}$$

with a positive continuous function $a(x)$, and the corresponding diffusion process are known to be obtained as the limit of appropriate random walks. These prelimiting random walks can be quite different, however. For example, approximating random walks in continuous time can be chosen as jump-type processes with the generators

$$L_h f(x) = \frac{1}{2h^2}[f(x+h\sqrt{a(x)}) + f(x-h\sqrt{a(x)}) - 2f(x)],$$

or as jump-type processes with the generators

$$\tilde{L}_h f(x) = \frac{a(x)}{2h^2}[f(x+h) + f(x-h) - 2f(x)],$$

since both L_h and $\tilde{L}_h$ tend to $(1/2)a(x)(d^2/dx^2)$, as $h \to 0$. In the first approximation, the diffusion coefficient $a(x)$ is responsible for the size of jumps (with constant intensity), and in the second approximation, it is responsible for the intensity of jumps (with constant sizes). The diffusion limit does not feel the difference. The situation changes if we model jump-type approximations via CTRW with non-exponential waiting times. If we make the diffusion coefficient responsible for the size of jumps and take waiting times from the domain of attraction of an α-stable law with a constant intensity α, then the standard scaling would lead (in the limit of small jumps and large intensities) to the most standard fractional diffusion equation

$$D^{\alpha}_{0+*}u(t,x) = \frac{1}{2}a(x)\frac{\partial^2 u}{\partial x^2}(t,x), \tag{2}$$

where D^{α}_{0+*} is the so-called Caputo–Dzerbashyan fractional derivative.

However, if we will use the CTRW approximations with fixed jump sizes and use $a(x)$ to distinguish intensities at different points, then (as will be shown) we get in the limit the equation with a variable position-dependent fractional derivative:

$$D^{\alpha a(x)}_{0+*}u(t,x) = \frac{1}{2}\frac{\partial^2 u}{\partial x^2}(t,x). \tag{3}$$

Of course, for any decomposition $a(x) = b(x)c(x)$ with positive $b(x)$, $c(x)$, one can use $b(x)$ as a function controlling the intensity of jumps and $c(x)$ as a function controlling the spread of the jumps. In this scenario, we get in the limit the equation

$$D_{0+*}^{\alpha b(x)} u(t, x) = \frac{1}{2} c(x) \frac{\partial^2 u}{\partial x^2}(t, x). \quad (4)$$

From this point of view, all Eqs. (4) are equally legitimate fractional extensions of the diffusion equation (1).

Similarly (though technically more cumbersome), for a time dependent diffusion equation

$$\frac{\partial u}{\partial t}(t, x) = \frac{1}{2} a(t, x) \frac{\partial^2 u}{\partial x^2}(t, x), \quad (5)$$

factorising $a(t, x) = b(t, x)c(t, x)$, we get by the CTRW approximation the fractional equation with variable order, which is time- and position-dependent:

$$D_{0+*}^{\alpha b(t,x)} u(t, x) = \frac{1}{2} c(t, x) \frac{\partial^2 u}{\partial x^2}(t, x). \quad (6)$$

In this chapter, we aim to develop these CTRW approximations in a general multidimensional case. Namely, let us look at the CTRW approximations and possible corresponding fractional limits for the diffusion processes generated by the operator

$$L_{dif} f(x) = \frac{1}{2} \mathrm{tr} \left(G(x) \frac{\partial^2 f}{\partial x^2}(x) \right) = \frac{1}{2} \sum_{ij} G_{ij}(x) \frac{\partial^2 f}{\partial x_i \partial x_j}(x), \quad x \in \mathbf{R}^d, \quad (7)$$

with $G(x)$ a positive $d \times d$-matrix, and stable-like processes generated by the operator

$$L_{st}^{\beta} f(x) = \int_{S^{d-1}} |(s, \nabla f(x))|^{\beta} \mu(x, s)\, ds, \quad x \in \mathbf{R}^d, \quad (8)$$

with $\beta \in (0, 2)$, $\mu(x, s)$ an even in s positive function on $\mathbf{R}^d \times S^{d-1}$ and ds Lebesgue measure on the unit sphere S^{d-1}. In particular, if $\mu(x, s) = 1$ (that is, the spectral measure $\mu(x, s)\, ds$ is uniform), then $L_{st}^{\beta} = \sigma |\Delta|^{\beta/2}$ (with σ a constant depending on d) becomes a standard fractional Laplacian. To avoid unnecessary repetitions, we shall speak about L_{st}^{β} with $\beta \in (0, 2]$ assuming that for $\beta = 2$ the diffusion operator (7) is meant.

We shall look at the CTRW approximations arising from the equations

$$\frac{\partial u}{\partial t}(t, x) = a(t, x) L_{st}^{\beta} u(t, x), \quad (9)$$

with some positive function $a(t,x)$. We stress that the separation of a multiplier $a(t,x)$ for a diffusion or a stable generator is not intrinsic, it can be done in an arbitrary way. Once the choice is made, we shall analyse the jump-type approximations to the processes governed by $a(t,x)L^{\beta}_{st}$, where $a(t,x)$ stands for the intensity of jumps and L^{β}_{st} stands for the distribution of jumps. We shall show that the limiting processes are governed by the equations of the type

$$D^{\alpha a(t,x)}_{0+*}u(t,x) = L^{\beta}_{st}u(t,x),$$

extending (6).

Our results are extendable to equations

$$\frac{\partial u}{\partial t}(t,x) = a(t,x)Lu(t,x), \tag{10}$$

with a generator of a general Feller process. We shall mostly work with (9), where concrete assumptions on coefficients are easy to check, and then comment on regularity assumptions required for a general L.

The CTRW models summing independent terms were analysed mathematically in many papers, see e.g. [13], following their development in physics literature, see e.g. [18]. The CTRW approximation for general models of Markov processes subordinated by their monotone coordinate turning in the scaling limit to a generalised fractional equation (of variable order) was seemingly first discussed mathematically in [7]. There is now a heavy body of literature on the variable order fractional equations. We are not trying to review it here, but rather refer to extensive reviews [2] and [17] dealing with both mathematical and applied questions. Let us mention specifically paper [16], where CTRW approximations to equation (3) were discussed in detail, and paper [15], with careful derivation of well-posedness for rather general variable order fractional equations. As basic references to general rules of fractional calculus and its applications, we mention books [1, 5, 14].

The rest of the chapter is organised as follows. In Sect. 2, we introduce carefully various CTRW approximations for diffusion or stable-like processes, when their coefficients can be used to distinguish either the sizes of jumps or their intensities. In Sect. 3, our main results are formulated. They are devoted to the convergence of the approximations to processes solving fractional equations of variable order and to the explicit representations of the solutions of these equations. The main result is actually Theorem 3.2 on the CTRW approximations and specific representation for their limit. The proof is based on the ideas from [7], where they were presented however rather sketchy. Theorem 3.3 is essentially known. Its detailed proof is given in [15]. We present here a different proof adding two details. Namely, our equation is a bit more general (variable fractional order depends on both space and time), and moreover we prove well-posedness of classical solutions (unlike only the mild solutions of [15]). In Sect. 4, we derive a handy general representation for the distributions of Markov processes subordinated by its monotone coordinate. It supplies an important ingredient to the proof of our main results but may be of independent interest. Section 5 presents a well-known result on the convergence of the standard CTRWs, but enhanced by the exact rates of (weak) convergence. Sects. 6–9 contain proofs of our main results.

Notice finally that similar approach to interacting particle systems leading to the new variable order fractional kinetic equations was developed in [11] and [12]. Quite different variable order kinetic equations are analysed in [3].

We shall use standard notations for the spaces of smooth function: $C(\mathbf{R}^d)$ the Banach space of bounded continuous functions equipped with sup-norm, $C^k(\mathbf{R}^d)$, $k \in \mathbf{N}$, the Banach space of k-times continuously differentiable bounded functions with bounded derivatives of order up to k, $C_\infty(\mathbf{R}^d)$ the closed subspace of $C(\mathbf{R}^d)$ consisting of functions vanishing at infinity and $C^k_\infty(\mathbf{R}^d)$ the closed subspace of $C^k(\mathbf{R}^d)$ consisting of functions such that all its derivatives up to order k belong to $C_\infty(\mathbf{R}^d)$.

2 CTRW Approximation

We are looking for the CTRW approximations to Eq. (9).

For the operator L^β_{st}, we choose standard random walk approximations

$$L^\beta_{st,\tau} f(x) = \frac{1}{\tau} \int (f(x+\tau^{1/\beta} z) - f(x)) p(x, dz), \tag{11}$$

with $p(x, dz)$ a symmetric probability kernel in $\mathbf{R}^d$ (i.e. $\int f(z) p(x, dz) = \int f(-z) p(x, dz)$), so that

$$\sup_x |(L^\beta_{st} - L^\beta_{st,\tau}) f(x)| \to 0, \quad \text{as} \quad \tau \to 0, \tag{12}$$

uniformly for f from any bounded subset of $C^2(\mathbf{R}^d)$.

The operator $L^\beta_{st,\tau}$ generates a jump-type Markov process with the intensity τ^{-1} and the distribution of jumps given by $p(x, dz)$.

Remark 1 Of course, such approximations are not unique. The existence is, however, more or less obvious. For the diffusion operator (7), the measures $p(x, dz)$ can be (and usually are) taken as discrete, for instance, of the type

$$\sum_i (\delta_{a_i e_i} + \delta_{-a_i e_i}) + \sum_{i>j} (\delta_{b_{ij}(e_i+e_j)} + \delta_{-b_{ij}(e_i+e_j)}),$$

with some numbers a_i, b_{ij}, where $\{e_i\}$ is the standard basis in $\mathbf{R}^d$. For the stable generator, one can take as $p(x, dz)$ a probability measure with an appropriate power tail, see e.g. [13] or Section 8.3 from [8].

We shall now construct a process with waiting times having power tails. For exponential waiting times, the intensity $a(s, x)$ governs the tail distribution $\mathbf{P}(T > t) = e^{-ta(s,x)\tau}$ with a scaling parameter τ. Hence, for the power tail, the analogous dependence will be the power law

$$\mathbf{P}(T > t) \sim \frac{1}{a(s, x)\alpha} t^{-a(s,x)\alpha}, \quad t \to \infty. \tag{13}$$

To simplify presentation, we shall make (13) more precise. Namely, we assume that these distributions have continuous densities $Q_{s,x}(r)$ such that

$$Q_{s,x}(r) = r^{-1-a(s,x)\alpha} \text{ for } r \geq B, \text{ and } Q_{s,x}(r) \leq 1 \text{ for all } r, \tag{14}$$

with some $B > 0$ uniformly for all s, x.

Remark 2 (i) The coefficients at the power of t in (13) are chosen for convenience. If not used here, then they would appear in (14) and in the limiting equation. (ii) Simplification arising from (14) is not essential and can be dispensed with.

As usual for CTRW modelling, we start by building an auxiliary scaled enhanced Markov chain with spatial jumps distributed according to $p(x, dz)$ and with an additional coordinate r specifying the total waiting times. In the usual CTRW scaling (see e.g. [13] and [8]), one scales discrete times by τ and the waiting times by the multiplier $\tau^{1/(\alpha a(s,x))}$ (see also (41)). Therefore, we define the Markov chain $(X^{\tau}_{x,s}, S^{\tau}_{x,s})(k\tau)$ evolving in discrete times τk, $k \in \mathbf{N}$, such that $(X^{\tau}_{x,s}, S^{\tau}_{x,s})(0) = (x, s)$, and, if a state of the chain in some discrete time $k\tau$ is (q, v), then in the next moment $(k + 1)\tau$ it will turn to the state

$$(q + \tau^{1/\beta} y, v + \tau^{1/(\alpha a(v,q))} r),$$

with y distributed according to $p(x, dy)$ and r distributed according to $Q_{v,q}$. We obtain the Markov chain with transition operators in time τ given by the formula

$$U^{\tau} F(x, s) = \int_{\mathbf{R}_+} \int_{\mathbf{R}^d} F(x+\tau^{1/\beta} y, s+\tau^{1/(\alpha a(s,x))} r) Q_{s,x}(r)\, dr\, p(x, dy). \tag{15}$$

We are interested in the value of the first coordinate $X^{\tau}_{x,s}$ evaluated at the random time $k\tau$ such that the total waiting time $S^{\tau}_{x,s}(k\tau)$ reaches t, that is, at the time

$$k\tau = T^{\tau}_{x,s}(t) = \inf\{m\tau : S^{\tau}_{x,s}(m\tau) \geq t\},$$

so that $T^{\tau}_{x,s}$ is the inverse process to $S^{\tau}_{x,s}$. We define the *scaled CTRW approximation for the process with the intensity of jumps governed by $a(s, x)$ and the distribution of jumps governed by $p(x, dy)$* as the (non-Markovian) process

$$\tilde{X}^{\tau}_{x,s}(t) = X^{\tau}_{x,s}(T^{\tau}_{x,s}(t)). \tag{16}$$

We are going to identify the weak limit of the process $\tilde{X}^{\tau}_{x,s}(t)$, as $\tau \to 0$, and to show that the limiting processes are governed by the equations extending (6). As an auxiliary step, we shall identify the weak limit of the Markov chain $(X^{\tau}_{x,s}, S^{\tau}_{x,s})$, that is, the weak limit of the chains with transitions $[U^{\tau}]^{[t/\tau]}$, where $[t/\tau]$ denotes the integer part of the number t/τ, as $\tau \to 0$. It is known (see e.g. Theorem 19.28 of [4] or Theorem 8.1.1 of [8]) that if such chain converges to a Feller process, then the generator of this limiting process can be obtained as the limit

$$\Lambda F = \lim_{\tau \to 0} \frac{1}{\tau}(U^{\tau} F - F). \tag{17}$$

3 Main Results

Let us assume the strict non-degeneracy condition:

$$g_1 \le G(x) \le g_2, \text{ or } m_1 \le \mu(x, s) \le m_2 \tag{18}$$

for the spatial part and

$$a_1 \le a(t, x) \le a_2 \tag{19}$$

for the temporal part of our stable generators, with some positive constants g_i, m_i, a_i such that $a_2\alpha < 1$, where the first inequality (18) is understood in the sense of matrix, and the smoothness condition:

$$a(.), G(.), \mu(., s) \in C^4(\mathbf{R}^d), \tag{20}$$

with $\mu(., s)$ having a bounded norm uniformly in s. Let us assume (12) for the approximating jumps and (14) for the distributions of waiting times.

Notice that we use the letter G to denote the diffusion coefficient $G(x)$ in the diffusion operator (7) and to denote the transition probability density $G(t; x, s; y, v)$ for our main processes below.

Theorem 3.1

(i) For $F \in C^2_{\infty}(R^{d+1})$, the limit in (17) exists and

$$\Lambda F(x, s) = \lim_{\tau \to 0} \frac{1}{\tau}(U^{\tau} F - F)(x, s)$$

$$= \int_0^{\infty} \frac{F(x, s + r) - F(x, s)}{r^{1+\alpha a(s,x)}} dr + L^{\beta}_{st} F(x, s). \tag{21}$$

(ii) *The operator* Λ *defined by the r.h.s. of* (21) *generates a (conservative time-homogeneous) Feller process* $(X_{x,s}, S_{x,s})(t)$ *in* $\mathbf{R}^{d+1}$ *(with the initial condition* $(X_{x,s}, S_{x,s})(0) = (x, s)$*) with continuous (in fact, even smooth) transition probability density* $G(t; x, s; y, v)$, $t > 0$, *and the corresponding Feller semigroup in* $C_\infty(\mathbf{R}^{d+1})$ *such that the space* $C^2_\infty(\mathbf{R}^{d+1})$ *is its invariant core. The process* $(X_{x,s}, S_{x,s})(t)$ *belongs to the class of processes usually referred to as stable-like processes.*

(iii) *The chains with transitions* $[U^\tau]^{[t/\tau]}$, *with* U^τ *given by* (15), *converge weakly to the process* $(X_{x,s}, S_{x,s})(t)$, *and the operators* $[U^\tau]^{[t/\tau]}$ *converge to the transition operators of the process* $(X_{x,s}, S_{x,s})(t)$ *strongly and uniformly for compact intervals of time.*

Theorem 3.2

(i) *The marginal distributions of the scaled CTRW* (16) *converge to the marginal distributions of the process*

$$\tilde{X}_{x,s}(t) = X_{x,s}(T_{x,s}(t)), \tag{22}$$

where

$$T_{x,s}(t) = \inf\{r : S_{x,s}(r) \geq t\},$$

that is, for a bounded continuous function $F(x)$, *it holds that*

$$\lim_{\tau \to 0} \mathbf{E} F(\tilde{X}^\tau_{x,s}(t)) = \mathbf{E} F(\tilde{X}_{x,s}(t)). \tag{23}$$

(ii) *For any* $F \in C_\infty(\mathbf{R}^d)$ *and arbitrary* $K > 0$,

$$\mathbf{E}[F(\tilde{X}_{x,s}(t))\mathbf{1}(T_{x,s}(t) \in [1/K, K])]$$

$$= \int dy \int_{1/K}^{K} du \int_s^t dv \frac{(t-v)^{-a(v,y)\alpha}}{a(v,y)\alpha} G(u; x, s; y, v) F(y). \tag{24}$$

This formula extends to infinite K, *that is,*

$$\mathbf{E}[F(\tilde{X}_{x,s}(t))] = \int dy \int_0^\infty du \int_s^t dv \frac{(t-v)^{-a(v,y)\alpha}}{a(v,y)\alpha} G(u; x, s; y, v) F(y). \tag{25}$$

Theorem 3.3 *For any* $F \in C^2(\mathbf{R}^d)$, *the evolution of averages* $F(x, s) = \mathbf{E} F(\tilde{X}_{x,s}(t))$ *represents the unique solution to the mixed fractional differential equation*

$$D^{\alpha a(s,x)}_{t-*} F(x, s) = L^\beta_{st} F(x, s), \quad s \in [0, t], \tag{26}$$

with the terminal condition $F(x,t) = F(x)$, where the right fractional derivative (of Caputo–Dzerbashyan type) acting on the variable $s \le t$ of $F(x,s)$ is defined as

$$D_{t-*}^{\alpha a(s,x)} g(s) = -\int_0^{t-s} \frac{g(s+r)-g(s)}{r^{1+\alpha a(s,x)}} dy - (g(t)-g(s)) \frac{(t-s)^{-\alpha a(s,x)}}{a(s,x)\alpha}. \tag{27}$$

Remark 3 With some ambiguity, we denote by the same letter the function $F(x)$ of one variable $x \in \mathbf{R}^d$ and the function $F(x,s)$ of two variables, $s \le t, x \in \mathbf{R}^d$, solving (26) with the boundary (actually terminal) condition $F(x,t) = F(x)$.

Remark 4 The limiting equation (27) is written in terms of the right fractional derivative because we have chosen the forward motion for the auxiliary coordinate $S_{s,x}$ of the Markov process $(X_{x,s}, S_{x,s})(t)$. The analogous equations with left derivatives, like in (4), can be obtained by changing the direction of $S_{s,x}$.

Remark 5 For a generator L of an arbitrary Feller process with a core $C_\infty^k(\mathbf{R}^d)$, where k equals one or two, one can use an arbitrary approximation L_τ of the type

$$L_\tau f(x) = \frac{1}{\tau} \int (f(x+z) - f(x)) p_\tau(x,dz), \tag{28}$$

with $p_\tau(x,dz)$ a family of probability kernels in $\mathbf{R}^d$ such that

$$\sup_x |(L - L_\tau) f(x)| \to 0, \quad \text{as} \quad \tau \to 0, \tag{29}$$

uniformly for f from any bounded subset of $C_\infty^k(\mathbf{R}^d)$. The corresponding Markov chain $(X_{x,s}^\tau, S_{x,s}^\tau)(k\tau)$ is defined by the transition operator

$$U^\tau F(x,s) = \int_{\mathbf{R}_+} \int_{\mathbf{R}^d} F(x+y, s+\tau^{1/(\alpha a(s,x))} r) Q_{s,x}(r)\, dr\, p_\tau(x,dy). \tag{30}$$

The results above remain true (with L instead of L_{st}^β) if we assume that the process generated by Λ from (21) has a continuous transition probability density, which is sufficiently smooth so that smooth functions form a core for its Feller semigroup and which has integrability condition for small and large times ensuring convergence of the integral in (25).

4 A General Representation for a Subordination

In this section, we derive a representation for Markov processes subordinated by its monotone coordinate. It supplies the proof for the part (ii) of Theorem 3.1 but may be of independent interest.

Lemma 4.1 *Let Y_s be an adapted process on a stochastic basis $(\Omega, \mathcal{F}, \mathcal{F}_t, P)$ and Σ a stopping time. Let the pairs (Y_s, Σ) have joint densities $g_s(y, \sigma)$ for $s > 0$ such that, for s from any bounded interval separated from zero, $g_s(y, \sigma)$ is bounded and right continuous in s. Then, for any $K > 0$ and a continuous bounded function $F(y)$,*

$$\mathbf{E}[F(Y_\Sigma)\mathbf{1}(\Sigma \in [1/K, K])] = \int dy \int_{1/K}^{K} d\sigma\, g_\sigma(y, \sigma)F(y). \tag{31}$$

Proof Let us choose the discrete approximations $\Sigma_\tau = \tau[\Sigma/\tau+1]$ (square brackets denoting the integer part), so that $\Sigma_\tau = k\tau$ when $(k-1)\tau \le \Sigma < k\tau$. Thus Σ_τ are right continuous stopping times such that $\Sigma_\tau \ge \Sigma$ for all τ, and they depend monotonically on τ and $\Sigma_\tau \to \Sigma$, as $\tau \to 0$. Then, we can write

$$\mathbf{E}[F(Y_\Sigma)\mathbf{1}(\Sigma \in [1/K, K])] = \lim_{\tau\to 0} \mathbf{E}[F(Y_{\Sigma_\tau})\mathbf{1}(\Sigma \in [1/K, K])]$$

$$= \lim_{\tau\to 0} \sum_{k=1}^{\infty} \mathbf{E}[F(Y_{\Sigma_\tau})\mathbf{1}(\Sigma \in [1/K, K])\mathbf{1}(\Sigma_\tau = k\tau)]$$

$$= \lim_{\tau\to 0} \sum_{k=1}^{\infty} \mathbf{E}\,[F(Y_{k\tau})\mathbf{1}(\Sigma \in [1/K, K])\mathbf{1}(\Sigma \in [(k-1)\tau, k\tau))]$$

$$= \lim_{\tau\to 0} \int dy \int_{1/K}^{K} d\sigma\, g_{\sigma_\tau}(y, \sigma)F(y).$$

And therefore

$$\mathbf{E}[F(Y_\Sigma)\mathbf{1}(\Sigma \in [1/K, K])] = \int dy \int_{1/K}^{K} d\sigma\, g_\sigma(y, \sigma)F(y)$$

$$+ \lim_{\tau\to 0} \int dy \int_{1/K}^{K} d\sigma\, (g_{\sigma_\tau} - g_\sigma)(y, \sigma)F(y).$$

Since the last term tends to zero (by the right continuity of g and dominated convergence), we obtain (31). □

Remark 6 (i) To get the same without the restrictions $[1/K, K]$, one needs some integrability conditions for g_s for small and large s. (ii) The existence of the density $g_s(y, \sigma)$ can be weakened to the condition

$$\mathbf{P}(Y_s \in dy, \Sigma \in d\sigma) = g_s(\sigma, dy)d\sigma$$

such that $g_s(\sigma, dy)$ is weakly right continuous in s uniformly in σ (on bounded intervals).

Lemma 4.2 *Let* (Y_s, V_s), $Y \in \mathbf{R}^d$, $V \in \mathbf{R}$ *be a Markov process with the generator* $L + A$*, where* L *acts on the first variable and does not depend on the second one (so that* Y_s *is itself a Markov process), and* A *acts on the second variable but may have coefficients depending on the first variable. Let the process* (Y_s, V_s) *have a continuous (for* $s > 0$*) transition density* $G(s; y_0, v_0; y, v)$*, and let* $G_Y(s, y_0, y)$ *be the transition density for the process* Y *so that*

$$G_Y(u, y_0, y) = \int G(u; y_0, v_0; y, v)dv$$

for any v_0.

Then, the random vector (Y_s, V_u) *has a density* $\phi_{y_0,v_0}(s, u; y, v)$ *such that, for* $s > u > 0$,

$$\phi_{y_0,v_0}(s, u; y, v) = \int G_Y(s-u, p, y)G(u; y_0, v_0; p, v)\, dp. \tag{32}$$

Moreover,

$$\frac{\partial}{\partial u}\phi_{y_0,v_0}(s, u; y, v) = \int G_Y(s-u, p, y)A^*_{(v;p)}G(u; y_0, v_0; p, v)\, dp, \tag{33}$$

where A^* *denotes the adjoint operator to* A *and the lower index* $(v; p)$ *indicates that this operator acts on the variable* v*, while the value of the spatial variable is* p.

Remark 7 Generally speaking, formula (33) may hold only in the sense of generalised functions. To ensure that it holds as an equation for functions, some additional mild regularity for transitions must be assumed.

Proof For a bounded continuous function F, we have

$$\mathbf{E}F(Y_s, V_u) = \int \mathbf{E}(F(Y_s, V_u)|Y_u = p, V_u = v)G(u; y_0, v_0; p, v)\, dpdv$$

$$= \int \mathbf{E}(F(Y_s, v)|Y_u = p)G(u; y_0, v_0; p, v)\, dpdv$$
$$= \int F(y, v)G_Y(s-u, p, y)G(u; y_0, v_0; p, v)\, dydpdv,$$

implying (32).

Next,

$$\frac{\partial}{\partial u}\phi_{y_0,v_0}(s, u; y, v) = -\int (L_{(p)}G_Y(s-u, p, y))G(u; y_0, v_0; p, v)\, dp$$

$$+\int G_Y(s-u, p, y))(L^*_{(p)} + A^*_{(v;p)})G(u; y_0, v_0; p, v)\, dp.$$

The key observation is that the terms with $L_{(p)}$ cancel implying (33). □

Lemma 4.3 *Under the assumption of Lemma 4.2, let*

$$A\phi(v) = \int_0^\infty (\phi(v+w) - \phi(v))\nu(v, w; p)\, dw, \tag{34}$$

where p is the value of the spatial variable, where ν is a strictly positive function for $w > 0$ such that

$$\sup_{v,p} \int \min(1, w)\nu(v, w; p)\, dw < \infty.$$

Then, the coordinate V_t is a strictly increasing process a.s., so that its generalised inverse (or hitting times) process

$$Z_t = \sup\{u \geq 0 : V(u) \leq t\} = \inf\{u \geq 0 : V(u) > t\}, \quad t > v_0,$$

is continuous, and for the density $g_{y_0,v_0}(s, t; y, z)$ of the pair (Y_s, Z_t), $s > 0, t > v_0$, the following formula holds:

$$g_{y_0,v_0}(s, t; y, z) = \frac{\partial}{\partial z} \int_t^\infty \phi_{y_0,v_0}(s, z; y, v)\, dv$$

$$= \int_t^\infty dv \int_{\mathbf{R}^d} dp\, G_Y(s - z, p, y) A^*_{(v;p)} G(z; y_0, v_0; p, v). \tag{35}$$

In particular,

$$\lim_{z \to s} g_{y_0,v_0}(s, t; y, z) = \int_t^\infty A^*_{(v;y)} G(s; y_0, v_0; y, v)\, dv. \tag{36}$$

Proof The fact that V_t is strictly increasing follows from (34), which in turn implies the continuity of Z_t.

Consequently, since $(Z_t \leq z) = (V_z \geq t) = (V_z > t)$ a.s., it follows that

$$g_{y_0,v_0}(s, t; y, z) = \frac{\partial^2}{\partial y \partial z} \mathbf{P}(Y_s \leq y, V_z \geq t)$$

$$= \frac{\partial^2}{\partial z \partial y} \int_{-\infty}^y \int_t^\infty \phi_{y_0,v_0}(s, z; \tilde{y}, v)\, dv dy = \frac{\partial}{\partial z} \int_t^\infty \phi_{y_0,v_0}(s, z; y, v)\, dv.$$

Substituting (33), we get (35). □

Theorem 4.1 *Under the assumptions of Lemma 4.2,*

$$\mathbf{E}[F(Y_{Z_t})\mathbf{1}(\mathbf{Z}_t \in [1/K, K])] = \int dy \int_{1/K}^{K} ds \int_t^{\infty} dv\, A^*_{(v;y)} G(s; y_0, v_0; y, v) F(y), \tag{37}$$

and also

$$\begin{aligned}\mathbf{E}[F(Y_{Z_t})\mathbf{1}(\mathbf{Z}_t \in [1/K, K])] \\ = \int dy \int_{1/K}^{K} ds \int_0^t dv\, (A_{(v;y)}\theta_{\geq t})(v) G(s; y_0, v_0; y, v) F(y)\end{aligned}$$

$$= \int dy \int_{1/K}^{K} ds \int_{v_0}^{t} dv \int_{t-v}^{\infty} \nu(v, w; y)\, dw\, G(s; y_0, v_0; y, v) F(y), \tag{38}$$

where $\theta_{\geq t}$ is the indicator function of the interval $[t, \infty)$.

Remark 8

(i) Formula (37) in this general form was seemingly first derived in [7] (though the proof was a bit sketchy there) and was reproduced in monograph [8]. The advantage of its modification (38) is that for its validity only the continuity of the density $G(s; y_0, v_0; y, v)$ is required (and even this condition can be weakened) and not any smoothness that may be needed to make sense out of $A^*_{(v;y)} G(s; y_0, v_0; y, v)$.
(ii) It follows that the density of the random variable Y_{Z_t} equals

$$\int_0^{\infty} ds \int_{v_0}^{t} dv\, (A_{(v;y)}\theta_{\geq t})(v)\, G(s; y_0, v_0; y, v), \tag{39}$$

whenever this integral is well defined.

Proof Formula (37) follows from (36) and (31). Next, by duality (effectively by the shift of the variable of integration),

$$\int_t^{\infty} A^*_{(v;y)} G(s; y_0, v_0; y, v)\, dv = \int_{-\infty}^{\infty} (\theta_{\geq t})(v) A^*_{(v;y)} G(s; y_0, v_0; y, v)\, dv$$

$$\begin{aligned}&= \int_{-\infty}^{\infty} (A_{(v;y)}\theta_{\geq t})(v) G(s; y_0, v_0; y, v)\, dv \\ &= \int_{v_0}^{t} dv \int_{t-v}^{\infty} \nu(v, w; y)\, dw\, G(s; y_0, v_0; y, v),\end{aligned}$$

implying (38). □

5 On the Rate of Convergence for the Standard CTRW

It is well known (see e.g. [13] and [8]) that the Markov chains with jumps distributed like (13) converge after appropriate scaling to stable subordinators. In the following result, we give exact rates of weak convergence for the corresponding generators under (13).

Proposition 5.1 *Let $p(y)$ be a probability density on $\mathbf{R}_+$ such that $p(y) = y^{-1-\alpha}$ for $y \geq B$ with some $\alpha \in (0, 1)$ and $B > 0$. Then, for any continuous f on $\mathbf{R}_+$ such that $f(0) = f(\infty) = 0$ and f is Lipschitz at zero so that $|f(y)| \leq Ly$ for $y \in [0, Bh]$ and some constant L, it follows that*

$$\left| h^{-\alpha} \int_0^\infty f(hy)p(y)dy - \int_0^\infty \frac{f(y)dy}{y^{1+\alpha}} \right| \leq C_B L h^{1-\alpha}, \tag{40}$$

with

$$C_B = \frac{B^{1-\alpha}}{1-\alpha} + \int_0^B yp(y)dy.$$

Remark 9 If, instead of assuming Lipschitz at zero, one assumes Hölder continuity at zero, that is, $|f(y)| \leq Ly^\beta$ for $y \in [0, Bh]$ and some $L > 0$, $\beta > \alpha$, then inequality (40) holds with the r.h.s.

$$\left(\frac{B^{\beta-\alpha}}{\beta - \alpha} + \int_0^B y^\beta p(y)dy \right) L h^{\beta-\alpha}.$$

Proof Let $f = \mathbf{1}_{\geq C}$ with $C \geq Bh$. Then,

$$h^{-\alpha} \int_0^\infty f(hy)p(y)dy = h^{-\alpha} \int_{C/h}^\infty p(y)dy$$

$$= h^{-\alpha}\alpha^{-1}(C/h)^{-\alpha} = \alpha^{-1}C^{-\alpha} = \int_C^\infty \frac{dy}{y^{1+\alpha}} = \int_0^\infty \frac{f(y)\, dy}{y^{1+\alpha}}.$$

Hence, by linearity and the density argument, it follows that

$$h^{-\alpha} \int_0^\infty f(hy)p(y)dy = \int_0^\infty \frac{f(y)dy}{y^{1+\alpha}},$$

for any bounded measurable f having support on $[Bh, \infty)$. Next, let f have support on $[0, Bh]$. Then,

$$\left| h^{-\alpha} \int_0^\infty f(hy)p(dy) \right| = \left| h^{-\alpha} \int_0^B f(hy)p(y)dy \right| \leq h^{1-\alpha} L \int_0^B yp(y)dy,$$

and

$$\left|\int_0^\infty \frac{f(y)dy}{y^{1+\alpha}}\right| = \left|\int_0^{Bh} \frac{f(y)dy}{y^{1+\alpha}}\right| \le L\int_0^{Bh} \frac{dy}{y^\alpha} = \frac{B^{1-\alpha}}{1-\alpha}Lh^{1-\alpha},$$

implying (40). □

In particular, setting $\tau = h^\alpha$, it follows that

$$\begin{aligned}&\left|\tau^{-1}\int_0^\infty (f(x\pm\tau^{1/\alpha}y) - f(x))p(y)dy - \int_0^\infty \frac{(f(x\pm y)-f(x))dy}{y^{1+\alpha}}\right| \\ &\quad \le C_B L\tau^{(1/\alpha)-1},\end{aligned} \tag{41}$$

where L is the sup of the derivative of f near x.

6 Proof of Theorem 3.1

(i) We have

$$\frac{1}{\tau}(U^\tau F - F)(x,s)$$

$$= \frac{1}{\tau}\int_{\mathbf{R}_+}\int_{\mathbf{R}^d} [F(x+\tau^{1/\beta}y, s+\tau^{1/(\alpha a(s,x))}r) - F(x,s)]Q_{s,x}(r)\,dr\; p(x,dy).$$

$$\begin{aligned}&= \tfrac{1}{\tau}\int_{\mathbf{R}_+}[F(x, s+\tau^{1/(\alpha a(s,x))}r) - F(x,s)]Q_{s,x}(r)\,dr \\ &\quad + \tfrac{1}{\tau}\int_{\mathbf{R}^d}[F(x+\tau^{1/\beta}y, s) - F(x,s)]p(x,dy) + R,\end{aligned} \tag{42}$$

where

$$R = \frac{1}{\tau}\int_{\mathbf{R}_+}\int_{\mathbf{R}^d} [g_y(x, s+\tau^{1/(\alpha a(s,x))}r) - g_y(x,s)]Q_{s,x}(r)\,dr\; p(x,dy),$$

with

$$g_y(x,s) = F(x+\tau^{1/\beta}y, s) - F(x,s).$$

By (41), the first term in (42) converges, as $\tau \to 0$, to

$$\int_0^\infty \frac{F(x,s+r) - F(x,s)}{r^{1+\alpha a(s,x)}}dr,$$

whenever F is continuously differentiable in s. By (12), the second term in (42) converges, as $\tau \to 0$, to $L^{\beta}_{st} F(x, s)$. Again by (41) and (12), it follows that $R \to 0$, as $\tau \to 0$, implying (21).

(ii) The operator L is the sum of two terms, and each one of them generates a Feller semigroup on $C_\infty(\mathbf{R}^{d+1})$ such that the spaces $C^2_\infty(\mathbf{R}^{d+1})$ and $C^4_\infty(\mathbf{R}^{d+1})$ represent its invariant core. The fact that L generates a Feller semigroup on $C_\infty(\mathbf{R}^{d+1})$ with an invariant core $C^2_\infty(\mathbf{R}^{d+1})$ follows from Theorems 5.2.1 and 5.2.2 of [10]. On the other hand, there is a well-known technique of building transition probabilities for stable-like processes with the generators of type (21). Though the author is unaware of any result for generator (21) exact, technique of [6] (see also Chapter 7 of [8]) is fully applicable yielding the existence of 4 times differentiable transition probabilities $G(t; x, s; y, v)$, $t > 0$, for the process $(X_{x,s}, S_{x,s})(t)$. This provides also an alternative proof of the statement about the cores.

(iii) It is a direct consequence of (i) and (ii) and the standard general result on the convergence of Markov chains, see e.g. Theorem 19.28 of [4].

7 Subordination for Discrete Markov Chains

As a starting point for the proof of Theorem 3.2, we give here some preliminary calculations for the subordinated Markov chains in discrete times.

Let $Y_{k\tau}$ be an adapted process on a stochastic basis $(\Omega, \mathcal{F}, \mathcal{F}_t, P)$ and Σ a stopping time with values in $\{k\tau\}$, $k \in \mathbf{N}$. Let the random variables $Y_{m\tau}\mathbf{1}(\Sigma = k\tau))$ have densities $g_{m\tau}(y, k\tau)$. Then, for any $K > 0$ and a continuous bounded function $F(y)$,

$$\mathbf{E}[F(Y_\Sigma)\mathbf{1}(\Sigma \in [1/K, K])] = \sum_{k\tau \in [1/K, K]} \int g_{k\tau}(y, k\tau) F(y) dy. \tag{43}$$

Let $(Y_{m\tau}, V_{m\tau})$, $Y \in \mathbf{R}^d$, $V \in \mathbf{R}$, be a Markov chain with the transition operator U^τ:

$$U^\tau F(y_0, v_0) = \int F(y, v) P_\tau(y_0, v_0; y, v)\, dy dv = \int F(y, v) P_\tau(y_0, v_0; dy, dv),$$

such that $Y_{m\tau}$ is a Markov chain on its own with the transition U^τ_Y.

Then, the density of the random vector $(Y_{k\tau}, V_{k\tau}, V_{(k-1)\tau})$ is

$$\int P_\tau(p, v; y, w) P_{(k-1)\tau}(y_0, v_0; p, v)\, dp \tag{44}$$

because

$$\mathbf{E}F(Y_{k\tau}, V_{k\tau}, V_{(k-1)\tau})$$

$$= \int \mathbf{E}(F(Y_{k\tau}, V_{k\tau}, v)|Y_{(k-1)\tau} = p, V_{(k-1)\tau} = v)P_{(k-1)\tau}(y_0, v_0; p, v)\,dpdv$$

$$= \int F(y, w, v)P_\tau(p, v; y, w)P_{(k-1)\tau}(y_0, v_0; p, v)\,dpdvdwdy.$$

Let the coordinate $V_{k\tau}$ be strictly increasing, and let

$$Z_t^\tau = \sup\{m\tau : V(m\tau) \le t\} = \inf\{m\tau : V(m\tau) > t\}, \quad t > v_0,$$

be its (generalised) inverse (or hitting times) process.

Then,

$$\begin{aligned}\mathbf{E}[F(Y_{Z_t^\tau})\mathbf{1}(Z_t^\tau \in [1/K, K])] &= \textstyle\sum_{k\tau\in[1/K,K]} F(Y_{k\tau})\mathbf{1}(Z_t^\tau = k\tau)\\ &= \textstyle\sum_{k\tau\in[1/K,K]} F(Y_{k\tau})\mathbf{1}(V_{(k-1)\tau} < t \le V_{k\tau})\end{aligned}$$

$$= \int F(y) \sum_{k\tau\in[1/K,K]} \int \mathbf{1}(v < t \le w)$$

$$P_\tau(p, v; y, w)P_{(k-1)\tau}(y_0, v_0; p, v)\,dydpdvdw. \tag{45}$$

This can be rewritten as

$$\begin{aligned}&\mathbf{E}[F(Y_{Z_t^\tau})\mathbf{1}(Z_t^\tau \in [1/K, K])]\\ &= \int \sum_{k\tau\in[1/K,K]} (U^\tau F_v)(p, v)P_{(k-1)\tau}(y_0, v_0; p, v)\,dpdv,\end{aligned} \tag{46}$$

with $F_v(y, w) = F(y)\mathbf{1}(v < t \le w)$.

8 Proof of Theorem 3.2

Theorem 4.1 implies formula (24):

$$\mathbf{E}[F(\tilde{X}_{x,s}(t))\mathbf{1}(T_{x,s}(t) \in [1/K, K])]$$

$$= \int dy \int_{1/K}^K du \int_s^t dv \frac{(t-v)^{-a(v,y)\alpha}}{a(v, y)\alpha} G(u; x, s; y, v)F(y). \tag{47}$$

To complete the proof of statement (ii) of Theorem 3.2, one has to note that, due to the asymptotic behavior of the transition probabilities $G(u; x, s; y, v)$ of the stable-like process $(X_{x,s}, S_{x,s})(t)$ (see [6] or [8]), the function $G(u; x, s; y, v)$ is integrable for small and large u, so that one can pass to the limit $K \to \infty$ in (47) to obtain (25).

Next, by (46),

$$\begin{aligned} &\mathbf{E}[F(\tilde{X}^\tau_{x,s}(t))\mathbf{1}(T^\tau_{x,s}(t) \in [1/K, K])] \\ &\quad = \int dy \int_0^t dv \sum_{k\tau \in [1/K,K]} (U^\tau F_v)(y, v) P_{(k-1)\tau}(x, s; y, v), \end{aligned} \tag{48}$$

where $F_v(q, w) = F(q)\mathbf{1}(v < t \le w)$ and

$$U^\tau F(y, v) = \int_{\mathbf{R}_+} \int_{\mathbf{R}^d} F(y + \tau^{1/\beta} q, v + \tau^{1/(\alpha a(v,y))} r) Q_{v,y}(r)\, dr\, p(y, dq), \tag{49}$$

with $Q_{s,x}$ satisfying (14).

Therefore,

$$\begin{aligned} &\mathbf{E}[F(\tilde{X}^\tau_{x,s}(t))\mathbf{1}(T^\tau_{x,s}(t) \in [1/K, K])] \\ &= \int dy \int_0^t dv \sum_{k\tau \in [1/K,K]} P_{(k-1)\tau}(x, s; y, v) F(y + \tau^{1/\beta} q) p(y, dq) \\ &\quad \mathbf{1}(v < t \le v + \tau^{1/(\alpha a(v,y))} r) Q_{v,y}(r)\, dr. \end{aligned} \tag{50}$$

Since the probability of small and large jumps is small both for $S_{x,s}$ and $S^\tau_{x,s}$, it follows that, in order to prove Statement (i) of Theorem 3.2, it is sufficient to show that expression (50) tends to expression (47), as $\tau \to 0$, for any finite K and any $F \in C_\infty(\mathbf{R}^d)$. By the density argument, this can be further reduced to the case of $F \in C^2_\infty(\mathbf{R}^d)$.

The key point for the argument is that, by Theorem 3.1, the sums

$$\tau \int dy \int_0^t dv \sum_{k\tau \in [1/K,K]} P_{(k-1)\tau}(x, s; y, v) \Omega(y, v)$$

approximate uniformly the Riemannian sums for the integral

$$\int dy \int_0^t dv \int_{[1/K,K]} du\, G(u; x, s; y, v) \Omega(y, v)$$

and thus converge to this integral, at least for $\Omega \in C_\infty(\mathbf{R}^{d+1})$.

Turning concretely to (50), we note firstly that the limit must be the same for the expression

$$\int dy \int_0^t dv \sum_{k\tau \in [1/K, K]}$$
$$P_{(k-1)\tau}(x, s; y, v) F(y) \mathbf{1}(v < t \leq v + \tau^{1/(\alpha a(v,y))} r) Q_{v,y}(r)\, dr. \tag{51}$$

In fact, expressions (50) and (51) differ by $\int (F(y + \tau^{1/\beta} q) - F(y)) p(y, dq)$, which tends to zero, as $\tau \to 0$, uniformly in y (for $F \in C^2_\infty(\mathbf{R}^d)$). Hence, if functions (51) have a finite limit (as we are going to show), then the same limit has functions (50).

Since

$$\begin{aligned} &\tfrac{1}{\tau} \int_0^\infty \mathbf{1}(v < t \leq v + \tau^{1/(\alpha a(v,y))} r) Q_{v,y}(r)\, dr \\ &\quad = \tfrac{1}{\tau} \int_0^\infty [\theta_{\geq t}(v + \tau^{1/a(v,y)\alpha} r) - \theta_{\geq t}(v)] Q_{v,y}(r)\, dr, \end{aligned} \tag{52}$$

it tends to $A\theta_{\geq t}(v)$, by (41), as claimed. Therefore, we just need to show that this limit goes through the Riemannian sum approximation (51). For this end, we note that the limit of (51) remains the same if we slightly decrease the domain of integration, namely if we choose the expression

$$\begin{aligned} &\int dy \int_{[0, t - B\tau^{1/a(v,y)\alpha}]} dv \sum_{k\tau \in [1/K, K]} \\ &\quad P_{(k-1)\tau}(x, s; y, v) F(y) \mathbf{1}(v < t \leq v + \tau^{1/(\alpha a(v,y))} r) Q_{v,y}(r)\, dr. \end{aligned} \tag{53}$$

In fact, by (52) and (14), the difference between (51) and (53) is bounded in magnitude by the expression

$$\tau \int dy \int_{[t - B\tau^{1/a(v,y)\alpha}, t]} dv \sum_{k\tau \in [1/K, K]} P_{(k-1)\tau}(x, s; y, v) F(y) \Omega_\tau(y, v),$$

with $\Omega(y, v) \leq 1/\tau$. And this expression tends to zero, as $\tau \to 0$, because

$$\int_{[t - B\tau^{1/a(v,p)\alpha}, t]} \Omega_\tau(p, v)\, dv$$

is bounded by $B\tau^{(1/a(v,y)\alpha) - 1}$, which is uniformly small.

Next, for $v > B\tau^{1/a(v,y)\alpha}$,

$$\frac{1}{\tau} \int_0^\infty \mathbf{1}(v < t \leq v + \tau^{1/(\alpha a(v,y))} r) Q_{v,y}(r)\, dr = A\theta_{\geq t}(v)$$

(see the proof of Proposition 5.1 for this simple calculation).

Hence, it remains to find the limit of the expression

$$\tau \int dy \int_{[0,t-A\tau^{1/a(v,y)\alpha}]} dv \sum_{k\tau\in[1/K,K]} P_{(k-1)\tau}(x,s;y,v)F(y)A\theta_{\geq t}(v). \tag{54}$$

Since $G(u;x,s;y,v)$ is continuous in y,v and $A\theta_{\geq t}(v)$ has an integrable singularity, it follows that this expression tends to (47) completing the proof of part (i) of Theorem 3.2.

9 Proof of Theorem 3.3

The statement on the probabilistic representation of the solutions to (26) follows from the standard argument (see e.g. [9]) with Dynkin's martingale. Namely, for a function F from the domain of the generator Λ of the process $(X_{x,s}, S_{x,s})(t)$, the process

$$M_F^t = F(X_{x,s}(t), S_{x,s}(t)) - F(x,s) - \int_0^t \Lambda F(X_{x,s}, S_{x,s})(r)\,dr$$

is a martingale, called Dynkin's martingale. We are now looking at the modification of the process $(X_{x,s}, S_{x,s})(t)$ obtained by stopping it at the attempt to cross the boundary $S_{x,s} = t$. This modification specifies a process on $\mathbf{R}^d \times (-\infty, t]$ with the generator

$$\tilde{\Lambda}(x,s) = L_{st}^{\beta} F(x,s)$$

$$+ \int_0^{t-s} \frac{F(x,s+r) - F(x,s)}{r^{1+\alpha a(s,x)}} dr + (F(x,t) - F(x,s)) \frac{(t-s)^{-a(s,x)\alpha}}{a(s,x)\alpha}, \tag{55}$$

obtained formally from (21) by reducing it to the set of functions such that $F(x,r) = F(x,t)$ for all $r \geq t$.

Applying Doob's optional sampling theorem to the martingale M_F^t and the stopping time

$$T_{x,s}(t) = \inf\{r : S_{x,s}(r) \geq t\},$$

we can conclude that, if $\tilde{\Lambda} F = 0$, which is equivalent to equation (26), then

$$F(x,s) = \mathbf{E}F[\tilde{X}_{x,s}(t)] = \mathbf{E}F[X_{x,s}(T_{x,s}(t))]. \tag{56}$$

Thus, formula (56) gives the probabilistic representation for a solution of the fractional boundary problem (26) implying also the uniqueness of such a solution.

To complete the proof of well-posedness, it remains to check that formula (56) really solves the problem. The fact that it satisfies the boundary condition $F(x,t) = F(x)$ is a direct consequence of (56). Let us check that it satisfies the equation $\tilde{\Lambda} F = 0$. To this end, we shall use the representation (25):

$$F(x,s) = \mathbf{E}[F(Y_{Z_t})] = \int dy \int_0^\infty du \int_0^t dv\, (A_{(v;y)}\theta_{\geq t})(v) G(u; x, s; y, v) F(y).$$

We know that $F(x,t) = F(x) = \lim_{s\to t} F(x,s)$. Setting $F(x,s) = F(x)$ for $s \geq t$, we make the function $F(x,s)$ continuous everywhere.

We want to show that

$$L^{\beta}_{st} F(x,s) + A_{(s;x)} F(s,x) = 0. \tag{57}$$

It is straightforward from (25) that $F(x,s)$ is smooth for $s < t$, so that all terms in this equation are well defined. Let

$$f(x,s;u,v) = \int dy\, G(u; x, s; y, v) F(y) (A_{(v;y)}\theta_{\geq t})(v), \quad u > 0,$$

so that $f(x,s;u,v) = 0$ for $s \geq v$ and any $u > 0$. By Kolmogorov's equation for the transition density,

$$\frac{\partial G}{\partial u} = L^{\beta}_{st} G + A_{(s;x)} G.$$

Therefore,

$$\frac{\partial f}{\partial u} = L^{\beta}_{st} f + A_{(s;x)} f.$$

Let

$$g(x,s;v) = \int_0^\infty f(x,s;u,v)\, du \text{ for } s < v \text{ and } g(x,s;v) = 0 \text{ for } s > v.$$

Integrating the equation for f over u and using that $G(0; x, s; y, v) = \delta(x-y)\delta(s-v)$ yield the equation

$$(L^{\beta}_{st} + A_{(s;x)}) g(x,s;v) + F(x)(A_{(v;x)}\theta_{\geq t})(v)\delta(v-s) = 0.$$

Since

$$F(x,s) = \int_s^t g(x,s;v) dv, \quad s \leq t \text{ and } F(x,s) = F(x) \text{ for } s \geq t,$$

it follows that

$$L^{\beta}_{st}F(x,s)+\int_s^t A_{(s;x)}g(x,s;v)dv+F(x)(A_{(s;x)}\theta_{\geq t})(s)=0.$$

Here

$$\int_s^t A_{(s;x)}g(x,s;v)dv+F(x)(A_{(s;x)}\theta_{\geq t})(s)$$

$$=\int_s^t dv\int_0^{t-s}dr(g(x,s+r;v)-g(x,s;v))r^{-1-a(s,x)\alpha}$$
$$+(A_{(s;x)}\theta_{\geq t})(s)(F(x)-F(x,s)).$$

Consequently, in order to establish (57), it remains to show that

$$\begin{aligned}\int_s^t dv\int_0^{t-s}dr(g(x,s+r;v)-g(x,s;v))r^{-1-a(s,x)\alpha}\\ =\int_0^{t-s}dr(F(x,s+r)-F(x,s))r^{-1-a(s,x)\alpha}.\end{aligned} \tag{58}$$

Let $\chi_\epsilon(s,r)$ be positive functions that coincide with $r^{-1-a(s,x)\alpha}$ for $r\leq\epsilon$, do not exceed these functions everywhere and are continuous and bounded. Then, to prove (58), it is sufficient to show

$$\begin{aligned}\int_s^t dv\int_0^{t-s}dr(g(x,s+r;v)-g(x,s;v))\chi_\epsilon(s,r)\\ =\int_0^{t-s}dr(F(x,s+r)-F(x,s))\chi_\epsilon(s,r),\end{aligned} \tag{59}$$

for any ϵ. The second terms cancel, and this equation turns to

$$\int_s^t dv\int_0^{t-s}drg(x,s+r;v)\chi_\epsilon(s,r)=\int_0^{t-s}drF(x,s+r)\chi_\epsilon(s,r). \tag{60}$$

And this holds because the l.h.s. is

$$\int_0^{t-s}dr\int_{s+r}^t dv\,g(x,s+r;v)\chi_\epsilon(s,r)=\int_0^{t-s}dr\,F(x,s+r)\chi_\epsilon(s,r).$$

Acknowledgments The author is grateful to the organisers of the 50th Memorial Barrett Lectures in the University of Tennessee, May 2021, for inviting him to give a plenary talk, which strongly stimulated his further research in fractional calculus, and to the anonymous referees for useful comments on the initial draft of the manuscript.

The author is grateful to the Simons foundation for the support of his residence at INI in Cambridge during the program Fractional Differential Equations January April 2022 and to the Newton Institute for the hospitality in Cambridge during the program, where the work was essentially written.

The work was supported by the Russian Science Foundation under grant no. 20-11-20119.

References

1. D. Baleanu, K. Diethelm, E. Scalas and J.J. Trujillo. Fractional calculus: Models and numerical methods: Second edition. World Scientific Publishing, Singapore, 2017. Series on Complexity, Nonlinearity and Chaos, **5**.
2. W. Ding et al. Applications of distributed-order fractional operators: a review. Entropy 23 (2021), no. 1, Paper No. 110.
3. S. Fedotov, D. Han, A Zubarev, M. Johnston and V. J. Allan. Variable-order fractional master equation and clustering of particles: non-uniform lysosome distribution. Philos. Trans. Roy. Soc. A 379 (2021), no. 2205, Paper No. 20200317.
4. O. Kallenberg. Foundations of Modern Probability. Second ed., Springer 2002.
5. V. Kiryakova, *Generalized fractional calculus and applications*, (Longman Scientific, Harlow. Copublished in the United States with John Wiley and Sons, New York, 1994. Pitman Research Notes in Mathematics Series, **301**.
6. V.N. Kolokoltsov. Symmetric Stable Laws and Stable-like Jump-Diffusions. Proc. London Math. Soc. **3 (80)** (2000), 725–768.
7. V. N. Kolokoltsov. Generalized Continuous-Time Random Walks (CTRW), Subordination by Hitting Times and Fractional Dynamics. Theory Probab. Appl. **53** (4), 594–609 (2009).
8. V. N. Kolokoltsov. *Markov processes, semigroups and generators*. DeGruyter Studies in Mathematics v. 38, DeGruyter, 2011.
9. V. N. Kolokoltsov. The probabilistic point of view on the generalized fractional PDEs. FCAA 22:3 (2019), 543-600 (open access mode).
10. V. N. Kolokoltsov, *Differential equations on measures and functional spaces*, (Birkhäuser, Cham, 2019) Birkhäuser Advanced Texts Basler Lehrbücher.
11. V. N. Kolokoltsov and O. A. Malafeyev. Many Agent Games in Socio-economic Systems: Corruption, Inspection, Coalition Building, Network Growth, Security. Springer Series in Operations Research and Financial Engineering, Springer Nature, 2019.
12. V. Kolokoltsov and M. Troeva. A New Approach to Fractional Kinetic Evolutions. Fractal Fract. 2022, 6, 49. https://doi.org/10.3390/fractalfract6020049
13. M. M. Meerschaert, E. Nane and Y. Xiao. Correlated continuous time random walks. Statist. Probab. Lett. 79:9 (2009), 1194–1202.
14. I. Podlubny, *Fractional differential equations, An introduction to fractional derivatives, fractional differential equations, to methods of their solution and some of their applications* (Academic Press, Inc., San Diego, 1999), Mathematics in Science and Engineering **198**.
15. M. Savov and B. Toaldo. Semi-Markov processes, integro-differential equations and anomalous diffusion-aggregation. Ann. Inst. Henri Poincaré Probab. Stat. 56:4 (2020), 2640–2671.
16. Peter Straka. Variable Order Fractional Fokker-Planck Equations derived from Continuous Time Random Walks. Phys. A 503 (2018), 451–463.
17. H. Sun et al. A review on variable-order fractional differential equations: Mathematical foundations, physical models, numerical methods and applications. Fract. Calc. Appl. Anal. 2019, 22, 27–59.
18. G.M. Zaslavsky. Fractional kinetic equation for Hamiltonian chaos. Physica D 76 (1994), 110–122.

Fractional Elliptic Problems on Lipschitz Domains: Regularity and Approximation

Juan Pablo Borthagaray, Wenbo Li, and Ricardo H. Nochetto

Abstract This survey hinges on the interplay between regularity and approximation for linear and quasilinear fractional elliptic problems on Lipschitz domains. For the linear Dirichlet integral Laplacian, after briefly recalling Hölder regularity and applications, we discuss novel optimal shift theorems in Besov spaces and their Sobolev counterparts. These results extend to problems with finite horizon and are instrumental for the subsequent error analysis. Moreover, we dwell on extensions of Besov regularity to the fractional p-Laplacian and review the regularity of fractional minimal graphs and stickiness. We discretize these problems using continuous piecewise linear finite elements and derive global and local error estimates for linear problems, thereby improving some existing error estimates for both quasi-uniform and graded meshes. We also present a BPX preconditioner which turns out to be robust with respect to both the fractional order and the number of levels. We

JPB has been supported in part by Fondo Vaz Ferreira grant 2019-068.
WL has been supported in part by NSF grant DMS-2111228.
RHN has been supported in part by NSF grant DMS-1908267.

J. P. Borthagaray (✉)
Departamento de Matemática y Estadística del Litoral, Universidad de la República, Salto, Uruguay

Instituto de Matemática y Estadística, Facultad de Ingeniería, Universidad de la República, Montevideo, Uruguay
e-mail: jpborthagaray@fing.edu.uy

W. Li
Department of Mathematics, University of Tennessee, Knoxville, TN, USA
e-mail: wli50@utk.edu

R. H. Nochetto
Department of Mathematics and Institute for Physical Science and Technology, University of Maryland, College Park, MD, USA
e-mail: rhn@math.umd.edu

T. Mengesha, A. J. Salgado (eds.), *A^3N^2M: Approximation, Applications, and Analysis of Nonlocal, Nonlinear Models*, The IMA Volumes in Mathematics and its Applications 165, https://doi.org/10.1007/978-3-031-34089-5_2

conclude with the discretization of fractional quasilinear problems and their error analysis. We illustrate the theory with several illuminating numerical experiments.

Keywords Finite elements · Fractional diffusion · Quasilinear problems · Regularity · Rate of convergence

1 Introduction and Motivation

Let $s \in (0, 1)$ and $u \colon \mathbb{R}^d \to \mathbb{R}$ be a sufficiently smooth function. The *integral fractional Laplacian* of order s of u, which we will denote by $(-\Delta)^s$, is given by

$$(-\Delta)^s v(x) := C_{d,s} \, \text{p.v.} \int_{\mathbb{R}^d} \frac{v(x) - v(y)}{|x - y|^{d+2s}} \, dy, \quad C_{d,s} := \frac{2^{2s} s \Gamma(s + \frac{d}{2})}{\pi^{d/2} \Gamma(1 - s)}. \tag{1.1}$$

This is a nonlocal operator: evaluation of $(-\Delta)^s v$ at some point x involves a weighted and regularized average of the values of v over the whole space $\mathbb{R}^d$. Equivalently, the fractional Laplacian is the pseudodifferential operator with symbol $|\xi|^{2s}$: for sufficiently smooth $v \colon \mathbb{R}^d \to \mathbb{R}$, it holds that

$$\mathcal{F}((-\Delta)^s v)(\xi) = |\xi|^{2s} \mathcal{F} v(\xi) \quad \forall \xi \in \mathbb{R}^d.$$

This characterization makes apparent the fact that $(-\Delta)^s$ approaches the classical Laplacian (resp., identity operator) as $s \to 1$ (resp., $s \to 0$); we point out that the asymptotic behavior of the constant in (1.1) $C_{d,s} \simeq s(1 - s)$ is crucial for this to hold.

This work is a survey of theoretical and computational aspects of boundary value problems involving the fractional Laplacian (1.1) and related fractional-order operators on bounded domains. Our main goal is to emphasize the interplay between regularity and approximation with continuous piecewise linear finite element methods (FEMs). The presence of algebraic boundary layers in the solution of both linear and nonlinear fractional elliptic PDEs—and even of discontinuities in minimal graph problems (stickiness)—limits the convergence rates achievable by the FEMs discussed below. We shall not dwell on the *spectral* fractional Laplacian; we refer to the surveys [19, 45, 81] for a comparison between such an operator and the integral fractional Laplacian (1.1) and a review of other discretization techniques.

After a thorough discussion on regularity and approximation of linear equations, we turn to the regularity and numerical treatment of certain quasilinear problems. Concretely, we focus on nonlocal minimal graph problems and on the Dirichlet problem for the (p, s)-Laplacian. Further discussion on these and related nonlinear, nonlocal equations can be found in [25, 35, 102].

1.1 Examples

We present three key examples that constitute the basis for our discussion below.

Random Walk with Long Jumps As a first motivation for the operator (1.1), we follow [35]. Given $s \in (0, 1)$, we consider the probability mass function on the nonzero natural numbers

$$\mathrm{P}(k) := \frac{c_s}{|k|^{1+2s}}, \qquad c_s := \left(\sum_{k=1}^{\infty} \frac{1}{|k|^{1+2s}} \right)^{-1}.$$

We consider a particle moving in the space $\mathbb{R}^d$ according to a time- and space-discrete process. Namely, every $\tau > 0$ units of time (here, τ is a time step), the particle draws a direction $\mathbf{n} \in \mathbb{S}^{d-1}$ according to a uniform probability distribution and a positive integer number k according to P. Then, the particle takes a step $kh\mathbf{n}$ (here, $h > 0$ is a length scale): if at time t the particle is at a point $x \in \mathbb{R}^d$, then at time $t + \tau$ it jumps to $x + kh\mathbf{n} \in \mathbb{R}^d$; the step is discrete but x is continuous.

Let $u(x, t)$ be the probability density of the particle to be at the location $x \in \mathbb{R}^d$ at time $t > 0$. At time $t+\tau$, such a probability equals the sum of the probabilities of finding the particle somewhere else, say at $y = x - kh\mathbf{n} \in \mathbb{R}^d$, times the probability of jumping from y to x:

$$u(x, t + \tau) = \frac{c_s}{\omega_{d-1}} \sum_{k=1}^{\infty} \int_{\mathbb{S}^{d-1}} \frac{u(x - kh\mathbf{n}, t)}{|k|^{1+2s}} \, d\sigma(\mathbf{n}),$$

where $\omega_{d-1} = |\mathbb{S}^{d-1}|$. At this point, it is important that τ and h have the correct scaling, namely $\tau = h^{2s}$. Subtracting $u(x, t)$ on both sides of the identity above and dividing by $\tau = h^{2s}$, we obtain

$$\frac{u(x, t + \tau) - u(x, t)}{\tau} = \frac{c_s h}{\omega_{d-1}} \sum_{k=1}^{\infty} \int_{\mathbb{S}^{d-1}} \frac{u(x - kh\mathbf{n}, t) - u(x, t)}{|hk|^{1+2s}} \, d\sigma(\mathbf{n}).$$

Since the right-hand side is a Riemann sum corresponding to an integral in polar coordinates over $\mathbb{R}^d$, taking the limit $\tau \to 0$ (or, equivalently, $h \to 0$) we formally obtain

$$u_t(x, t) = \frac{c_s}{\omega_{d-1}} \int_{\mathbb{R}^d} \frac{u(y, t) - u(x, t)}{|y - x|^{d+2s}} \, dy = C(-\Delta)^s u(x, t).$$

We observe that a *fractional heat equation* arises as a formal limit of the random walk with jumps.

There are extensive reports in the literature of natural phenomena being successfully modeled by processes of this type. For instance, in biology, a "hit-and-run" hunting strategy consists in the following [75, 93, 99]: a predator moves according to the random walk we discussed above, searches prey in its surroundings, and then goes on. For non-destructive foraging, that is, assuming the prey is distributed in patches and being only temporarily depleted, reference [104] shows that, whenever no a priori information about the surroundings is available, the value $s = 1/2$ delivers an optimal searching strategy.

In contrast, for destructive foraging (namely, in the case the target resource is depleted once found), the optimal search pattern corresponds to the limit $s \to 0$. This is consistent with data gathered from in situ observations; reference [99] compares the behavior of diverse marine vertebrates and shows that the best fitting for the conduct of such species corresponds to values of s between 0.35 and 0.7.

From a mathematical viewpoint, [78] studies a model of two competing species that have the same population dynamics but different dispersal strategies: one species moves according to a classical random walk, while the other adopts a nonlocal dispersal strategy. Supported both by a local stability analysis and numerical simulations, the authors conjecture that nonlocal dispersal is always preferred over random dispersal.

In nature, it may be unrealistic to allow arbitrarily long jumps. Instead, there may well be a maximal jump length δ. For a given function $\phi\colon [0,\infty) \to [0,\infty)$ supported in the unit interval $[0,1]$, we can consider a *linear fractional diffusion operator with finite horizon* δ,

$$\mathcal{A}_{\delta,s} v(x) := \int_{\mathbb{R}^d} \phi\left(\frac{|x-y|}{\delta}\right) \frac{(v(x)-v(y))}{|x-y|^{d+2s}}\, dy. \tag{1.2}$$

This operator localizes the interactions built in (1.1) to a ball of radius δ centered at $x \in \Omega$. Formally, the operator above recovers (up to a constant) the fractional Laplacian in the limit $\delta \to \infty$. Operators of this type become *local* in either limit $s \to 1$ or $\delta \to 0$. The behavior of operators of this type in the limit $\delta \to 0$ (with a suitable scaling with respect to δ) is explored in [52, 101], for example.

Fractional Perimeters and Fractional Minimal Graphs Suppose we want to measure the perimeter of a region E by using an image of it. In particular, let us assume the figure Ω is composed of square pixels with length h and the region is a tilted square with sides at 45° with respect to the orientation of the pixels. Then, independently of the image resolution (namely, of the pixel sizes), the approximation of the perimeter of E by the perimeter of E_h always produces an error by a factor of $\sqrt{2}$, see Fig. 1.

As an alternative, one can consider the fractional s-perimeter of E in Ω (see Definition 3.1 below),

$$P_s(E;\Omega) := |\chi_E|_{\widetilde{W}_1^{2s}(\Omega)}. \tag{1.3}$$

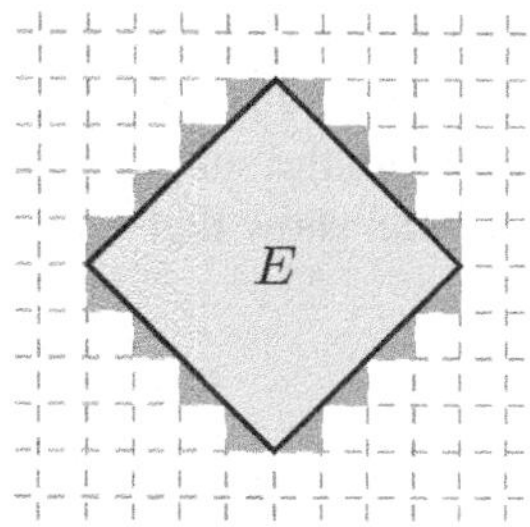

Fig. 1 Discrepancy between the perimeter of a tilted square E and its pixel representation. Independently of the grid size h, the perimeter of the dark gray approximation E_h is off by a factor of $\sqrt{2}$

Estimation of the s-perimeter of E in Ω by using E_h has an error of the order of h^{1-2s}. Moreover, because fractional s-perimeters recover the classical perimeter as $s \to 1/2$ (after a suitable normalization), they can provide a consistent mean to estimate the perimeter of E.

A classical result [88] states that classical minimal graphs arise as Γ-limits of certain Ginzburg–Landau energies. More precisely, given $\Omega \subset \mathbb{R}^d$ a bounded domain with Lipschitz boundary and $\varepsilon > 0$, we consider the double-well potential $W(t) = \frac{1}{2}(1-t^2)^2$ and the energy

$$E_{1,\varepsilon}[v;\Omega] = \frac{\varepsilon^2}{2}|v|^2_{H^1(\Omega)} + \int_\Omega W(v).$$

Then, for every sequence $\{u_\varepsilon\}$ of minimizers of the rescaled energy $\varepsilon^{-1}E_{1,\varepsilon}[\cdot;\Omega]$ with uniformly bounded energies, there exists a subsequence that converges in $L^1(\Omega)$ to $\chi_E - \chi_{E^c}$, where E is a set of minimal perimeter in Ω.

Instead, when one considers the following fractional energy:

$$E_{s,\varepsilon}[v;\Omega] = \frac{\varepsilon^{2s}}{2}\iint_{Q_\Omega} \frac{|v(x)-v(y)|^2}{|x-y|^{d+2s}}\,dydx + \int_\Omega W(v), \quad s \in (0,1),$$

defined in $Q_\Omega := (\mathbb{R}^d \times \mathbb{R}^d) \setminus (\Omega^c \times \Omega^c)$ with $\Omega^c := \mathbb{R}^d \setminus \overline{\Omega}$, and the Γ-limit of the corresponding rescaled energy $\gamma_\varepsilon E_{s,\varepsilon}[\cdot;\Omega]$ with scaling parameter

$$\gamma_\varepsilon = \begin{cases} \varepsilon^{-2s} & \text{if } s < 1/2, \\ \varepsilon^{-1}|\log\varepsilon|^{-1} & \text{if } s = 1/2, \\ \varepsilon^{-1} & \text{if } s > 1/2, \end{cases}$$

then a remarkable phenomenon arises [97]. Indeed, in the limit with $\varepsilon \to 0$, one also obtains convergence in $L^1(\Omega)$ to $\chi_E - \chi_{E^c}$; the set $E \subset \Omega$ has minimal perimeter if $s \geq 1/2$, but it is a minimizer of the s-perimeter $P_s(\cdot;\Omega)$ (cf. (1.3)) in case $s < 1/2$.

Let us now consider fractional minimal sets over cylinders of the form $\Omega \times \mathbb{R} \subset \mathbb{R}^{d+1}$ that, outside the cylinder, correspond to the subgraph of a given bounded function g. In that case, the solution of the Plateau problem of finding a set E that minimizes $P_s(\cdot; \Omega\times\mathbb{R})$ among all sets that coincide with the subgraph of g in $\Omega^c\times\mathbb{R}$ is in turn the subgraph of a function, say u. After some technical considerations [83], one can show that such a function u can be found by minimizing the energy

$$I_s[v] = \iint_{Q_\Omega} F_s\left(\frac{v(x)-v(y)}{|x-y|}\right)\frac{1}{|x-y|^{d+2s-1}}\,dydx,$$

where $F_s\colon \mathbb{R}\to\mathbb{R}$ is a suitable convex and nonnegative function; see (3.1.2) below. This energy is a fractional-order analogue of the classical graph area functional

$$I_1[v] = \int_\Omega \sqrt{1+|\nabla v|^2}.$$

The Euler–Lagrange equation corresponding to a minimizer of I_s turns out to be quasilinear and elliptic, although not *uniformly* elliptic.

Quasilinear Operators The third class of fractional operators is given by energies of the form

$$\mathcal{F}(u) = \iint_{Q_\Omega} G\left(x, y, \frac{u(x)-u(y)}{|x-y|^s}\right)\frac{1}{|x-y|^d}\,dydx - \int_\Omega fu,$$

where the function $G\colon \mathbb{R}^d\times\mathbb{R}^d\times\mathbb{R}\to\mathbb{R}$ is required to satisfy suitable assumptions described in detail in [23]. For the sake of clarity, in this chapter, we shall restrict our attention to the case $G(x,y,\rho) = \frac{C_{d,s,p}}{2p}|\rho|^p$ with $1<p<\infty$ and $C_{d,s,p}$ a suitable normalization constant; see (3.3) below. The resulting energy is $\frac{1}{p}|v|^p_{\widetilde{W}^s_p(\Omega)}$ and gives rise to the *fractional* (p,s)*-Laplacian operator*:

$$(-\Delta)^s_p u(x) := C_{d,s,p}\int_{\mathbb{R}^d}\frac{|u(x)-u(y)|^{p-2}(u(x)-u(y))}{|x-y|^{d+sp}}\,dy.$$

This generalizes (1.1) to $1<p<\infty$. We can rewrite the operator above as

$$(-\Delta)^s_p u(x) = C_{d,s,p}\int_{\mathbb{R}^d}\left(\frac{|u(x)-u(y)|}{|x-y|^s}\right)^{p-2}\frac{(u(x)-u(y))}{|x-y|^{d+2s}}\,dy.$$

which suggests that, heuristically, one can understand the fractional (p,s)-Laplacian as a weighted fractional Laplacian of order s, with a weight $\left(\frac{|u(x)-u(y)|}{|x-y|^s}\right)^{p-2}$. This is analogous to the local case, for which the p-Laplacian $(-\Delta)_p u = \mathrm{div}(|\nabla u|^{p-2}\nabla u)$ can be regarded as a Laplacian with weight $|\nabla u|^{p-2}$. The Dirichlet problem for the local p-Laplacian arises in a number of models of physical processes, including non-Newtonian fluids [12] and turbulent flows in porous media [46].

2 Linear Problems

This section deals with the homogeneous Dirichlet problem for the fractional Laplacian (1.1). We shall make use of fractional-order Sobolev spaces and Besov spaces. We employ the notation from [27] and refer to that work for elementary properties of these spaces.

Given $f \in H^{-s}(\Omega)$, we set $\Omega^c := \mathbb{R}^d \setminus \overline{\Omega}$ and look for a function $u \in \widetilde{H}^s(\Omega)$, the space of functions in $H^s(\mathbb{R}^d)$ that vanish in Ω^c, such that

$$\begin{cases} (-\Delta)^s u = f & \text{in } \Omega, \\ u = 0 & \text{in } \Omega^c. \end{cases} \tag{2.1}$$

To formulate (2.1) weakly, we introduce the following inner product in $\widetilde{H}^s(\Omega)$:

$$(v, w)_s := \frac{C_{d,s}}{2} \iint_{Q_\Omega} \frac{(v(x) - v(y))(w(x) - w(y))}{|x - y|^{d+2s}} \, dx \, dy, \quad \|v\|_{\widetilde{H}^s(\Omega)} := (v, v)_s^{1/2}, \tag{2.2}$$

and recall the notation $Q_\Omega = (\mathbb{R}^d \times \mathbb{R}^d) \setminus (\Omega^c \times \Omega^c)$. We denote by $\langle \cdot, \cdot \rangle$ the duality pairing between $\widetilde{H}^s(\Omega)$ and its dual $H^{-s}(\Omega)$. The constant $C_{d,s}$, defined in (1.1), enforces consistency between (2.2) and the definition of fractional Laplacian via Fourier transform

$$(v, w)_s = \langle (-\Delta)^s v, w \rangle = \int_{\mathbb{R}^d} |\xi|^{2s} \mathcal{F}[v](\xi) \overline{\mathcal{F}[w]}(\xi) d\xi.$$

The weak formulation of (2.1) reads: find $u \in \widetilde{H}^s(\Omega)$ such that

$$(u, v)_s = \langle f, v \rangle \quad \forall v \in \widetilde{H}^s(\Omega). \tag{2.3}$$

Moreover, the following Poincaré inequality is valid with a constant $C(\Omega)$ uniform in $s \in (0, 1)$:

$$\|v\|_{L^2(\Omega)} \leq C(\Omega) \|v\|_{\widetilde{H}^s(\Omega)} \quad \forall v \in \widetilde{H}^s(\Omega) \tag{2.4}$$

and implies that $(\cdot, \cdot)_s$ is indeed an inner product. By the Lax–Milgram theorem, one immediately deduces that for all $f \in H^{-s}(\Omega)$ there exists a unique $u \in \widetilde{H}^s(\Omega)$ satisfying (2.3) and that the solution map is continuous,

$$\|u\|_{\widetilde{H}^s(\Omega)} \leq \|f\|_{H^{-s}(\Omega)}. \tag{2.5}$$

2.1 Regularity

We now review some results regarding regularity of solutions to problem (2.3). It is natural to expect that if the right-hand side f is smoother than $H^{-s}(\Omega)$, then some additional regularity is inherited by the weak solution u, which a priori is only known to belong to $\widetilde{H}^s(\Omega)$.

By using the definition of $(-\Delta)^s$ as a pseudodifferential operator of order $2s$ and the characterization of fractional Sobolev spaces as Bessel potential spaces for $p = 2$, it is clear that if $u \in L^2(\mathbb{R}^d)$ is such that $(-\Delta)^s u \in H^t(\mathbb{R}^d)$, then $u \in H^{t+2s}(\mathbb{R}^d)$. A much less trivial question is whether this result can be used to derive interior regularity estimates for problems posed on a bounded domain. In this regard, we point out interesting estimates in [17, 42, 61].

Let us next discuss regularity of solutions to (2.1) up to the boundary of the domain Ω. For problems posed on one-dimensional domains, one can take advantage of explicit expression of $\mathrm{dist}(x, \partial\Omega)$ to obtain sharp regularity results for linear operators. Indeed, for the fractional Laplacian (1.1), a certain kind of Jacobi polynomials, the *Gegenbauer polynomials*, play a crucial role in the analysis of (2.1) [5, 57, 84]. On the interval $[-1, 1]$, these polynomials are L^2-orthogonal with respect to the weight $\omega(x) = (1 - x^2)^s$. Special polynomials are also a crucial tool in the analysis of problems with advection and/or reaction terms [56, 73] and are instrumental in the development of spectral methods for these problems in 1D.

For problems based on domains in $\mathbb{R}^d$ with $d \geq 1$, techniques based on Fourier analysis, such as the ones employed by Višik and Èskin [58, 103] or Grubb [71], allow for a full characterization of mapping properties of the integral fractional Laplacian of functions supported in Ω. However, such arguments typically require the domain Ω to be C^∞; the recent work by Abels and Grubb [2] introduces a method to handle nonsmooth coordinate changes that leads to regularity results for domains with $C^{1+\beta}$ boundary with $\beta > 2s$. The maximal regularity reads [2, 58, 71, 103]

$$u \in H^{s+\frac{1}{2}-\varepsilon}(\Omega), \tag{2.6}$$

for any $0 < \varepsilon < s+\frac{1}{2}$ regardless of the regularity of f. The following exact solution u of (2.1) in $\Omega = B(0, 1) \subset \mathbb{R}^d$ with $f \equiv 1$ exhibits this extreme behavior:

$$u(x) = C_s(1 - |x|^2)^s_+, \tag{2.7}$$

where $t_+ = \max\{t, 0\}$ and $C_s = \frac{\Gamma(\frac{d}{2})}{2^{2s}\Gamma(\frac{d+2s}{2})\Gamma(1+s)}$; in fact, $u \notin H^{s+\frac{1}{2}}(\Omega)$, see Example 2.1. Therefore, it stems from this example that the boundary behavior of the solution u to (2.1) is expected to be

$$u(x) \approx d(x, \partial\Omega)^s, \tag{2.8}$$

for every x close to $\partial\Omega$ irrespective of the regularity of $\partial\Omega$. This is in striking contrast with the boundary behavior of the classical Laplacian that inherits the regularity of $\partial\Omega$.

2.1.1 Hölder Regularity

Using potential theory tools, references [1, 94] studied problem (2.1) and related problems for integral operators with translation-invariant kernels. These references obtain a fine characterization of boundary Hölder regularity of solutions. More precisely, [94, Proposition 1.1] proves that the solution u satisfies

$$\|u\|_{C^s(\mathbb{R}^d)} \leq C(\Omega)\|f\|_{L^\infty(\Omega)}, \tag{2.9}$$

provided Ω is a Lipschitz domain that satisfies an *exterior ball condition* property. This estimate is also a consequence of [1, Theorem 1.4] in case Ω is of class $C^{1,\beta}$ and $f \in C^{\beta-s}(\overline{\Omega})$ for $s < \beta < 1$. Estimate (2.9) does not seem to be valid uniformly in s for polygonal domains in 2D with reentrant corners [62, 67]. Once the s-Hölder continuity (2.9) of u is established, references [1, 94] derive higher order estimates in case f possesses certain additional regularity. In fact, if

$$\delta(x, y) := \min\left\{\operatorname{dist}(x, \partial\Omega), \operatorname{dist}(y, \partial\Omega)\right\} \quad \forall x, y \in \overline{\Omega},$$

then a typical weighted estimate from [1, 94] reads

$$\sup_{x,y\in\overline{\Omega}} \left\{\delta(x, y)^{\beta+s} \frac{|u(x) - u(y)|}{|x - y|^{\beta+2s}}\right\} \leq C(f, u), \tag{2.10}$$

where $\beta > 0$ and $C(f, u)$ depends on suitable regularity of f and (2.9).

Estimates such as (2.10) are not immediately useful in the analysis of finite element methods, which rely on Sobolev estimates. To capture the boundary behavior of u in terms of Sobolev norms, [4] converts estimates such as (2.9) into (weighted) Sobolev estimates for u in terms of Hölder norms of f. To make this precise, given $\ell = k + s$, with $k \in \mathbb{N}$ and $s \in (0, 1)$, and $\kappa \geq 0$, we define the weighted norm

$$\|v\|^2_{H^\ell_\kappa(\Omega)} := \|v\|^2_{H^k(\Omega)} + \sum_{|\beta|=k} \iint_{\Omega\times\Omega} \frac{|D^\beta v(x) - D^\beta v(y)|^2}{|x - y|^{d+2s}} \delta(x, y)^{2\kappa} \, dy \, dx$$

and the associated weighted space

$$H^\ell_\kappa(\Omega) := \left\{v \in H^\ell(\Omega) : \|v\|_{H^\ell_\kappa(\Omega)} < \infty\right\}. \tag{2.11}$$

The regularity estimate in the weighted Sobolev scale (2.11) for $d = 2$ reads as follows [4].

Theorem 2.1 (Weighted Sobolev Estimate) *Let Ω be a bounded Lipschitz domain in $\mathbb{R}^2$, let the solution u of (2.3) satisfy (2.9), and $f \in C^{1-s}(\overline{\Omega})$. Then, for all $\varepsilon > 0$ sufficiently small, $u \in \widetilde{H}^{1+s-2\varepsilon}_{1/2-\varepsilon}(\Omega)$ satisfies the estimate*

$$\|u\|_{\widetilde{H}^{1+s-2\varepsilon}_{1/2-\varepsilon}(\Omega)} \leq \frac{C(\Omega, s)}{\varepsilon} \|f\|_{C^{1-s}(\overline{\Omega})}.$$

Compared with (2.6), we observe a gain of about $1/2$-order differentiability at the expense of about $1/2$-order weight.

We explore next an alternative to L^2-based weighted estimates that consists of reducing the integrability index [28]. We consider $\Omega = (0, \infty)$ to be the half line in 1D and interpret the behavior (2.8) as $u(x) = x_+^s$. We then wonder under what conditions this function belongs to a Sobolev space with differentiability index r and integrability index p. For that purpose, let us compute (Riemann–Liouville) derivatives of order $r > 0$ of the function $v : \mathbb{R} \to \mathbb{R}$, $v(x) = x_+^s$:

$$\partial^r v(x) \simeq x_+^{s-r}, \quad r > 0. \tag{2.12}$$

We realize that $\partial^r v$ is L^p-integrable near $x = 0$ if and only if $p(s - r) > -1$, namely, if $r < s + \frac{1}{p}$. This heuristic discussion illustrates the natural interplay between the differentiability order r and integrability index p for membership of solutions to (2.1) in the class W^r_p, at least for dimension $d = 1$. It turns out that the restriction $r < s + \frac{1}{p}$ is needed irrespective of d [28, Theorem 3.7].

To figure out the optimal choice of indices r, p compatible with nonlinear approximation theory for $d = 2$, we inspect the DeVore diagram; see Fig. 2. Recall the definition of Sobolev number $\text{Sob}(W^r_p) := r - \frac{d}{p}$ and the Sobolev line corresponding to the nonlinear approximation scale of H^s,

$$\left\{\left(\frac{1}{p}, r\right) : \text{Sob}(W^r_p) = \text{Sob}(H^s)\right\} = \left\{\left(\frac{1}{p}, r\right) : r = s - 1 + \frac{2}{p}\right\}.$$

In order to have a compact embedding $W^r_p(\Omega) \subset H^s(\Omega)$, we require that $\text{Sob}(W^r_p) > \text{Sob}(H^s)$ or equivalently that $(1/p, r)$ lies above the Sobolev line, as well as $r > s$. In addition, this line intersects the regularity line $r = s + \frac{1}{p}$ at $(1, 1 + s)$, which is not an admissible pair (see Fig. 2). Letting $p = 1 + \varepsilon$ and $r = s + 1 - \varepsilon$ with $\varepsilon > 0$ arbitrarily small, we have

$$\text{Sob}(W^{s+1-\varepsilon}_{1+\varepsilon}) = s - 1 + \varepsilon\left(\frac{1-\varepsilon}{1+\varepsilon}\right) > \text{Sob}(H^s),$$

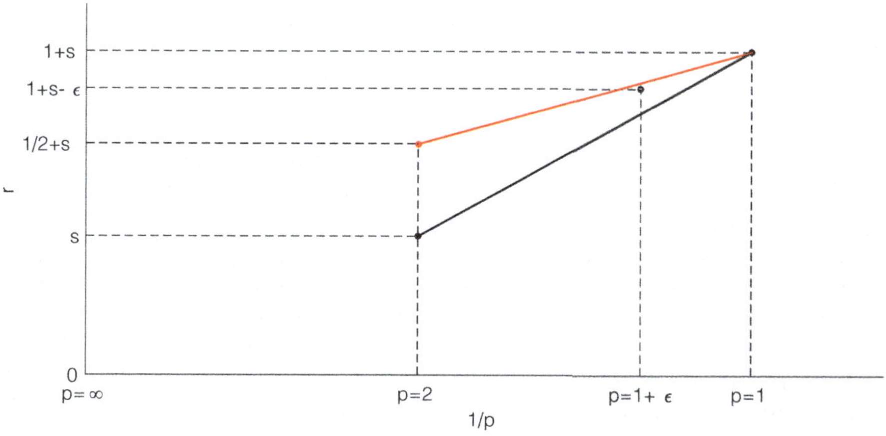

Fig. 2 DeVore diagram for $d = 2$. Solutions to (2.1) are of class W_p^r, where $(1/p, r)$ must be below the regularity line $\{(1/p, r) \colon r = s + \frac{1}{p}\}$ (red). The Sobolev line $\{(1/p, r) \colon r = s - 1 + \frac{2}{p}\}$ (black) connects the spaces W_p^r and H^s and corresponds to the nonlinear approximation scale of H^s. Both lines intersect at $(1, 1 + s)$

while $\frac{1}{p} = \frac{1}{1+\varepsilon} > 1 - \varepsilon = r - s$ is also satisfied. This is thus an optimal choice of parameters for $d = 2$. The Hölder regularity of [1, 94] leads to the following Sobolev estimate [28, Corollary 3.8].

Theorem 2.2 (Differentiability vs Integrability) *Let Ω be a bounded Lipschitz domain in $\mathbb{R}^2$, let the solution u of (2.3) satisfy (2.9), and $f \in C^{1-s}(\overline{\Omega})$. Then, for ε sufficiently small, the function $u \in \widetilde{W}_{1+\varepsilon}^{s+1-\varepsilon}(\Omega)$ satisfies the a priori bound*

$$\|u\|_{\widetilde{W}_{1+\varepsilon}^{s+1-\varepsilon}(\Omega)} \le \frac{C(\Omega, s)}{\varepsilon^3} \|f\|_{C^{1-s}(\overline{\Omega})}. \tag{2.13}$$

We recall that (2.9) hinges on the exterior ball condition [1, 94]. In finite element applications, where Ω is usually a polygon for $d = 2$, such a condition implies Ω must be convex, which is too restrictive in practice. We next discuss regularity estimates that solely require Ω to be Lipschitz.

2.1.2 Besov Regularity

We follow [23, 27], which are in turn inspired in [96] for second-order linear elliptic problems. This is a functional analysis approach to regularity and starts with the observation that the solution u of (2.1) is the minimizer of the quadratic functional $v \mapsto \mathcal{F}_2(v) - \mathcal{F}_1(v)$, where

$$\mathcal{F}_1(v) := \int_\Omega f v, \quad \mathcal{F}_2(v) := \frac{1}{2}\|v\|_{\widetilde{H}^s(\Omega)}^2 \quad \forall\, v \in \widetilde{H}^s(\Omega),$$

whence it satisfies the stationarity condition

$$\frac{1}{2}\|u-v\|^2_{\widetilde{H}^s(\Omega)} = \big(\mathcal{F}_2(v)-\mathcal{F}_2(u)\big)-\big(\mathcal{F}_1(v)-\mathcal{F}_1(u)\big) \quad \forall\, v\in \widetilde{H}^s(\Omega). \tag{2.14}$$

If we now think of v as a suitable translation of u, then (2.14) reveals that further regularity of u beyond $\widetilde{H}^s(\Omega)$ could be inferred from regularity of the functionals $\mathcal{F}_1, \mathcal{F}_2$. However, to carry out this program, we face two important difficulties: first, we need to localize the translations because global translations are not admissible for bounded domains and second instead of all possible directions. we need to deal with a convex cone of directions dictated by the Lipschitz regularity of $\partial\Omega$. We briefly describe both issues now and refer to [27, 96] for details.

We start with the definition of Besov spaces by real interpolation. Given a pair of compatible Banach spaces (X_0, X_1), $u \in X_0 + X_1$, and $t > 0$, we set the K-functional [86]

$$K(t,u) := \inf\Big\{\big(\|u_0\|^2_{X_0} + t^2\|u_1\|^2_{X_1}\big)^{1/2} : u = u_0 + u_1,\ u_0 \in X_0,\ u_1 \in X_1\Big\}. \tag{2.15}$$

For $\theta \in (0,1)$ and $q \in [1,\infty]$, let us define interpolation spaces

$$\big[X_0, X_1\big]_{\theta,q} := \{u \in X_0 + X_1 \colon \|u\|_{(X_0,X_1)_{\theta,q}} < \infty\},$$

where

$$\|u\|_{[X_0,X_1]_{\theta,q}} = \begin{cases} \Big[q\theta(1-\theta)\int_0^\infty t^{-(1+\theta q)}|K(t,u)|^q\,dt\Big]^{1/q} & \text{if } 1 \le q < \infty,\\ \sup_{t>0} t^{-\theta}|K(t,u)| & \text{if } q=\infty. \end{cases} \tag{2.16}$$

The normalization factor $q\theta(1-\theta)$ in the norm (2.16) guarantees the correct scalings in the limits $\theta \to 0$, $\theta \to 1$, and $q \to \infty$ for norm continuity; see [86, Appendix B]. In our setting, we let $X_0 := L^2(\Omega)$, $X_1 := H^2(\Omega)$, $\sigma \in (0,2)$, and $q \in [1,\infty]$ to obtain

$$B^\sigma_{2,q}(\Omega) := \big[L^2(\Omega), H^2(\Omega)\big]_{\sigma/2,q}, \quad \dot{B}^\sigma_{2,q}(\Omega) := \{v \in B^\sigma_{2,q}(\Omega) : \operatorname{supp} v \subset \overline{\Omega}\}.$$

In particular, we have that $B^\sigma_{2,2}(\Omega) = H^\sigma(\Omega)$ for all $\sigma \in (0,2)$. If $\sigma \in (0,1)$, then we can also set $\dot{B}^\sigma_{2,q}(\Omega) := [L^2(\Omega), H^1_0(\Omega)]_{\sigma,q}$ and point out that the spaces coincide but the corresponding norms exhibit a different scaling as $\sigma \to 1$; we will quantify this discrepancy in (2.19) below. Besov spaces can also be characterized in terms of first and second difference operators $\delta_1(h)$ and $\delta_2(h)$:

$$\delta_1(h)v(x) := v(x+h) - v(x), \qquad \delta_2(h)v(x) := v(x+h) + v(x-h) - 2v(x),$$

for $v \in L^2(\Omega)$ and $x \in \Omega_{|h|} := \{y \in \Omega : d(y, \partial\Omega) > |h|\}$. In fact, if $D = D_\rho(0)$ is the ball in $\mathbb{R}^d$ of radius ρ centered at the origin and $\sigma \in (0, 2)$, we define the seminorms for $q \in [1, \infty)$

$$|v|_{B^\sigma_{2,q}(\Omega)} := \left(q\sigma(2-\sigma) \int_D \frac{\|\delta_2(h)v\|^q_{L^2(\Omega_{|h|})}}{|h|^{d+q\sigma}} dh, \right)^{1/q}, \tag{2.17}$$

and $q = \infty$

$$|v|_{B^\sigma_{2,\infty}(\Omega)} := \sup_{h\in D} \frac{\|\delta_2(h)v\|_{L^2(\Omega_{|h|})}}{|h|^\sigma} \tag{2.18}$$

and observe that $\|v\|_{L^2(\Omega)} + |v|_{B^\sigma_{2,q}(\Omega)}$ is equivalent to the norm induced by interpolation and is robust with respect to σ, q, and ρ [6, Theorem 7.47]. It turns out that $q = \infty$ is the most significant index for us, whence we focus on it and state the *Marchaud inequality* for $\sigma \in (0, 1)$ (cf. [51])

$$\sup_{h\in D} \frac{\|\delta_1(h)v\|_{L^2(\Omega_{|h|})}}{|h|^\sigma} \lesssim \|v\|_{L^2(\Omega)} + \frac{1}{\sqrt{1-\sigma}} \sup_{h\in D} \frac{\|\delta_2(h)v\|_{L^2(\Omega_{|h|})}}{|h|^\sigma} \qquad \forall v \in \dot{B}^\sigma_{2,\infty}(\Omega), \tag{2.19}$$

which quantifies the precise blow-up on first differences relative to second differences as $\sigma \to 1$. This behavior is sharp as we discuss next with an explicit example.

Example 2.1 (Regularity of Explicit Solution) Our goal now is to justify that $q = \infty$ is the most adequate choice in the present setting. To this end, we examine first and second differences for the 1D function $v(x) = x_+^s$ for $x \in [0, 1]$, $v(x) = 0$ for $x \in (-1, 0)$, and $v(x) = 1$ for $x \in (1, 2)$, which possesses a profile at $x = 0$ consistent with (2.7) and (2.8). For $h \in (0, \rho)$, we decompose $\|\delta_1(h)v\|^2_{L^2(\Omega_h)}$ into integrals over $(-h, 0)$, $(0, h)$, $(h, 1-h)$, and $(1-h, 1)$ which are all of order h^{1+2s} for $s \in (0, 1/2)$. To estimate the most delicate integral over the interval $(h, 1-h)$, we use a dyadic partition $(2^k h, 2^{k+1}h)$ with $k \le K < \frac{|\log h|}{\log 2}$

$$\int_h^{1-h} \left[(x+h)^s - x^s\right]^2 dx \approx \sum_{k=0}^{K} \int_{2^k h}^{2^{k+1}h} x^{2s}\left[(1+2^{-k})^s - 1\right]^2 dx$$

$$\approx s^2 h^{1+2s} \sum_{k=0}^{K} 2^{k(2s-1)} \approx (1-2s)^{-1} h^{1+2s}.$$

This implies

$$\int_0^\rho \frac{\|\delta_1(h)v\|^q_{L^2(\Omega_h)}}{h^{1+q(s+1/2)}} dh \approx \int_0^\rho \frac{dh}{h} = \infty, \tag{2.20}$$

for any $q \in [1, \infty)$, whereas for $q = \infty$

$$\sup_{h \in (0,\rho)} \frac{\|\delta_1(h)v\|_{L^2(\Omega_h)}}{h^{s+1/2}} \approx \frac{1}{\sqrt{1-2s}},$$

which shows a blow-up as $s \to \frac{1}{2}$. Moreover, if $s = 1/2$, the calculation above becomes

$$\begin{aligned} \|\delta_1(h)v\|^2_{L^2(\Omega_h)} &\approx \sum_{k=0}^{K} \int_{2^k h}^{2^{k+1}h} x\big[(1+2^{-k})^{\frac{1}{2}} - 1\big]^2 dx \\ &\approx h^2 \sum_{k=1}^{K} 1 = h^2 K \approx h^2 |\log h|. \end{aligned}$$

This shows that first differences are not adequate for $s = \frac{1}{2}$. In contrast, utilizing second differences yields a critical term of the form

$$\int_h^{1-h} \big[(x+h)^s + (x-h)^s - 2x^s\big]^2 dx \approx h^{1+2s} \quad \Rightarrow \quad \sup_{h \in (0,\rho)} \frac{\|\delta_2(h)v\|_{L^2(\Omega_h)}}{h^{s+1/2}} \approx 1$$

for all $s \in (0, 1/2]$ with a uniform constant as $s \to \frac{1}{2}$. This derivation reveals that the constant $(1-\sigma)^{-\frac{1}{2}}$ in (2.19) is sharp. Moreover, if $s \in (1/2, 1)$, then $v \in H^1(\Omega)$, whence computing the first weak derivative $v'(x) = s x_+^{s-1}$ and repeating the previous calculation yield

$$\|\delta_1(h)v'\|_{L^2(\Omega_h)} \approx \frac{h^{s-1/2}}{\sqrt{2s-1}}, \qquad \|\delta_2(h)v'\|_{L^2(\Omega_h)} \approx h^{s-\frac{1}{2}}.$$

In view of our definition (2.18) involving second differences, we deduce that $v(x) = x_+^s$ satisfies

$$v \in B^{s+\frac{1}{2}}_{2,\infty}(\Omega) \quad \forall s \in (0,1)$$

and motivates the use of the Besov space $\dot{B}^{s+\frac{1}{2}}_{2,\infty}(\Omega)$ in our regularity theory for Lipschitz domains. In contrast, (2.20) implies that $v \notin B^{s+\frac{1}{2}}_{2,q}(\Omega)$ for any $1 \le q < \infty$, and in particular $v \notin H^{s+\frac{1}{2}}(\Omega)$.

In light of Example 2.1, we next discuss the key steps leading to the *optimal shift property*

$$\|u\|_{\dot{B}^{s+1/2}_{2,\infty}(\Omega)} \le C(\Omega, d)\|f\|_{B^{-s+1/2}_{2,1}(\Omega)}.$$

To guarantee that local translations of u are admissible test functions, we need to restrict the set of *admissible directions* h to a convex cone related to the local Lipschitz structure of $\partial\Omega$: there exist $\rho > 0$, $\theta \in (0, \pi]$, and a map $\mathbf{n} : \Omega \to \mathbb{S}^{d-1}$ such that for all $x \in \Omega$ the cone $\mathcal{C}_\rho(\mathbf{n}(x), \theta)$ with height ρ, aperture θ, apex 0, and axis $\mathbf{n}(x)$ gives admissible outward vectors in the sense that

$$h \in \mathcal{C}_\rho(\mathbf{n}(x), \theta) \quad \Rightarrow \quad D_{3\rho}(x) \cap \Omega^c + th \subset \Omega^c \quad \forall t \in [0, 1]. \tag{2.21}$$

Such cone $D = \mathcal{C}_\rho(\mathbf{n}(x), \theta)$ has the property that generates $\mathbb{R}^d$ in the sense that for all $h \in D_\rho(0)$, there exist $\{h_j\}_{j=1}^d \subset D \cup (-D)$ and a constant $c > 0$ only depending on θ such that

$$h = \sum_{j=1}^d h_j, \qquad \sum_{j=1}^d |h_j| \le c|h|.$$

It thus follows that restricting the ball $D_\rho(0)$ to the cone D in the definitions of Besov seminorms (2.17) and (2.18) yields equivalent seminorms [27, Proposition 2.2].

Given $x_0 \in \Omega$, let $\phi \in C_0^\infty(D_{2\rho}(x_0))$ be a cut-off function satisfying $0 \le \phi(x) \le 1$ for all $x \in D_{2\rho}(x_0)$ and $\phi(x) = 1$ for all $x \in D_\rho(x_0)$. Given an admissible direction $h \in D = \mathcal{C}_\rho(\mathbf{n}(x_0), \theta)$, we define the *localized translation operator* $T_h : \widetilde{H}^s(\Omega) \to \widetilde{H}^s(\Omega)$ to be

$$T_h v(x) := v\big(x + h\phi(x)\big) \quad \forall x \in \Omega. \tag{2.22}$$

The operator T_h translates v along the direction $h \in D$ and coincides with the identity in $D_{2\rho}(x_0)^c$. Moreover, if $v \in \widetilde{H}^s(\Omega)$, then $T_h v \in \widetilde{H}^s(\Omega)$ is an admissible test function to insert in (2.14) and probe the behavior of the functionals $\mathcal{F}_1$ and $\mathcal{F}_2$. Combining a reiteration property of Besov seminorms (cf. [6, Theorem 7.21],[27, Proposition 2.1]) with (2.14) yields for $\sigma \in (0, 2)$

$$\begin{aligned}
|u|^2_{B^{s+\sigma/2}_{2,\infty}(D_\rho(x_0))} &\lesssim \sup_{h\in D} \frac{|\delta_1(h)u|^2_{H^s(D_\rho(x_0))}}{|h|^\sigma} \\
&\lesssim \sup_{h\in D} \frac{\|T_h u - u\|^2_{\widetilde{H}^s(\Omega)}}{|h|^\sigma} \\
&\lesssim \sup_{h\in D} \frac{|\mathcal{F}_1(T_h u) - \mathcal{F}_1(u)|}{|h|^\sigma} + \sup_{h\in D} \frac{|\mathcal{F}_2(T_h u) - \mathcal{F}_2(u)|}{|h|^\sigma}
\end{aligned}$$

and shows that bounding the right-hand side induces further local regularity of u. The following estimates for $\mathcal{F}_1$ and $\mathcal{F}_2$ are derived in [23]: for $\sigma \in (0,1]$ and $t \in (-1, 1+\sigma)$, we have

$$\sup_{h\in D} \frac{|\mathcal{F}_1(T_h v) - \mathcal{F}_1(v)|}{|h|^\sigma} \lesssim \|f\|_{B^t_{2,1}(D_{2\rho}(x_0)\cap\Omega)} |v|_{B^{\sigma-t}_{2,\infty}(D_{2\rho}(x_0))} \qquad \forall v \in \dot{B}^{\sigma-t}_{2,\infty}(D_{2\rho}(x_0)); \tag{2.23}$$

$$\sup_{h\in D} \frac{|\mathcal{F}_2(T_h v) - \mathcal{F}_2(v)|}{|h|^\sigma} \lesssim \int\int_{Q_{D_{2\rho}(x_0)}} \frac{|v(x)-v(y)|^2}{|x-y|^{d+2s}} dydx \qquad \forall v \in \widetilde{H}^s(\Omega). \tag{2.24}$$

These estimates improve upon [27, Propositions 3.1 and 3.2] due to the special structure of the translation operator T_h in (2.22). We illustrate this point with the first estimate, which is the trickiest and most insightful. We first observe that the estimate

$$\|v - T_h v\|_{B^r_{2,q}(D_{2\rho}(x_0))} \lesssim |h|^\sigma \|v\|_{B^{r+\sigma}_{2,q}(D_{2\rho}(x_0))} \quad \forall\, v \in B^{r+\sigma}_{2,q}(D_{2\rho}(x_0)) \tag{2.25}$$

is valid for $r \in (-1,1)$ with constants independent of $\sigma \in [0,1]$ and $q \in [1,\infty]$, with possible blow up as $r \to \pm 1$; we will come back to this below. Applying (2.25) for $q = \infty$ readily implies

$$\big|\mathcal{F}_1(T_h v) - \mathcal{F}_1(v)\big| \lesssim |h|^\sigma \|f\|_{B^{-r}_{2,1}(\Omega\cap D_{2\rho}(x_0))} \|v\|_{B^{r+\sigma}_{2,\infty}(D_{2\rho}(x_0))}.$$

We next rewrite $T_h v$ in (2.22) as $T_h v = v \circ S_h$ with $S_h = I + h\phi$, whence $\int_\Omega f T_h v = \int_{S_h(\Omega)} f \circ S_h^{-1} v|J|$ with $J = \det \nabla S_h^{-1}$. Exploiting this along with the fact that S_h is close to the identity, and using (2.25) with $q = 1$, yields

$$\big|\mathcal{F}_1(T_h v) - \mathcal{F}_1(v)\big| \lesssim |h|^\sigma \|f\|_{B^{r+\sigma}_{2,1}(\Omega\cap D_{2\rho}(x_0))} \|v\|_{B^{-r}_{2,\infty}(D_{2\rho}(x_0))}.$$

Since the map $(v, f) \mapsto \mathcal{F}_1(T_h v) - \mathcal{F}_1(v)$ is bilinear, operator interpolation theory gives the asserted estimate (2.23). Consequently, by this estimate and (2.24), we arrive at the local estimate for a generic point $x_0 \in \Omega$,

$$\begin{aligned} |u|^2_{B^{s+\sigma/2}_{2,\infty}(D_\rho(x_0))} &\lesssim \|f\|_{B^t_{2,1}(D_{2\rho}(x_0)\cap\Omega)} |u|_{B^{\sigma-t}_{2,\infty}(D_{2\rho}(x_0))} \\ &\quad + \iint_{Q_{D_{2\rho}(x_0)}} \frac{|u(x)-u(y)|^2}{|x-y|^{d+2s}} dydx. \end{aligned}$$

To apply this estimate, we consider a finite covering of $\{x \in \mathbb{R}^d : \operatorname{dist}(x,\Omega) < \rho/2\}$ with balls $D_j = D(x_j, \rho)$ centered at $x_j \in \Omega$ and radius ρ, $j \leq J$ and recall the *norm localization property*

$$|v|^2_{\dot{B}^r_{2,q}(\Omega)} \approx \sum_{j=1}^{J} |v|^2_{B^r_{2,q}(D_j)} \quad \forall r > 0,\ q \in [1, \infty]. \tag{2.26}$$

This, in conjunction with the previous estimate and (2.5), gives the *fundamental recursion formula*

$$\|u\|^2_{\dot{B}^{s+\sigma/2}_{2,\infty}(\Omega)} \lesssim \|f\|_{B^t_{2,1}(\Omega)} \|u\|_{\dot{B}^{\sigma-t}_{2,\infty}(\Omega)} + \|f\|^2_{H^{-s}(\Omega)} \qquad \forall \sigma \in (0, 1],\ t \in (-1, 1+\sigma). \tag{2.27}$$

We finally discuss two direct consequences of (2.27). The next result is proven in [23].

Theorem 2.3 (Optimal Shift Property) *Let Ω be a bounded Lipschitz domain and $f \in B^{-s+1/2}_{2,1}(\Omega)$. Then, the solution u to (2.1) belongs to the Besov space $\dot{B}^{s+1/2}_{2,\infty}(\Omega)$ and satisfies*

$$\|u\|_{\dot{B}^{s+1/2}_{2,\infty}(\Omega)} \le C(\Omega, d) \|f\|_{B^{-s+1/2}_{2,1}(\Omega)}, \tag{2.28}$$

with a constant $C(\Omega, d)$ uniform with respect to $s \in (0, 1)$.

Proof We proceed by iteration of (2.27). We fix $t = \frac{1}{2} - s$ and note that, according to (2.5),

$$\|u\|_{\dot{B}^s_{2,\infty}(\Omega)} \lesssim \|u\|_{\widetilde{H}^s(\Omega)} \le \|f\|_{H^{-s}(\Omega)} \lesssim \|f\|_{B^{-s+1/2}_{2,1}(\Omega)} \quad \Rightarrow \quad \|u\|_{\dot{B}^s_{2,\infty}(\Omega)} \le \Lambda_0 \|f\|_{B^{-s+1/2}_{2,1}(\Omega)}, \tag{2.29}$$

where Λ_0 depends on Ω, d, and s. We rewrite (2.27) as

$$\|u\|^2_{\dot{B}^{\sigma_{k+1}}_{2,\infty}(\Omega)} \le \Big(C_1 \|f\|_{B^{-s+1/2}_{2,1}(\Omega)} + C_2 \|u\|_{\dot{B}^{\sigma_k}_{2,\infty}(\Omega)} \Big) \|f\|_{B^{-s+1/2}_{2,1}(\Omega)}, \tag{2.30}$$

with constants $C_1, C_2 > 0$ depending on Ω, d, and s and parameters

$$\sigma_{k+1} := s + \sigma/2, \quad \sigma_k := \sigma - t = \sigma - \frac{1}{2} + s,$$

for $k \ge 0$. We now set $\sigma_0 = s$ and prove by induction that $\sigma_k = s + \frac{1}{2}\big(1 - \frac{1}{2^k}\big)$ and

$$\|u\|_{\dot{B}^{\sigma_k}_{2,\infty}(\Omega)} \le \Lambda_k \|f\|_{B^{-s+1/2}_{2,1}(\Omega)}, \tag{2.31}$$

with uniformly bounded constants $\Lambda_k \le M := \max\{\Lambda_0, (2C_1 + C_2^2)^{\frac{1}{2}}\}$. We first observe that $\sigma = 1 - \frac{1}{2^{k+1}} \in (0, 1]$ and $-1 < t = \frac{1}{2} - s < 1 + \sigma$ are within the range of validity of (2.27). For $k = 0$, formula (2.31) is true in view of (2.29) and $\sigma_0 = s$, whereas for $k \ge 0$ the expression for σ_{k+1} is easy to verify and (2.30) gives

$$\|u\|^2_{\dot{B}^{\sigma_{k+1}}_{2,\infty}(\Omega)} \le \big(C_1 + C_2\Lambda_k\big)\|f\|^2_{B^{-s+1/2}_{2,1}(\Omega)} = \Lambda^2_{k+1}\|f\|^2_{B^{-s+1/2}_{2,1}(\Omega)}.$$

It remains to show that $\Lambda_k \le M$ for all $k \ge 0$. This is clearly true for $k = 0$, and for $k \ge 0$, we see that if $\Lambda_k \le M$, then

$$\Lambda^2_{k+1} = C_1 + C_2\Lambda_k \le C_1 + C_2 M \le \frac{1}{2}\big(2C_1 + C_2^2 + M^2\big) \le M^2.$$

We finally replace Λ_k by M and take $k \to \infty$ in (2.31) to obtain (2.28). □

Remark 2.1 (Intermediate Estimate) Interpolating between (2.28) and (2.29) yields

$$\|u\|_{\dot{B}^{s+r/2}_{2,\infty}(\Omega)} \le C(\Omega, d, s)\|f\|_{B^{-s+r/2}_{2,1}(\Omega)}$$

for all $r \in (0, 1)$, with a constant independent of r. This estimate will be useful in Proposition 2.2.

We now state the second consequence of (2.27), which was originally derived in [27].

Theorem 2.4 (Regularity with L^2-Data) *Let Ω be a bounded Lipschitz domain and $f \in L^2(\Omega)$. If $s \neq \frac{1}{2}$, then the solution u to* (2.1) *belongs to the Besov space $\dot{B}^{s+\alpha}_{2,\infty}(\Omega)$ with $\alpha = \min\{s, 1/2\}$ and satisfies*

$$\|u\|_{\dot{B}^{s+\alpha}_{2,\infty}(\Omega)} \le \frac{C(\Omega, d)}{\sqrt{|1-2s|}}\|f\|_{L^2(\Omega)}. \tag{2.32}$$

Instead, if $s = \frac{1}{2}$, then for any $\varepsilon > 0$ sufficiently small there holds

$$\|u\|_{\dot{B}^{1-\varepsilon}_{2,\infty}(\Omega)} \le \frac{C(\Omega, d)}{\sqrt{\varepsilon}}\|f\|_{L^2(\Omega)}. \tag{2.33}$$

Proof We proceed in two steps. We consider first the case $s \le \frac{1}{2}$. To account for the Marchaud inequality (2.19), we let $\sigma \in (0, 1)$ and modify (2.25) as follows:

$$\|v - T_h v\|_{L^2(D_{2\rho}(x_0))} \lesssim \frac{|h|^\sigma}{\sqrt{1-\sigma}}\|u\|_{B^\sigma_{2,\infty}(D_{2\rho}(x_0))}.$$

Correspondingly, we change the estimate of the functional $\mathcal{F}_1$ to read

$$\left|\mathcal{F}_1(T_h v) - \mathcal{F}_1(v)\right| \lesssim \frac{|h|^\sigma}{\sqrt{1-\sigma}} \|f\|_{L^2(D_{2\rho}(x_0))} \|u\|_{B^\sigma_{2,\infty}(D_{2\rho}(x_0))}.$$

The bound (2.27) changes accordingly and becomes

$$\|u\|^2_{\dot{B}^{s+\sigma/2}_{2,\infty}(\Omega)} \lesssim \frac{1}{\sqrt{1-\sigma}} \|f\|_{L^2(\Omega)} \|u\|_{B^\sigma_{2,\infty}(\Omega)} + \|f\|^2_{L^2(\Omega)}.$$

We now set $\sigma_0 = s$ and convert this expression into the recursion formula

$$\|u\|^2_{\dot{B}^{\sigma_{k+1}}_{2,\infty}(\Omega)} \le \left(\frac{C_1}{\sqrt{1-\sigma_k}} \|u\|_{B^{\sigma_k}_{2,\infty}(\Omega)} + C_2 \|f\|_{L^2(\Omega)} \right) \|f\|_{L^2(\Omega)},$$

with $\sigma_{k+1} = s + \frac{\sigma_k}{2}$. We again proceed by induction to show that $\sigma_k = 2s\left(1 - 2^{-(k+1)}\right)$ and $\Lambda_k \le M\left(1-2s\right)^{-1/2}$ with M depending on Ω and d, and

$$\|u\|_{\dot{B}^{\sigma_k}_{2,\infty}(\Omega)} \le \Lambda_k \|f\|_{L^2(\Omega)}. \tag{2.34}$$

This is valid for $k = 0$ and for $k \ge 0$ can be derived along the lines of the proof of Theorem 2.3. For $s < \frac{1}{2}$, we realize that $\sigma_k \to 2s$ and (2.32) follows immediately. For $s = \frac{1}{2}$ and $0 < \varepsilon < \frac{1}{2}$, let $k_\varepsilon \in \mathbb{N}$ satisfy $2^{-k_\varepsilon} < \varepsilon \le 2^{-(k_\varepsilon - 1)}$. Using (2.34) for this k_ε yields (2.33).

It remains to tackle the case $s > \frac{1}{2}$. In view of(2.28), we only need to worry about the dependence in s of the constant in the inequality $\|f\|_{B^{-s+1/2}_{2,1}(\Omega)} \lesssim \|f\|_{L^2(\Omega)}$ as $s \to 1/2$. We aim to prove that such a constant scales as $(s-1/2)^{-1/2}$. By duality, it suffices to show

$$\|v\|_{L^2(\Omega)} \lesssim \frac{1}{\sqrt{2s-1}} \|v\|_{B^{s-1/2}_{2,\infty}(\Omega)} \quad \forall\, v \in B^{s-1/2}_{2,\infty}(\Omega).$$

This follows from a careful manipulation of the K-functional (2.15). To do this, we regard $L^2(\Omega)$ and $B^{s-1/2}_{2,\infty}(\Omega)$ as interpolation spaces between $X_0 = H^{-1}(\Omega)$ and $X_1 = H^1_0(\Omega)$ with

$$L^2(\Omega) = \left[H^{-1}(\Omega), H^1_0(\Omega)\right]_{1/2,2}, \quad B^{s-1/2}_{2,\infty}(\Omega) = \left[H^{-1}(\Omega), H^1_0(\Omega)\right]_{\theta,\infty},$$

where $\theta = \frac{1}{2}\left(s + \frac{1}{2}\right)$. This choice of spaces (X_0, X_1) guarantees that θ is uniformly far from 0, 1, and the norms in (2.16) are robust. Using (2.16) for $q = 2$, we deduce that for any $N \ge 1$ to be found

$$\|v\|^2_{L^2(\Omega)} \lesssim \int_0^N t^{-2} \left|K(t,v)\right|^2 dt + \int_N^\infty t^{-2} \left|K(t,v)\right|^2 dt.$$

Moreover, exploiting again (2.16) but now for $q = \infty$ yields

$$\int_0^N t^{-2}\left|K(t,v)\right|^2 dt \le \sup_{t>0}\left(t^{-2\theta}\left|K(t,v)\right|^2\right)\int_0^N t^{2\theta-2}dt = \frac{N^{s-\frac{1}{2}}}{s-\frac{1}{2}}\|v\|_{B^{s-1/2}_{2,\infty}(\Omega)}.$$

On the other hand, since Ω is Lipschitz, combining a duality argument with the standard Poincaré inequality in $H^1_0(\Omega)$ gives $\left|K(t,v)\right| \le \|v\|_{H^{-1}(\Omega)} \le C(\Omega)\|v\|_{L^2(\Omega)}$ and

$$\int_N^\infty t^{-2}\left|K(t,v)\right|^2 dt \le C(\Omega)^2\|v\|^2_{L^2(\Omega)}\int_N^\infty t^{-2}dt = \frac{C(\Omega)^2}{N}\|v\|^2_{L^2(\Omega)}.$$

Finally, choosing $N = 2C(\Omega)^2$ leads to the desired estimate and concludes the proof. □

We are now in a position to compare the estimate (2.28) for $s = \frac{1}{2}$ and (2.33). We see again the special role played by the Besov space $B^0_{2,1}(\Omega)$ which is a strict subspace of $L^2(\Omega)$.

We next convert the Besov estimates in Theorems 2.3 and 2.4 into Sobolev estimates. Such estimates can be compared with those in the literature and, more importantly, used to improve the existing error analyses in Sects. 2.3 and 2.4.

Corollary 2.1 (Sobolev Estimates) *Let Ω be a bounded Lipschitz domain, $\alpha = \min\{s, \frac{1}{2}\}$, and u be the solution to (2.1). For any $\varepsilon > 0$ sufficiently small, the following estimates are valid:*

$$\|u\|_{\widetilde{H}^{s+1/2-\varepsilon}(\Omega)} \le \frac{C(\Omega,d)}{\sqrt{\varepsilon}}\|f\|_{B^{-s+1/2}_{2,1}(\Omega)} \quad s \in (0,1), \tag{2.35}$$

$$\|u\|_{\widetilde{H}^{s+\alpha-\varepsilon}(\Omega)} \le \frac{C(\Omega,d)}{\sqrt{\varepsilon}|1-2s|}\|f\|_{L^2(\Omega)} \quad s \neq \frac{1}{2}, \tag{2.36}$$

$$\|u\|_{\widetilde{H}^{1-\varepsilon}(\Omega)} \le \frac{C(\Omega,d)}{\varepsilon}\|f\|_{L^2(\Omega)} \quad s = \frac{1}{2}. \tag{2.37}$$

Proof The three estimates follow by combining, respectively, (2.28), (2.32), or (2.33), with the embedding $\dot{B}^\sigma_{2,\infty}(\Omega) \subset \widetilde{H}^{\sigma-\varepsilon}(\Omega)$ for any $\sigma \in (0,2)$ and the ε-dependent estimate for $\varepsilon > 0$ small

$$\|v\|_{\widetilde{H}^{\sigma-\varepsilon}(\Omega)} \lesssim \frac{1}{\sqrt{\varepsilon}}\|v\|_{B^\sigma_{2,\infty}(\Omega)} \quad \forall v \in B^\sigma_{2,\infty}(\Omega). \tag{2.38}$$

To derive (2.38), we proceed as in the proof of Theorem 2.4 and consider the K-functional (2.15) for the pair of spaces $X_0 = L^2(\Omega)$ and $X_1 = H^2(\Omega)$ and $\theta = \frac{\sigma-\varepsilon}{2}$. We see that

$$\|v\|_{H^{\sigma-\varepsilon}(\Omega)}^2 \lesssim \int_0^\infty t^{-1-\sigma+\varepsilon}\left|K(t,v)\right|^2 dt \le \frac{1}{\varepsilon}\sup_{t>0}\left(t^{-\sigma}\left|K(t,v)\right|^2\right) + \frac{1}{\sigma-\varepsilon}\|v\|_{L^2(\Omega)}^2 \lesssim \frac{1}{\varepsilon}\|v\|_{B^\sigma_{2,\infty}(\Omega)}^2$$

is the asserted estimate. □

The estimates (2.35) and (2.37) are consistent with the maximal regularity (2.6) derived in [2, 58, 71, 103] under assumptions on $\partial\Omega$ stronger than Lipschitz continuity.

2.1.3 Linear Problems with Finite Horizon

We now extend the preceding results to operators of the form (1.2). Indeed, if ϕ is bounded, then we immediately have that the energy norm

$$|||v||| := \left(\frac{C_{d,s}}{2}\iint_{\mathbb{R}^d\times\mathbb{R}^d}\phi\left(\frac{|x-y|}{\delta}\right)\frac{|v(x)-v(y)|^2}{|x-y|^{d+2s}}\,dydx\right)^{\frac{1}{2}}$$

with $C_{d,s} \approx s(1-s)$ defined in (1.1) satisfies

$$|||v||| \le \|\phi\|_{L^\infty(\mathbb{R}^d)}^{\frac{1}{2}}\|v\|_{\widetilde{H}^s(\Omega)} \qquad \forall v \in \widetilde{H}^s(\Omega), \tag{2.39}$$

for all $s \in (0,1)$. Additionally, if $\phi \ge \phi_0 > 0$ on an interval $[0,r]$ for some $r \in (0,1)$, then the localization estimate (cf., for example, [55, Lemma 7])

$$\iint_{D_R(0)\times D_R(0)}\frac{|v(x)-v(y)|^2}{|x-y|^{d+2s}}\,dydx \le \left(\frac{3R}{r\delta}\right)^{2(1-s)}\iint_{D_R(0)\times D_R(0)}\frac{|v(x)-v(y)|^2}{|x-y|^{d+2s}}\,\chi_{\{|x-y|\le r\delta\}}\,dydx,$$

valid for all $R > r\delta > 0$ and $s \in (0,1)$, implies

$$\iint_{D_R(0)\times D_R(0)}\frac{|v(x)-v(y)|^2}{|x-y|^{d+2s}}\,dydx \le \frac{2(3R)^{2(1-s)}}{C_{d,s}(r\delta)^{2(1-s)}\phi_0}|||v|||^2$$

regardless of whether or not the support of ϕ is compact. Given any $v \in \widetilde{H}^s(\Omega)$, we use this bound with $R > 0$ sufficiently large so that $\Omega \subset D_R(0)$ and $\operatorname{dist}(\Omega, \partial D_R(0)) \ge \frac{R}{2}$, exploit the fact that $|x-y| \ge \frac{R}{2}$ for all $x \in \Omega$ and $y \in D_R(0)^c$, and integrate in polar coordinates to get

$$
\begin{aligned}
\|v\|^2_{\widetilde{H}^s(\Omega)} &= \frac{C_{d,s}}{2} \iint_{D_R(0)\times D_R(0)} \frac{|v(x)-v(y)|^2}{|x-y|^{d+2s}}\,dydx \\
&\quad + C_{d,s} \iint_{D_R(0)\times D_R(0)^c} \frac{|v(x)|^2}{|x-y|^{d+2s}}\,dydx \\
&\le \frac{(3R)^{2(1-s)}}{(r\delta)^{2(1-s)}\phi_0} |||v|||^2 + \frac{2^{2s-1}C_{d,s}\omega_{d-1}}{s\,R^{2s}} \|v\|^2_{L^2(\Omega)},
\end{aligned}
$$

where $\omega_{d-1} = |\mathbb{S}^{d-1}|$ denotes the $(d-1)$-dimensional measure of the unit sphere $\mathbb{S}^{d-1} = \partial D_1(0)$ in $\mathbb{R}^d$. We next resort to the Poincaré inequality (2.4) and fix $R > 0$ in such a way that

$$
\frac{2^{2s-1}C_{d,s}\omega_{d-1}}{s\,R^{2s}} \|v\|^2_{L^2(\Omega)} \le \frac{1}{2}\|v\|^2_{\widetilde{H}^s(\Omega)} \quad \forall v \in \widetilde{H}^s(\Omega),
$$

to obtain a constant $C(d,\Omega,r,\phi_0,\delta)$ uniform in $s \in (0,1)$ such that

$$
\|v\|_{\widetilde{H}^s(\Omega)} \le C(d,\Omega,r,\phi_0,\delta)\,|||v||| \quad \forall v \in \widetilde{H}^s(\Omega). \tag{2.40}
$$

By combining (2.39) and (2.40), we deduce that the energy norm $|||\cdot|||$ is equivalent to the $\|\cdot\|_{\widetilde{H}^s(\Omega)}$-norm. Consequently, the Dirichlet problem for the operator $\mathcal{A}_{\delta,s}$ defined in (1.2) but scaled by $C_{d,s}$ is well-posed in $\widetilde{H}^s(\Omega)$ uniformly in s: if $f \in H^{-s}(\Omega)$, then there exists a unique $u \in \widetilde{H}^s(\Omega)$ so that

$$
((u,v))_{\delta,s} = \langle \mathcal{A}_{\delta,s}u, v\rangle = \langle f, v\rangle \quad \forall v \in \widetilde{H}^s(\Omega), \tag{2.41}
$$

and it satisfies $\|u\|_{\widetilde{H}^s(\Omega)} \lesssim \|f\|_{H^{-s}(\Omega)}$; hereafter, $((\cdot,\cdot))_{\delta,s}$ is the scalar product associated with $|||\cdot|||$.

Another consequence of the equivalence between the energy norm $|||\cdot|||$ and the $\widetilde{H}^s(\Omega)$ norm is that we can adapt the technique employed in the proof of Theorem 2.3 (optimal shift property) to this finite-horizon problem. We refer to [23] for further details.

Corollary 2.2 (Optimal Shift Property for Finite-Horizon Operators) *Let $\phi\colon [0,\infty) \to [0,\infty)$ be a bounded function of class C^β on a neighborhood of the origin, for some $\beta \in (0,1]$, and satisfy $\phi(0) > 0$. Let Ω be a bounded Lipschitz domain, $s \in (0,1)$, and $f \in B^{-s+\beta/2}_{2,1}(\Omega)$. Then, the unique function $u \in \widetilde{H}^s(\Omega)$ that solves* (2.41) *belongs to the Besov space $\dot{B}^{s+\beta/2}_{2,\infty}(\Omega)$ and*

$$
\|u\|_{\dot{B}^{s+\beta/2}_{2,\infty}(\Omega)} \le C(\Omega,d)\|f\|_{B^{-s+\beta/2}_{2,1}(\Omega)}. \tag{2.42}
$$

Proof It suffices to realize that the proof of Theorem 2.3 hinges on the action of the localized translation operator T_h on generic balls $D_{2\rho}(x_0)$ of radius 2ρ and center x_0 via (2.23) and (2.24); the only change is that (2.24) is now valid for $\sigma \in (0, \beta]$ due the presence of ϕ. Consequently, if ρ is sufficiently small so that $\phi \in C^\beta([0, 4\rho])$ and $\phi \geq \phi_0 > 0$ on $[0, 4\rho]$, then we end up with the following variant of (2.30) regardless of the regularity of ϕ outside $[0, 4\rho]$ and the size of its support:

$$\|u\|^2_{\dot{B}^{\sigma_{k+1}}_{2,\infty}(\Omega)} \leq \left(C_1 \|f\|_{B^t_{2,1}(\Omega)} + C_2 \|u\|_{\dot{B}^{\sigma_k}_{2,\infty}(\Omega)}\right) \|f\|_{B^t_{2,1}(\Omega)},$$

with

$$\sigma_{k+1} := s + \sigma/2, \quad \sigma_k := \sigma - t, \quad t := \frac{\beta}{2} - s.$$

If $\sigma_0 = s$, then we can prove by induction that $\sigma_k = s + \frac{\beta}{2}\left(1 - \frac{1}{2^k}\right)$ and $\sigma = \beta\left(1 - \frac{1}{2^{k+1}}\right) \in (0, \beta]$. The rest of the proof follows that of Theorem 2.3. □

Remark 2.2 (Sobolev Estimates for Finite-Horizon Operators) In the same fashion as in Corollary 2.1 (Sobolev estimates), but with $\alpha := \{s, \frac{\beta}{2}\}$, we can convert (2.42) into the Sobolev regularity estimates for all $\varepsilon > 0$ sufficiently small

$$\|u\|_{\widetilde{H}^{s+\beta/2-\varepsilon}(\Omega)} \leq \frac{C(\Omega, d)}{\sqrt{\varepsilon}} \|f\|_{B^{-s+\beta/2}_{2,1}(\Omega)}, \tag{2.43}$$

and

$$\|u\|_{\widetilde{H}^{s+\alpha-\varepsilon}(\Omega)} \leq \frac{C(\Omega, d, s)}{\varepsilon^\kappa} \|f\|_{L^2(\Omega)}; \tag{2.44}$$

here $\kappa = \frac{1}{2}$ for $s \neq \frac{1}{2}$ and $\kappa = 1$ for $s = \frac{1}{2}$ and $C(\Omega, d, s) \approx |1 - 2s|^{-\frac{1}{2}}$ for $s \neq \frac{1}{2}$ provided $\beta = 1$, whereas $\kappa = \frac{1}{2}$ and the constant does not blow up in s provided $\beta < 1$. The latter is a consequence of $\sigma \leq \beta < 1$ in the proof of Theorem 2.4 due to the functional $\mathcal{F}_2$. These estimates are instrumental for the finite element error analysis of linear problems with finite horizon.

Remark 2.3 (Assumptions on the Diffusivity) We stress that (2.42), (2.43), and (2.44) do not require global Hölder-β regularity of the diffusivity function ϕ but just in a neighborhood of the origin. Thus, this result applies to the case of *truncated Laplacians* (which correspond to $\phi(\rho) = \chi_{[0,1]}(\rho)$) and thereby extends the estimates from [36] to Lipschitz domains. To the best of our knowledge, these are the first Besov or Sobolev regularity estimates for finite-horizon operators with non-constant diffusivity ϕ, whose support can even extend beyond $[0, 1]$ provided ϕ is globally bounded, and valid on Lipschitz domains.

2.2 Direct Finite Element Discretization

We next consider a direct finite element discretization of (2.3) by using piecewise linear continuous functions. Given $h_0 > 0$, for $h \in (0, h_0]$, we let $\mathcal{T}_h$ denote a triangulation of Ω, i.e., $\mathcal{T} = \{T\}$ is a conforming partition of Ω into simplices T of diameter h_T. We assume the family $\{\mathcal{T}\}$ to be shape-regular, namely,

$$\sigma := \sup_{\mathcal{T}} \max_{T \in \mathcal{T}} \frac{h_T}{\rho_T} < \infty,$$

where ρ_T is the diameter of the largest ball contained in T [34]. As usual, the parameter h denotes the mesh size, $h = \max_{T \in \mathcal{T}} h_T$; moreover, we take elements to be closed sets.

Let $\mathcal{N}_h$ be the set of interior vertices of $\mathcal{T}$, N be its cardinality, and $\{\varphi_i\}_{i=1}^N$ be the standard piecewise linear Lagrangian basis, with φ_i associated with the node $x_i \in \mathcal{N}_h$ and star $S_i = \operatorname{supp} \varphi_i$. The finite element space is the set of continuous piecewise linear functions over $\mathcal{T}_h$,

$$\mathbb{V}_h := \left\{ v \in C_0(\Omega) : v = \sum_{i=1}^N v_i \varphi_i \right\}. \tag{2.45}$$

It is clear that $\mathbb{V}_h \subset \widetilde{H}^s(\Omega)$ for all $s \in (0, 1)$, and therefore we have a conforming discretization.

With the notation described above, the discrete weak formulation reads as follows: find $u_h \in \mathbb{V}_h$ such that

$$(u_h, v_h)_s = \langle f, v_h \rangle \quad \forall v_h \in \mathbb{V}_h. \tag{2.46}$$

We now briefly describe some practical aspects of the implementation and solution process of (2.46). In view of (2.45), the linear system associated with (2.46) can be expressed as $\mathbf{AU} = \mathbf{F}$, where $\mathbf{U} = (u_i)_{i=1}^N$ and $u_h = \sum_{i=1}^N u_i \varphi_i$, and the entries of the stiffness matrix $\mathbf{A} = (a_{ij})_{ij=1}^N$ and right-hand side vector $\mathbf{F} = (f_i)_{i=1}^N$ read

$$a_{ij} = (\varphi_i, \varphi_j)_s, \quad f_i = \langle f, \varphi_i \rangle.$$

Computation of the Stiffness Matrix There are two issues in dealing with the entries $\mathbf{A}_{ij}$ $(i, j = 1, \ldots, N)$. The first one is that the integration domain in the bilinear form $(\varphi_i, \varphi_j)_s$ is unbounded if $S_i \cap S_j \neq \emptyset$. Namely, one has to compute integrals of the form

$$\int_{S_i \cap S_j} \varphi_i(x) \varphi_j(x) \int_{S_i^c} \frac{dx}{|x - y|^{d+2s}} dy dx.$$

In the experiments we display below, we have used auxiliary exterior domains and Dirichlet data truncation as proposed in [3]. We point out that—at least for homogeneous problems—an efficient alternative to compute the integrals over $\Omega \times \Omega^c$ is to transform them into integrals over $\Omega \times \partial\Omega$ by means of the Divergence Theorem [7].

The second issue is that the singular (non-integrable) kernel $|x - y|^{-d-2s}$ offers significant difficulties when computing stiffness matrix entries corresponding to nodal basis functions with supports close to each other. Enforcing a degree of cancellation compatible with the accuracy of (2.46) requires suitable quadrature rules to compute such entries. Reference [3] adopts techniques from the boundary element method [39, 95].

Matrix Compression The finite element spaces (2.45) give rise to *full* stiffness matrices regardless of the value of $s \in (0, 1)$. Indeed, if i and j are such that $S_i \cap S_j = \emptyset$, then

$$a_{ij} = -\frac{C_{d,s}}{2} \iint_{S_i \times S_j} \frac{\varphi_i(x)\, \varphi_j(y)}{|x - y|^{d+2s}}\, dy\, dx < 0.$$

Thus, in a naive implementation, most of the matrix assembly time is devoted to the computation of elements a_{ij} for $\mathtt{x}_i$ and $\mathtt{x}_j$ far away from one another. However, these matrix elements should be significantly smaller than the ones that involve neighboring nodes. Therefore, in an efficient implementation of the finite element method, these far field contributions can be replaced by computationally cheaper low-rank blocks. The cluster paneling method from the boundary element literature has been considered in [7, 106], resulting in a data-sparse representation with $\mathcal{O}(N \log^\alpha N)$ complexity for some $\alpha \geq 0$; see also [16]. We finally remark that in [79] it is shown that the inverse of $\mathbf{A}$ can be represented using the same block structure as employed to compress $\mathbf{A}$.

Preconditioning The use of matrix factorization techniques to solve a dense matrix equation $\mathbf{AU} = \mathbf{F}$ has complexity $\mathcal{O}(N^3)$. A common alternative is the use of the conjugate gradient method, for which the number of iterations needed for a fixed tolerance scales like $\sqrt{\kappa(\mathbf{A})}$, where $\kappa(\mathbf{A})$ is the condition number of $\mathbf{A}$. Reference [9] shows that

$$\kappa(\mathbf{A}) = \mathcal{O}\left(N^{2s/d} \left(\frac{h_{max}}{h_{min}} \right)^{d-2s} \right), \tag{2.47}$$

where h_{max} and h_{min} denote the maximum and minimum element sizes, respectively. Therefore, on quasi-uniform meshes, we have $\kappa(\mathbf{A}) = \mathcal{O}(h^{-2s})$, where $h \simeq N^{-1/d}$ is a mean element diameter. As we discuss in Sect. 2.3.3, the use of graded meshes yields a higher convergence rate with respect to the number of degrees of freedom. However, as (2.47) illustrates, such an improvement comes at

Table 1 Number of iterations needed when using a PCG method with BPX preconditioner without ($\widetilde{\gamma} = 0$) and with ($\widetilde{\gamma} = \frac{1}{2}$) a correction factor. We display results on a family of uniformly refined grids (left panel) and on a sequence of suitably graded bisection grids (right panel)

Uniform grids				
DOFs	$s = 10^{-1}$		$s = 10^{-2}$	
	$\tilde{\gamma} = 0$	$\tilde{\gamma} = \frac{1}{2}$	$\tilde{\gamma} = 0$	$\tilde{\gamma} = \frac{1}{2}$
225	14	10	16	10
961	17	10	18	10
3969	19	10	21	10
16,129	20	10	23	9

Graded bisection grids				
DOFs	$s = 10^{-1}$		$s = 10^{-2}$	
	$\tilde{\gamma} = 0$	$\tilde{\gamma} = \frac{1}{2}$	$\tilde{\gamma} = 0$	$\tilde{\gamma} = \frac{1}{2}$
161	13	10	15	11
853	17	12	19	13
2265	20	12	22	14
9397	22	13	25	14

the expense of poorer conditioning; a simple diagonal rescaling restores the same condition number as for uniform meshes [9].

There have been some recent progress in the development of preconditioners for fractional elliptic problems. Reference [7] mentions the use of multigrid preconditioners, while [68] analyzes an operator preconditioner based on an explicit representation of Green's function for the integral fractional Laplacian on a ball (Boggio's formula).

Additive Schwarz preconditioners of BPX-type have been considered in [30, 63]. Reference [63] develops a local multilevel diagonal preconditioner, while [30] introduces a scaling for coarse spaces that yields condition numbers uniformly bounded with respect to both the number of levels and the order $s \in (0, 1)$. We briefly discuss the latter next.

Since the s-Laplacian (1.1) tends to the classical Laplacian as $s \to 1$ and to the identity as $s \to 0$, due to the suitable s-depending scaling built into $C_{d,s}$, the challenge is to design a preconditioner $\mathbf{B}$ such that $\kappa(\mathbf{BA}) \leq C$ uniformly in s as well as in the number of levels J. The standard BPX preconditioners cannot do this because $\mathbf{A}$ is not a good approximation of the identity. The key idea of [30] is an s-dependent scaling that captures the transition to the identity as $s \to 0$. Let $V = \mathbb{V}_h$ and consider the space decomposition $\{V_j\}_{j=0}^J$ of V with nested spaces V_j, $V = \sum_{j=1}^J V_j$, and the L^2-projection operators $Q_j : V \to V_j$. The following *s-uniform decomposition* is valid for all $\gamma, \widetilde{\gamma} \in (0, 1)$, $s > 0$, and for every $v \in V$,

$$\sum_{j=0}^{J} \gamma^{-2sj} \|(Q_j - Q_{j-1})v\|_0^2 = \inf_{\substack{v_j \in V_j \\ \sum_{j=0}^J v_j = v}} \left(\gamma^{-2sJ} \|v_J\|_0^2 + \sum_{j=0}^{J-1} \frac{\gamma^{-2sj}}{1 - \widetilde{\gamma}^{2s}} \|v_j\|_0^2 \right). \tag{2.48}$$

The critical role of the weight $1 - \widetilde{\gamma}^{2s}$ is documented in Table 1 for a simple example $\Omega = (0, 1)^2$ and $f = 1$ with $s = 10^{-1}, 10^{-2}$: the number of iterations is not uniform for $\widetilde{\gamma} = 0$!

Consider first the case of *quasi-uniform meshes* $\mathcal{T}_j$ with mesh size h_j and corresponding spaces V_j. If $I_j : V_j \to V$ is the injection operator, so that $Q_j = I_j^T$, then the additive preconditioner B in operator form reads

Table 2 Quasi-uniform meshes: iterations of CG and PCG with BPX preconditioner (2.49), parameter $\tilde{\gamma} = 0.5$, and fractional order $s = 0.9, 0.5, 0.1$

			$s = 0.9$		$s = 0.5$		$s = 0.1$	
J	h_J	DOFs	CG	PCG	CG	PCG	CG	PCG
1	2^{-1}	9	4	4	4	4	4	4
2	2^{-2}	49	12	12	8	8	8	9
3	2^{-3}	225	25	16	11	10	8	10
4	2^{-4}	961	46	19	17	11	8	10
5	2^{-5}	3969	84	21	24	12	8	10
6	2^{-6}	16,129	157	22	32	13	8	10

$$B := I_J h_J^{2s} Q_J + (1 - \tilde{\gamma}^s) \sum_{j=0}^{J-1} I_j h_j^{2s} Q_j. \tag{2.49}$$

The performance of B, showing robustness with respect to J and s, is displayed in Table 2.

Consider now *graded bisection meshes* $\mathcal{T}_j = \mathcal{T}_{j-1} + b_j = \mathcal{T}_0 + \{b_1, \cdots, b_j\}$ obtained from an initial mesh $\mathcal{T}_0$ by successive compatible bisections b_j. Associated with each b_j, there is a triplet of nodes, namely the bisection node x_j and the parent nodes of x_j at the end of the bisection edge containing x_j; h_j is the local mesh size. The local space V_j is the span of the three hat functions corresponding to this triplet on the mesh $\mathcal{T}_j$. Let $\mathcal{P}$ be the set of interior nodes of the finest graded mesh $\mathcal{T}_J$ and let V_p be the one dimensional space spanned by the hat function $\phi_p \in V$ associated with $p \in \mathcal{P}$ and local mesh size h_p. Given the space decomposition

$$V = \sum_{p \in \mathcal{P}} V_p + \sum_{j=1}^{J} V_j,$$

the additive preconditioner B has a similar structure to (2.49) and reads

$$B = \sum_{p \in \mathcal{P}} I_p h_p^{2s} Q_p + (1 - \tilde{\gamma}^s) \sum_{j=0}^{J} I_j h_j^{2s} Q_j. \tag{2.50}$$

Table 3 documents the robust performance of B with respect to J and s.

The robust performance of B is supported by theory. The following result is shown in [30]. The proof for quasi-uniform meshes $\mathcal{T}_j$ hinges on (2.48) as well as s-uniform interpolation and local inverse estimates. However, for graded meshes, there is no global notion of scale and the subspaces V_j are local and non-nested, so (2.48) cannot be used directly. The proof relies on the geometric structure of bisection grids.

Theorem 2.5 (Uniform Preconditioning) *The BPX preconditioner of* (2.49) *and* (2.50) *satisfies* $\kappa(\mathbf{BA}) \le C$, *where the constant* C *is independent of the number of levels* J *and fractional order* s.

Table 3 Graded bisection meshes: iterations of CG and PCG with BPX preconditioner (2.50), parameter $\tilde{\gamma} = \sqrt{2}/2$, and fractional orders $s = 0.9, 0.5, 0.1$

$\bar{j}$	DOFs	$s = 0.9$		$s = 0.5$		$s = 0.1$	
		CG	PCG	CG	PCG	CG	PCG
7	61	10	10	10	7	13	8
8	153	15	13	15	9	21	10
9	161	15	14	15	9	21	10
10	369	20	16	20	11	34	11
11	405	21	16	19	11	31	12
12	853	26	18	26	12	48	12
13	973	30	19	26	12	47	12
14	1921	34	20	33	13	72	12
15	2265	40	21	32	13	65	12
16	4269	46	22	39	14	97	13
17	5157	55	22	40	14	92	12
18	9397	64	24	48	14	135	13

2.3 Global Error Estimates

In this section, we review global error estimates for the finite element discretization (2.46) in $\widetilde{H}^s(\Omega)$, $L^2(\Omega)$, and negative-order Sobolev norms.

The *energy norm* in $\widetilde{H}^s(\Omega)$ satisfies the best approximation property,

$$\|u - u_h\|_{\widetilde{H}^s(\Omega)} = \min_{v_h \in \mathbb{V}_h} \|u - v_h\|_{\widetilde{H}^s(\Omega)}, \tag{2.51}$$

because u_h is the projection of u onto $\mathbb{V}_h$ with respect to such a norm. Therefore, we must account for nonlocality of the fractional norm in $\widetilde{H}^s(\Omega)$ as well as the regularity of u. We present error estimates for quasi-uniform meshes and suitably graded meshes that compensate for the singular boundary behavior of u. The rates of convergence improve upon previous results because they hinge on the new regularity estimates of Theorem (2.3) (optimal shift property).

2.3.1 Localization and Interpolation Estimates

A fundamental tool in the error analysis is the use of interpolation estimates. In order to obtain estimates valid for arbitrary meshes, a typical approach is to derive them elementwise (or patchwise), thereby giving rise to local interpolation estimates. Since the energy norm is *nonlocal,* a localization procedure is thus required.

We start by defining the star (or patch) of a subset $A \subset \Omega$ by

$$S_A^1 := \bigcup \{T \in \mathcal{T}: T \cap A \neq \emptyset\}.$$

Given $T \in \mathcal{T}$, the star S_T^1 of T is the first ring of T and the star S_T^2 of S_T^1 is the second ring of T. References [59, 60] derive the following localized estimate:

$$|v|^2_{H^s(\Omega)} \leq \sum_{T\in\mathcal{T}} \left[\int_T \int_{S_T^1} \frac{|v(x)-v(y)|^2}{|x-y|^{d+2s}}\, dy\, dx + \frac{C(d,\sigma)}{s h_T^{2s}} \|v\|^2_{L^2(T)} \right] \quad \forall v \in H^s(\Omega). \tag{2.52}$$

Therefore, to estimate an $H^s(\Omega)$-seminorm, it suffices to compute local contributions on patches of the form $\{T \times S_T^1\}_{T\in\mathcal{T}}$ and elementwise L^2-contributions. The energy norm in our problem is not exactly the $H^s(\Omega)$ one but the $\widetilde{H}^s(\Omega)$ one instead. The latter involves integration over the space $\mathbb{R}^d$ even if the functions are supported in Ω, and thus we need to make some slight modifications on (2.52). Following [28], given $T \in \mathcal{T}$, we denote its barycenter by x_T and consider a ball $B_T = B(x_T, Ch_T)$ with center x_T and radius Ch_T, where $C = C(\sigma)$ is a shape-regularity-dependent constant such that $S_T^1 \subset B_T$. Then, we define the *extended* stars

$$\widetilde{S}_T^1 := \begin{cases} S_T^1 \text{ if } T \cap \partial\Omega = \emptyset, \\ B_T \text{ otherwise} \end{cases}$$

and the extended second ring $\widetilde{S}_T^2$

$$\widetilde{S}_T^2 := \bigcup \{\widetilde{S}_{T'}^1 : T' \in \mathcal{T}, \mathring{T}' \cap S_T^1 \neq \emptyset\}.$$

With these modifications, [28] proves the following estimate:

$$\|v\|^2_{\widetilde{H}^s(\Omega)} \leq \sum_{T\in\mathcal{T}} \left[\int_T \int_{\widetilde{S}_T^1} \frac{|v(x)-v(y)|^2}{|x-y|^{d+2s}}\, dy\, dx + \frac{C(d,\sigma)}{s h_T^{2s}} \|v\|^2_{L^2(T)} \right] \quad \forall v \in \widetilde{H}^s(\Omega). \tag{2.53}$$

Let us now mention some standard local quasi-interpolation estimates, involving a suitable quasi-interpolation operator that we shall denote by Π. Examples of possible choices for Π include the Scott–Zhang and the Clément interpolation operators [34, 41]. One can prove the following local estimates (see, for example, [4, 29, 41]) in either standard or weighted Sobolev spaces.

Lemma 2.1 (Local Interpolation Error Estimates) *Let $T \in \mathcal{T}$, $s \in (0,1)$, $r \in (s, 2]$, and Π be a suitable local quasi-interpolation operator. If $v \in W_p^r(\widetilde{S}_T^2)$, then*

$$\int_T \int_{\widetilde{S}_T^1} \frac{|(v-\Pi v)(x) - (v-\Pi v)(y)|^2}{|x-y|^{d+2s}}\, dy\, dx \leq C\, h_T^{2l} |v|^2_{W_p^r(\widetilde{S}_T^2)}, \tag{2.54}$$

where $l = Sob\,(W_p^r) - Sob\,(H^s) = r - s - d\left(\frac{1}{p} - \frac{1}{2}\right) > 0$ and $C = C(\Omega, d, s, \sigma, r)$.

Moreover, considering the weighted Sobolev scale (2.11)*, it holds that for all* $v \in H^r_\gamma(\widetilde{S}^2_T)$

$$\int_T \int_{\widetilde{S}^1_T} \frac{|(v - \Pi v)(x) - (v - \Pi v)(y)|^2}{|x - y|^{d+2s}} \, dy \, dx \le C h_T^{2(r-s-\gamma)} |v|^2_{H^r_\gamma(\widetilde{S}^2_T)}. \tag{2.55}$$

An important feature of either (2.52) and (2.53) is that when applied to an interpolation error—which typically has zero mean over elements—the scaled L^2-terms can be converted into H^s terms by means of a Poincaré inequality. This means that, when measuring interpolation errors, the $H^s(\Omega)$-seminorm effectively localizes as the sum of patchwise H^s-seminorms.

We finally point out that (2.53) avoids the need to use the Hardy inequality. This is particularly important in the case $s = 1/2$ because such a step gives rise to an additional logarithmic factor [22, 29].

2.3.2 Quasi-uniform Meshes

Combining the best approximation property (2.51) with regularity, localization, and local interpolation estimates, one can derive the following quasi-optimal a priori estimates in the energy norm. The first estimate is new and the second one improves upon [22] the power of the logarithmic factor for all $s \in (0, 1)$.

Theorem 2.6 (Energy Error Estimates) *Let* $\Omega \subset \mathbb{R}^d$ *be a bounded Lipschitz domain,* $0 < s < 1$*, and let* u *denote the solution of* (2.3) *and* $u_h \in \mathbb{V}_h$ *the solution of the discrete problem* (2.46)*, computed over a quasi-uniform mesh* $\mathcal{T}_h$*, where* $h = \max_{T \in \mathcal{T}} h_T$*. If* $f \in B^{1/2-s}_{2,1}(\Omega)$*, then there exists a constant* $C = C(\Omega, d, \sigma)$ *such that for all* $s \in (0, 1)$

$$\|u - u_h\|_{\widetilde{H}^s(\Omega)} \le C h^{\frac{1}{2}} |\log h|^{\frac{1}{2}} \, \|f\|_{B^{1/2-s}_{2,1}(\Omega)}. \tag{2.56}$$

If instead $f \in L^2(\Omega)$ *and* $\alpha = \min\{s, \frac{1}{2}\}$*, then for* $\kappa = \frac{1}{2}$ *if* $s \neq \frac{1}{2}$ *and* $\kappa = 1$ *if* $s = \frac{1}{2}$ *we have*

$$\|u - u_h\|_{\widetilde{H}^s(\Omega)} \le C h^{\alpha} |\log h|^{\kappa} \, \|f\|_{L^2(\Omega)}. \tag{2.57}$$

The constant $C = C(\Omega, d, s, \sigma)$ *blows up as* $s \to \frac{1}{2}$ *as* $|1 - 2s|^{-\frac{1}{2}}$.

Proof Employing (2.51) in conjunction with the interpolation estimates (2.53) and (2.54) and the Sobolev regularity estimate (2.35) yields

$$\|u - u_h\|_{\widetilde{H}^s(\Omega)} \le C h^{\frac{1}{2}-\varepsilon} \|u\|_{\widetilde{H}^{s+\frac{1}{2}-\varepsilon}(\Omega)} \le C h^{\frac{1}{2}} \frac{h^{-\varepsilon}}{\varepsilon^{\frac{1}{2}}} \|f\|_{B^{1/2-s}_{2,1}(\Omega)}.$$

Taking $\varepsilon = |\log h|^{-1}$ gives the desired estimate (2.56). Replacing (2.35) by (2.36) if $s \neq \frac{1}{2}$ or by (2.37) if $s = \frac{1}{2}$ leads to (2.57) and completes the proof. □

With error estimates in $\widetilde{H}^s(\Omega)$ at hand, one can perform an Aubin–Nitsche duality argument to derive convergence rates in weaker norms. To see this, let $g \in L^2(\Omega)$ or $g \in H^{s-\frac{1}{2}}(\Omega)$ in case $s \in (0, \frac{1}{2})$. Let $u_g \in \widetilde{H}^s(\Omega)$ solve $(v, u_g)_s = \langle v, g\rangle$ for all $v \in \widetilde{H}^s(\Omega)$. We then have

$$\begin{aligned}\langle u - u_h, g\rangle &= (u - u_h, u_g)_s \\ &= (u - u_h, u_g - \Pi u_g)_s \leq \|u - u_h\|_{\widetilde{H}^s(\Omega)} \|u_g - \Pi u_g\|_{\widetilde{H}^s(\Omega)}.\end{aligned} \tag{2.58}$$

Proposition 2.1 (L^2-Error Estimate) *Let Ω be a bounded Lipschitz domain, $s \in (0, 1)$, and $\alpha = \min\{s, \frac{1}{2}\}$. If $f \in B^{1/2-s}_{2,1}(\Omega)$, then for κ defined in Theorem 2.6 we have*

$$\|u - u_h\|_{L^2(\Omega)} \leq Ch^{\frac{1}{2}+\alpha} |\log h|^{\frac{1}{2}+\kappa} \, \|f\|_{B^{1/2-s}_{2,1}(\Omega)}.$$

If $f \in L^2(\Omega)$ instead, then we have

$$\|u - u_h\|_{L^2(\Omega)} \leq Ch^{2\alpha} |\log h|^{2\kappa} \, \|f\|_{L^2(\Omega)}.$$

Proof Choose $g \in L^2(\Omega)$ in (2.58) and use either (2.36) or (2.37) for u_g to get

$$\|u_g - \Pi u_g\|_{\widetilde{H}^s(\Omega)} \lesssim h^{\alpha} |\log h|^{\kappa} \|g\|_{L^2(\Omega)}.$$

Combining this estimate with either (2.56) or (2.57) gives the asserted error bounds. □

For classical second-order problems, the pivot space is $H^1_0(\Omega)$ and the pick-up regularity for piecewise linear elements is just 1. Therefore, the accessible regularity is $H^2(\Omega)$, and by duality the lowest order space is just $L^2(\Omega)$. Getting higher order estimates in negative Sobolev spaces entails increasing the polynomial degree beyond 1. For the fractional Laplacian, the situation is different: the pivot space is $\widetilde{H}^s(\Omega)$ and the pick-up regularity is $\frac{1}{2}$. Therefore, the accessible regularity is $H^{s+\frac{1}{2}-\varepsilon}(\Omega)$ and the lowest order space accessible by duality is $H^{s-\frac{1}{2}}(\Omega)$, which turns out to be a negative Sobolev space if $s < \frac{1}{2}$. The following convergence rates improve upon Proposition 2.1.

Proposition 2.2 (Error Estimate in Negative Sobolev Spaces) *Let Ω be a bounded Lipschitz domain and $s \in (0, \frac{1}{2})$. If $f \in B^{1/2-s}_{2,1}(\Omega)$, then we have*

$$\|u - u_h\|_{H^{s-\frac{1}{2}}(\Omega)} \leq Ch |\log h|^{\frac{3}{2}} \, \|f\|_{B^{1/2-s}_{2,1}(\Omega)}.$$

If $f \in L^2(\Omega)$ instead, then we have

$$\|u - u_h\|_{H^{s-\frac{1}{2}}(\Omega)} \le C h^{\frac{1}{2}+s} |\log h|^{\frac{3}{2}} \|f\|_{L^2(\Omega)}.$$

Proof We invoke the Sobolev estimate

$$\|v\|_{\dot{B}_{2,1}^{-s+1/2-\varepsilon}(\Omega)} \lesssim \varepsilon^{-\frac{1}{2}} \|v\|_{\widetilde{H}^{-s+1/2}(\Omega)} \quad \forall\, v \in \widetilde{H}^{-s+1/2}(\Omega),$$

which follows by duality and an argument with the K-functional and spaces $X_0 = H^{-1}(\Omega)$ and $X_1 = H_0^1(\Omega)$ similar to the second part of the proof of Theorem 2.4 and the proof of Corollary 2.1.

We take $g \in \widetilde{H}^{-s+1/2}(\Omega)$ in (2.58) and combine (2.35) with the preceding estimate to arrive at

$$\begin{aligned} \|u_g - \Pi u_g\|_{\widetilde{H}^s(\Omega)} &\lesssim h^{\frac{1}{2}-2\varepsilon} |u_g|_{\widetilde{H}^{s+\frac{1}{2}-2\varepsilon}(\Omega)} \lesssim \frac{h^{\frac{1}{2}-2\varepsilon}}{\varepsilon^{\frac{1}{2}}} \|g\|_{\dot{B}_{2,1}^{-s+1/2-\varepsilon}(\Omega)} \\ &\lesssim \frac{h^{\frac{1}{2}-2\varepsilon}}{\varepsilon} \|g\|_{\widetilde{H}^{-s+1/2}(\Omega)}. \end{aligned}$$

This, in conjunction with either (2.56) or (2.57) and $\varepsilon = |\log h|^{-1}$, gives the desired bounds. □

On the other hand, by combining inverse inequalities and interpolation estimates, one can also derive convergence estimates on *higher order* seminorms such as the $H^1(\Omega)$-seminorm [21]. We remark that the restriction $s > 1/2$ below is due to the fact that under such a condition one can guarantee that the solution u is actually in $H^1(\Omega)$. The condition on f is weaker than in [21] because here we are exploiting Theorem 2.3 (optimal shift property).

Proposition 2.3 **(H^1-Error Estimate)** *Let Ω be a bounded Lipschitz domain, $f \in B_{2,1}^{-s+1/2}(\Omega)$, and $s \in (1/2, 1)$. If h is sufficiently small, then we have*

$$\|u - u_h\|_{H^1(\Omega)} \le C h^{s-1/2} |\log h|^{1/2} \|f\|_{B_{2,1}^{-s+1/2}(\Omega)}.$$

Proof Let $\varepsilon \in (0, s - 1/2)$. Combining the interpolation error estimates (2.53) and (2.54) with the Sobolev regularity estimate (2.35), we obtain

$$\|u - \Pi u\|_{H^1(\Omega)} \lesssim \frac{h^{s-1/2-\varepsilon}}{\sqrt{\varepsilon}} \|f\|_{B_{2,1}^{-s+1/2}(\Omega)}.$$

It remains to bound $\|\Pi u - u_h\|_{H^1(\Omega)}$. By using the standard inverse inequality

$$\|v_h\|_{H^1(\Omega)} \le C h^{s-1} \|v_h\|_{\widetilde{H}^s(\Omega)} \quad \forall v_h \in \mathbb{V}_h$$

and the triangle inequality, it follows

$$\|\Pi u - u_h\|_{H^1(\Omega)} \leq Ch^{s-1}\left(\|\Pi u - u\|_{\widetilde{H}^s(\Omega)} + \|u - u_h\|_{\widetilde{H}^s(\Omega)}\right).$$

Finally, we resort again to (2.53), (2.54), and the energy error estimate (2.56), to deduce

$$\|u - u_h\|_{H^1(\Omega)} \lesssim \frac{h^{s-1/2-\varepsilon}}{\sqrt{\varepsilon}}\|f\|_{B^{-s+1/2}_{2,1}(\Omega)}, \quad \forall \varepsilon \in (0, s-1/2).$$

The proof is concluded upon setting $\varepsilon = |\log h|^{-1}$ in the estimate above. □

2.3.3 Graded Meshes

Prior convergence rates on quasi-uniform meshes suffer from boundary pollution. A natural remedy to improve upon them is to consider graded meshes adapted to the boundary behavior of the solution to problem (2.1) and to exploit the regularity estimates from Theorem 2.1 or Theorem 2.2 for their design and analysis. This is the objective of this section.

The construction of graded meshes à-la-Grisvard hinges on weighted Sobolev estimates [70]. We let the parameter h represent a local mesh size in the interior of Ω and assume that the family of meshes $\{\mathcal{T}\}$ is shape-regular and admits a parameter $\mu \geq 1$ such that for every $T \in \mathcal{T}$,

$$h_T \leq C(\sigma)\begin{cases} h^{\mu} \text{ if } T \cap \partial\Omega \neq \emptyset, \\ h\,\mathrm{dist}(T, \partial\Omega)^{(\mu-1)/\mu} \text{ if } T \cap \partial\Omega = \emptyset, \end{cases} \tag{2.59}$$

for $d \geq 1$. This yields mesh cardinality (see [13, 29])

$$\#\mathcal{T} \approx \dim \mathbb{V}_h \approx \begin{cases} h^{-d} \text{ if } \mu < \frac{d}{d-1}, \\ h^{-d}|\log h| \text{ if } \mu = \frac{d}{d-1}, \\ h^{(1-d)\mu} \text{ if } \mu > \frac{d}{d-1}, \end{cases} \tag{2.60}$$

for $d > 1$; for $d = 1$ the relation $\dim \mathbb{V}_h \approx h^{-1}$ is valid regardless of the value of μ. Although (2.59) provides a sufficient grading condition, it does not guarantee the existence of such meshes, especially for complicated geometries. We will discuss a constructive approach below (cf. Algorithm 2.1).

If $\mu \leq \frac{d}{d-1}$, then the interior mesh size h and $\#\mathcal{T}$ satisfy the optimal relation $h \simeq \#\mathcal{T}^{-1/d}$ (up to logarithmic factors if $\mu = \frac{d}{d-1}$). To derive optimal convergence rates for meshes satisfying (2.59), one needs to tune the parameter μ, and the optimal choice of μ depends on the dimension; we refer to [22] for details. In two dimensions, the optimal choice is $\mu = 2$. We combine (2.53) with either (2.55) or

(2.54), depending on whether S_T^2 intersects $\partial\Omega$ or not, and Theorem 2.1 (weighted Sobolev estimate), to derive the following result.

Theorem 2.7 (Energy Error Estimates on Graded Meshes) *Let $\Omega \subset \mathbb{R}^2$ be a bounded Lipschitz domain satisfying an exterior ball condition, $u \in \widetilde{H}^s(\Omega)$ denote the solution to* (2.3), *and $u_h \in \mathbb{V}_h$ denote the solution of the discrete problem* (2.46), *computed over a mesh $\mathcal{T}$ satisfying* (2.59) *with $\mu = 2$. If $f \in C^{1-s}(\overline{\Omega})$, then we have*

$$\|u - u_h\|_{\widetilde{H}^s(\Omega)} \le C(\Omega, s, \sigma) h |\log h| \|f\|_{C^{1-s}(\overline{\Omega})}. \tag{2.61}$$

Equivalently, in terms of mesh cardinality $\#\mathcal{T}$, the estimate above reads

$$\|u - u_h\|_{\widetilde{H}^s(\Omega)} \le C(\Omega, s, \sigma) (\#\mathcal{T})^{-\frac{1}{2}} |\log \#\mathcal{T}|^{\frac{3}{2}} \|f\|_{C^{1-s}(\overline{\Omega})}.$$

Since the practical implementation of meshes satisfying (2.59) might be problematic, [28] proposes a constructive algorithm based on the *bisection method*, which works as follows. Given a mesh $\mathcal{T}$, we assume every simplex $T \in \mathcal{T}$ has an edge $e(T)$ marked for refinement. To subdivide T into two children T_1, T_2 such that $T = T_1 \cup T_2$, one connects the midpoint of $e(T)$ with the vertices of T that do not lie in $e(T)$. If every simplex sharing $e(T)$ has $e(T)$ marked for refinement, then the patch is *compatible* and the refinement does not propagate beyond it. Otherwise, at least one element in the patch has an edge other than $e(T)$ marked for refinement, and the refinement procedure must go outside the patch to maintain conformity (namely, we have a nonlocal step). Therefore, two natural questions arise:

- *Completion*: How many elements other than T must be refined to keep the mesh conforming?
- *Termination*: Does this procedure terminate?

To guarantee termination, a special labeling of the initial mesh $\mathcal{T}_0$ is required (a suitable choice of the edge $e(T)$ for each element $T \in \mathcal{T}_0$). Completion is rather tricky to assess and was done by P. Binev, W. Dahmen, and R. DeVore for $d = 2$ [18] and R. Stevenson for $d > 2$ [100]; we refer to the surveys [91, 92] for a rather complete discussion.

Given the j-th refinement $\mathcal{T}_j$ of $\mathcal{T}_0$ and a subset $\mathcal{M}_j \subset \mathcal{T}_j$ of elements marked for bisection,

$$\mathcal{T}_{j+1} = \mathsf{REFINE}(\mathcal{T}_j, \mathcal{M}_j)$$

is a procedure that creates the smallest conforming refinement $\mathcal{T}_{j+1}$ of $\mathcal{T}_j$ upon bisecting all elements of $\mathcal{M}_j$ at least once and perhaps additional elements to keep conformity. We point out that it is simple to construct counterexamples to the estimate

$$\#\mathcal{T}_{j+1} - \#\mathcal{T}_j \le \Lambda \, \#\mathcal{M}_j,$$

where Λ is a universal constant independent of j; see [92, Section 1.3]. However, this can be repaired upon considering the cumulative effect of a sequence of conforming bisection meshes $\{\mathcal{T}_j\}_{j=0}^k$ for any k. In fact, the following crucial estimate is valid [18, 100] (see also [91, 92]):

$$\#\mathcal{T}_k - \#\mathcal{T}_0 \leq \Lambda \sum_{j=0}^{k-1} \#\mathcal{M}_j. \tag{2.62}$$

We propose a greedy algorithm to choose simplices for refinement [28]. For $d = 2$, the basic idea is to equidistribute the local H^s-interpolation errors, for which one assumes access to the quantities

$$E_T(u) = Ch_T^t R_T(u), \qquad R_T(u) = |u|_{W^{s+1-\varepsilon}_{1+\varepsilon}(\widetilde{S}_T^2)}, \tag{2.63}$$

with C being a constant depending on the mesh shape-regularity and $t = 2 - \varepsilon - \frac{2}{1+\varepsilon} > 0$. We point out that such regularity was stated in Theorem 2.2 (differentiability vs integrability). Given a tolerance $\delta > 0$ and a conforming mesh $\mathcal{T}_0$ with suitable labeling, Algorithm 2.1 finds a conforming refinement $\mathcal{T}$ of $\mathcal{T}_0$ by bisection such that $E_T(u) \leq \delta$ for all $T \in \mathcal{T}$.

Algorithm 2.1 GREEDY$(\mathcal{T}_0, \delta)$

```
Let T = T_0.
while M := {T ∈ T : E_T(u) > δ} ≠ ∅ do
    T := REFINE(T, M)
end while
return (T)
```

Theorem 2.8 (Quasi-Optimal Error Estimate on Bisection Meshes) *Let $\Omega \subset \mathbb{R}^2$ be a polygonal domain. If $u \in \widetilde{W}^{s+1-\varepsilon}_{1+\varepsilon}(\Omega)$ satisfies (2.13) with $\beta = 1 - s$, then GREEDY terminates in finite steps and the resulting isotropic mesh $\mathcal{T}$ satisfies*

$$\|u - u_h\|_{\widetilde{H}^s(\Omega)} \leq C(\Omega, s)\big(\#\mathcal{T}\big)^{-\frac{1}{2}} |\log \#\mathcal{T}|^3 \|f\|_{C^{1-s}(\overline{\Omega})}. \tag{2.64}$$

Proof We sketch the main steps and refer to [28, Theorem 4.5] for details. Finite termination is guaranteed by $t > 0$. To facilitate counting, we split the set of all marked elements $\mathcal{M} = \cup_{j=0}^{k-1} \mathcal{M}_j$ into the disjoint sets $\mathcal{P}_j$ of elements $T \in \mathcal{M}$ with size $h_T := |T|^{1/d}$ satisfying

$$2^{-(j+1)} < |T| \leq 2^{-j} \quad \Rightarrow \quad 2^{-(j+1)/d} < h_T \leq 2^{-j/d}.$$

Exploiting the definition of $\mathcal{P}_j$ and that $E_T(u) > \delta$ for all $T \in \mathcal{P}_j$, we deduce the key properties

$$\#\mathcal{P}_j \le |\Omega| 2^j, \qquad \#\mathcal{P}_j \le C\big(\delta^{-1}R(u)\big)^{1+\varepsilon} 2^{-jt(1+\varepsilon)/d}, \qquad R(u) = \|u\|_{\widetilde{W}^{s+1-\varepsilon}_{1+\varepsilon}(\Omega)}.$$

These complementary bounds allow us to split $\sum_j \#\mathcal{P}_j$ into $j \le j_0$ and $j > j_0$, where j_0 is the smallest index for which the second term is smaller than the first one. This simple trick minimizes the counting and yields $\#\mathcal{M} = \sum_j \#\mathcal{P}_j \lesssim \big(\delta^{-1}R(u)\big)^{1+\varepsilon}$. This in conjunction with (2.62) gives

$$\#\mathcal{T} - \mathcal{T}_0 \lesssim \big(\delta^{-1}R(u)\big)^{1+\varepsilon} \quad \Rightarrow \quad \delta \lesssim (\#\mathcal{T})^{-\frac{1}{1+\varepsilon}} R(u),$$

provided $\#\mathcal{T} \ge 2\#\mathcal{T}_0$. Upon termination of **GREEDY**, we have $E_T(u) \le \delta$ for all $T \in \mathcal{T}$, whence

$$\|u - \Pi_{\mathcal{T}} u\|^2_{\widetilde{H}^s(\Omega)} \lesssim \delta^2 \#\mathcal{T} \lesssim (\#\mathcal{T})^{-1+\frac{2\varepsilon}{1+\varepsilon}} R(u)^2$$

with $\Pi_{\mathcal{T}}$ a local quasi-interpolant. In view of (2.13), we see that $R(u) \lesssim \varepsilon^{-3}\|f\|_{C^{1-s}(\Omega)}$ and

$$\|u - \Pi_{\mathcal{T}} u\|_{\widetilde{H}^s(\Omega)} \lesssim (\#\mathcal{T})^{-\frac{1}{2}} \frac{(\#\mathcal{T})^{\frac{\varepsilon}{1+\varepsilon}}}{\varepsilon^3} \|f\|_{C^{1-s}(\Omega)}.$$

The desired estimate (2.64) follows upon choosing $\varepsilon = |\log \#\mathcal{T}|^{-1}$ and applying (2.51). □

Remark 2.4 (Practical Estimator) For computational purposes, the error estimator (2.63) is not practical. One can replace it with the geometric quantity

$$\mathcal{E}_T(u) = |T| \mathrm{dist}(x_T, \partial\Omega)^{-1};$$

we refer to [28, Section 5] for details. The subordinate **GREEDY** algorithm to the surrogate estimator $\mathcal{E}_T(u)$ exhibits similar convergence rates to the one in Theorem 2.7.

Remark 2.5 (Convergence in Dimensions $d \neq 2$) We point out that both Theorem 2.7 (energy error estimates on graded meshes) and Theorem 2.8 (quasi-optimal error estimate on bisection meshes) can also be extended to dimensions $d \neq 2$, cf. [22, Theorem 3.5] and [28, Theorem 4.5]. In the former, an optimal choice of the mesh grading parameter turns out to provide convergence with order $2 - s$ in $d = 1$ and $\frac{1}{2(d-1)}$ in $d \ge 2$ (up to logarithmic terms), with respect to $\#\mathcal{T}$. In the latter, **GREEDY** delivers the same convergence rates with respect to $\#\mathcal{T}$, although with a higher power in the logarithmic factor.

We conclude this section with error estimates in weaker norms than the energy norm. It is important to realize that we pick up additional powers of the global mesh size h rather than the local mesh size h_T. This is due to the fact that the duality argument is not completely local unless the mesh size changes slowly; see [34, Section 0.8] for details for $d = s = 1$.

Proposition 2.4 (Error Estimates in Weaker Norms) *Let $\Omega \subset \mathbb{R}^2$ be a bounded Lipschitz domain satisfying an exterior ball condition, $u \in \widetilde{H}^s(\Omega)$ denote the solution to* (2.3)*, and $u_h \in \mathbb{V}_h$ denote the solution of the discrete problem* (2.46)*, computed over a mesh $\mathcal{T}$ satisfying* (2.59) *with $\mu = 2$. If $f \in C^{1-s}(\overline{\Omega})$ and $\alpha = \min\{s, \frac{1}{2}\}$, then for $\kappa = \frac{1}{2}$ if $s \neq \frac{1}{2}$ and $\kappa = 1$ if $s = \frac{1}{2}$ we have*

$$\|u - u_h\|_{L^2(\Omega)} \leq C(\Omega, s, \sigma) h^{1+\alpha} |\log h|^{1+\kappa} \|f\|_{C^{1-s}(\overline{\Omega})},$$

and if $s < \frac{1}{2}$

$$\|u - u_h\|_{H^{s-\frac{1}{2}}} \leq C(\Omega, s, \sigma) h^{\frac{3}{2}} |\log h|^2 \|f\|_{C^{1-s}(\overline{\Omega})}.$$

Proof We simply carry out the duality argument (2.58) and estimate the adjoint solution $u_g \in \widetilde{H}^s(\Omega)$ as in the proofs of Proposition 2.1 (L^2-error estimate) and Proposition 2.2 (error estimate in negative Sobolev spaces). This, together with (2.61), gives the assertions. □

2.3.4 Problems with Finite Horizon

The conforming finite element discretization of the weak formulation (2.41) with continuous piecewise linear functions $\mathbb{V}_h$ is similar to (2.46), namely

$$u_h \in \mathbb{V}_h : \quad ((u_h, v_h))_{\delta,s} = \langle f, v_h \rangle \quad \forall v_h \in \mathbb{V}_h. \tag{2.65}$$

Since $((\cdot, \cdot))_{\delta,s}$ is a scalar product equivalent to $(\cdot, \cdot)_s$ in $\widetilde{H}^s(\Omega)$, Lax–Milgram guarantees the existence of a unique solution u_h. The following novel error estimates mimic those in Sect. 2.3.2.

Theorem 2.9 (Error Estimates for Quasi-uniform Meshes) *Let Ω be a bounded Lipschitz domain and $\delta > 0$ be the horizon. Let $\phi\colon [0, \infty) \to [0, \infty)$ be a bounded function of class C^β on a neighborhood of the origin, for some $\beta \in (0, 1]$, and satisfy $\phi(0) > 0$. If $\alpha = \min\{s, \frac{\beta}{2}\}$, then the solutions $u \in \widetilde{H}^s(\Omega)$ of* (2.41) *and $u_h \in \mathbb{V}_h$ of* (2.65) *satisfy the error estimates*

$$\|u - u_h\|_{\widetilde{H}^s(\Omega)} + h^{-\alpha} |\log h|^{-\kappa} \|u - u_h\|_{L^2(\Omega)} \leq C_1 h^{\frac{1}{2}} |\log h|^{\frac{1}{2}} \|f\|_{B^{1/2-s}_{2,1}(\Omega)},$$

$$\|u - u_h\|_{\widetilde{H}^s(\Omega)} + h^{-\alpha} |\log h|^{-\kappa} \|u - u_h\|_{L^2(\Omega)} \leq C_2 h^{\alpha} |\log h|^{\kappa} \|f\|_{L^2(\Omega)},$$

where $\kappa = \frac{1}{2}$ if $s \neq \frac{1}{2}$ and $\kappa = 1$ if $s = \frac{1}{2}$ provided $\beta = 1$, whereas $\kappa = \frac{1}{2}$ if $\beta < 1$. Moreover, we have $C_1 = C_1(\Omega, d, \sigma, \delta)$ and the constant $C_2 = C_2(\Omega, d, s, \sigma, \delta)$ blows up as $s \to \frac{1}{2}$ as $|1 - 2s|^{-\frac{1}{2}}$ for $\beta = 1$.

Proof Argue as in Theorem 2.6 (energy error estimates) and Proposition 2.1 (L^2-error estimate) but utilizing instead Corollary 2.2 (optimal shift property for finite-horizon operators) and Remark 2.2 (Sobolev estimates for finite-horizon operators). □

2.4 Local Error Estimates

The reduced convergence rates in Theorem 2.6 and Propositions 2.1 and 2.3 are essentially due to the boundary behavior of solutions. A natural question is whether it is possible to obtain better convergence rates in the interior of the domain. Such a question has been recently addressed in [22, 61]. We will discuss this next.

2.4.1 Caccioppoli Estimate

This estimate, well known for second-order PDEs, quantifies the property that solutions do not oscillate. Its derivation is particularly simple and revealing for harmonic functions. If $u \in H^1(\Omega)$ satisfies $\Delta u = 0$ in the ball $D_R = D_R(0) \subset \mathbb{R}^d$ of radius R centered at the origin, then

$$\int_{D_{R/2}} |\nabla u|^2 \lesssim \frac{1}{R^2} \int_{D_R} u^2.$$

To prove it, let $\eta \in C_0^\infty(D_R)$ be a cut-off function such that $\eta = 1$ in D_R, $\eta = 0$ in $D^c_{3R/4}$ and $|\nabla \eta| \lesssim R^{-1}$. Since $\int_{D_R} \nabla u \cdot \nabla v = 0$ for all $v \in H^1(D_R)$, taking $v = \eta^2 u \in H^1(D_R)$ yields

$$0 = \int_{D_R} \nabla u \cdot \nabla(\eta^2 u) = \int_{D_R} |\nabla(\eta u)|^2 - \int_{D_R} u^2 |\nabla \eta|^2,$$

whence

$$\int_{D_{R/2}} |\nabla u|^2 \le \int_{D_R} |\nabla(\eta u)|^2 = \int_{D_R} u^2 |\nabla \eta|^2 \lesssim R^{-2} \int_{D_R} u^2.$$

This simple estimate extends to local s-harmonic functions [42]. In fact, if u satisfies $(u, v)_s = 0$ for all $v \in \widetilde{H}^s(D_R)$ and $\int_{D_R^c} \frac{u(x)}{|x|^{d+2s}} dx < \infty$, then

$$|u|^2_{H^s(D_{R/2})} \lesssim R^{-2s} \|u\|^2_{L^2(D_R)} + R^{d+2s} \left(\int_{D_R^c} \frac{u(x)}{|x|^{d+2s}} dx \right)^2.$$

We consider now sets $\Omega_0 \Subset \Omega_1 \Subset \Omega$ such that $0 < R < \operatorname{dist}(\Omega_0, \partial\Omega_1)$ and a finite covering of Ω_0 with balls of radius $R/2$ centered at points in Ω_0. The localization property (2.26) enables us to extend the previous Caccioppoli estimate as

$$|u|^2_{H^s(\Omega_0)} \lesssim R^{-2s} \|u\|^2_{L^2(\Omega_1)} + R^{d+2s} \left(\int_{\Omega_1^c} \frac{u(x)}{|x|^{d+2s}} dx \right)^2,$$

with a hidden constant depending on the covering cardinality. Indeed, for any ball $D_{R/2} = D_{R/2}(x_j)$ in the covering, we split the integral over D_R^c and apply Hölder's inequality to deduce

$$\begin{aligned} \left(\int_{D_R^c} \frac{u(x)}{|x - x_j|^{d+2s}} dx \right)^2 &\le 2 \left(\int_{\Omega_1 \setminus D_R} \frac{u(x)}{|x - x_j|^{d+2s}} dx \right)^2 \\ &\quad + 2 \left(\int_{\Omega_1^c} \frac{u(x)}{|x - x_j|^{d+2s}} dx \right)^2 \\ &\le \frac{2\omega_{d-1}}{d + 4s} R^{-(d+4s)} \|u\|^2_{L^2(\Omega_1 \setminus D_R)} \\ &\quad + 2 \left(\int_{\Omega_1^c} \frac{u(x)}{|x - x_j|^{d+2s}} dx \right)^2. \end{aligned}$$

2.4.2 Local Energy Estimates

Local estimates for FEMs go back to the seminal paper by J. Nitsche and A. Schatz [90] for second-order linear PDEs; see the survey [105]. Such estimates are typically of the following form: the energy error in a subdomain Ω_0 is bounded by the interpolation error on a larger subdomain Ω_1 and a pollution (or slush) term that involves a lower order norm of the error. This structure is instrumental to characterize the pollution effect in Ω_0 due to singularities remote from Ω_0. It is thus natural to wonder whether such a localization is actually possible for the nonlocal problems at hand and, as a consequence, whether the boundary singularity propagates inside the domain Ω or not. This question has been recently studied in [22, 61].

We start with the estimates from [22], which measure the slush term with a global L^2-norm. Its proof is a nontrivial discrete version of the above Caccioppoli estimate.

Theorem 2.10 (Local Estimates with L^2-Slush Term) *Let Ω be a Lipschitz domain, $u \in \widetilde{H}^s(\Omega)$ be the solution of (2.3), and $u_h \in \mathbb{V}_h$ satisfy the local Galerkin orthogonality condition*

$$(u - u_h, v_h)_s = 0 \quad \forall\, v_h \in \mathbb{V}_h(D_R).$$

Table 4 The interior estimates exhibit an improvement rate $h^{\min\{s,1/2\}}$ over global estimates for $f \in B^{-s+1/2}_{2,1}(\Omega)$ and qusi-uniform meshes regardless of the regularity of Ω, according to Theorem 2.10

	Interior rates		Global rates	
	Ω-smooth	Ω-Lipschitz	Ω-smooth	Ω-Lipschitz
$s \le \frac{1}{2}$	$h^{s+\frac{1}{2}}$	$h^{s+\frac{1}{2}}$	$h^{\frac{1}{2}}$	$h^{\frac{1}{2}}$
$s > \frac{1}{2}$	h	h	$h^{\frac{1}{2}}$	$h^{\frac{1}{2}}$

Table 5 The interior estimates exhibit an improvement rate $h^{\min\{s,1-s\}}$ over global estimates for $f \in C^{1-s}(\overline{\Omega})$ and graded meshes for Ω smooth or Lipschitz satisfying the exterior ball condition (e.b.c), according to Theorem 2.10

	Ω-smooth or Lipschitz e.b.c.	
	Interior rates	Global rates
$s \le \frac{1}{2}$	h^{s+1}	h
$s > \frac{1}{2}$	h^{2-s}	h

If $\mathcal{T}_h$ is a shape-regular graded mesh such that $16h_T \le R$ for all $T \in \mathcal{T}_h$, $T \subset D_R$, then for all $v_h \in \mathbb{V}_h$ we have

$$|u-u_h|_{H^s(D_{R/2})} \lesssim \inf_{v_h \in \mathbb{V}_h} \Big(|u-v_h|_{H^s(D_R)} + R^{-s}\|u-v_h\|_{L^2(\Omega)} \Big) + R^{-s}\|u-u_h\|_{L^2(\Omega)}.$$

Combining this estimate with the localization property (2.26) yields the following local estimates for subdomains:

$$|u-u_h|_{H^s(\Omega_0)} \lesssim \inf_{v_h \in \mathbb{V}_h} \Big(|u-v_h|_{H^s(\Omega_1)} + R^{-s}\|u-v_h\|_{L^2(\Omega)} \Big) + R^{-s}\|u-u_h\|_{L^2(\Omega)}.$$

We are now in a position to compare the global error $\|u - u_h\|_{\widetilde{H}^s(\Omega)}$ and the local error $|u - u_h|_{H^s(\Omega_0)}$, where Ω_0 is an interior subdomain of Ω. For *quasi-uniform meshes* and $f \in B^{-s+1/2}_{2,1}(\Omega)$ or smoother, the interior estimates exhibit an improvement rate $h^{\min\{s,1/2\}}$ regardless of the regularity of Ω. We summarize this in Table 4, which neglects logarithmic factors for clarity.

In two dimensions, for *graded meshes* satisfying the condition $h_T \approx h\,\mathrm{dist}(T, \partial\Omega)^{1/2}$, as discussed in (2.59), the interior estimates exhibit an improvement rate $h^{\min\{s,1-s\}}$ for Ω either smooth or Lipschitz with an exterior ball condition (e.b.c.). This is documented in Table 5.

In contrast to [22], the estimates of [61] express the slush term in the $H^{s-1/2}$-norm. This is of interest in case $s < 1/2$,and is actually optimal, because the duality argument used to estimate the $H^{s-1/2}$-norm exploits the maximal regularity of the dual problem; see Proposition 2.2 (error estimate in negative Sobolev spaces). Another relevant difference between [22] and [61] regards mesh grading: there is no restriction in [22], which might in turn be of independent interest, whereas the assumption $h_{\max} h_{\min}^{-2} \lesssim 1$ is required in [61] and agrees with Remark 3.3. The latter is consistent with the mesh grading (2.59) with $\mu = 2$, which is optimal for $d = 2$. We now present a scaled variant of the local estimate from [61, Theorem 2.3].

Proposition 2.5 (Local Estimates with $H^{s-1/2}$-Slush Term) *Let $\Omega \subset \mathbb{R}^d$ be a domain that satisfies the shift property $\|u\|_{\widetilde{H}^{1/2+s-\varepsilon}(\Omega)} \le C\|f\|_{H^{1/2-s-\varepsilon}(\Omega)}$ with a constant $C = C(\Omega, d, s, \varepsilon)$. On shape-regular meshes satisfying $h_{max}h_{min}^{-2} \lesssim 1$ and $h_{max} \le 10R$, we have*

$$|u - u_h|_{H^s(\Omega_0)} \lesssim \left(\inf_{v_h \in \mathbb{V}_h} |u - v_h|_{H^s(\Omega_1)} + R^{-s}\|u - v_h\|_{L^2(\Omega_1)}\right) + R^{-\frac{1}{2}}\|u - u_h\|_{H^{s-1/2}(\Omega)}.$$

Moreover, in case $u \in H^1(\Omega_1)$, we also have

$$|u - u_h|_{H^1(\Omega_0)} \lesssim \inf_{v_h \in \mathbb{V}_h}\left(|u - v_h|_{H^1(\Omega_1)} + \|u - v_h\|_{L^2(\Omega_1)}\right) + R^{s-\frac{3}{2}}\|u - u_h\|_{H^{s-1/2}(\Omega)}.$$

We observe that the above shift property is a consequence of Remark 2.1 (intermediate estimate) with $r = \frac{1}{2} - 2\varepsilon$ for Ω Lipschitz. We also point out that, ignoring logarithmic factors, the interior rates improve by a power $h^{\frac{1}{2}-s}$ for $s < \frac{1}{2}$, namely they become h for quasi-uniform meshes instead of $h^{s+\frac{1}{2}}$ and $h^{\frac{3}{2}}$ instead of h^{1+s} for graded meshes with respect to Tables 4 and 5.

2.5 Computational Examples

We now present two numerical experiments that explore further two important points discussed earlier, namely that the boundary layer (2.8) is generic irrespective of the forcing f and domain regularity and the performance of GREEDY established in Theorem 2.8 (quasi-optimal error estimate on bisection meshes).

Example 2.2 (Boundary Layer Effect) We asserted that the boundary behavior of the special solution (2.7) with $f = 1$ on the unit sphere is generic. We now probe this assertion in the square $\Omega = (-1, 1)^2$ with eccentric right-hand side compactly supported in Ω

$$f(x_1, x_2) = -(r^2 - (x_1 - a)^2 - x_2^2)_+ \tag{2.66}$$

with $a = r = 0.25$. In Fig. 3, we plot the numerical solutions for $s = 0.2, 0.8$ on a uniform mesh with size $h = 2^{-6}$. To numerically study the boundary behavior, we assume the solution can be approximated by

$$u(x) \approx \mathrm{dist}(x, \partial\Omega)^{\alpha(s)},$$

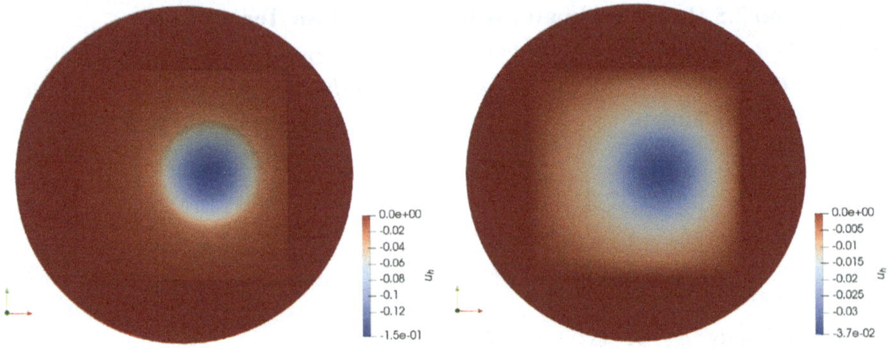

Fig. 3 Numerical solutions of Example 2.2 for $s = 0.2$ (left panel) and for $s = 0.8$ (right panel). The former is much less diffusive than the latter

Table 6 Example 2.2: exponents for different values of s. We observe the boundary behavior (2.8), which is intrinsic to the integral fractional Laplacian

Value of s	0.1	0.2	0.3	0.4	0.5	0.6	0.7	0.8	0.9
Value of $\alpha(s)$	0.118	0.216	0.315	0.414	0.513	0.612	0.709	0.807	0.903

where $\alpha(s)$ is to be determined, and consider mesh points near the boundary point $(-1, 0)$ along the slice $\{x_2 = 0\}$. We use numerical solutions to fit the power $\alpha(s)$ and report the results we obtain in Table 6. Even though f vanishes in a neighborhood of $\partial\Omega$, we observe a good agreement of $\alpha(s)$ with the boundary behavior (2.8).

Example 2.3 (Quasi-optimal Convergence with GREEDY *Algorithm)* We present an example from [28] to illustrate Theorem 2.8 (quasi-optimal error estimate on bisection meshes). We solve (2.1) on the L-shaped domain $\Omega = (-1, 1)^2 \setminus ([0, 1) \times (-1, 0])$ with $f = 1$ and $s \in \{0.25, 0.5, 0.75\}$, using the finite element setting from Sect. 2.2 and the MATLAB code from [3] to assemble the resulting stiffness matrices. We construct a family of bisection grids by employing an adaptive mesh refinement algorithm with a greedy marking strategy based on the package provided in [66]. As an error estimator, we use the surrogate quantity $\mathcal{E}_{\mathcal{T}}(u)$ described in Remark 2.4 (practical estimator).

We run the GREEDY algorithm with tolerance $\delta_k = 2^{-k} \cdot 10^{-2}$, $k = 2, \ldots 5$ in order to construct meshes $\mathcal{T}_2, \ldots \mathcal{T}_5$ and examine the error decay $\|u - \Pi_{\mathcal{T}_k} u\|_{\widetilde{H}^s(\Omega)}$ in terms of $\#\mathcal{T}_k$. We emphasize that the marking strategy $\mathcal{E}_{\mathcal{T}}(u) > \delta_k$ is independent of s and insensitive to the presence of reentrant corners. This is reflected in Fig. 4 whose left and middle panels depict meshes with $\#\mathcal{T} = 9504$ elements and $\#\mathcal{T} = 15118$ elements, respectively.

To compute the error $\|u - \Pi_{\mathcal{T}_k} u\|_{\widetilde{H}^s(\Omega)}$, we resort to a solution on a highly refined mesh because of the lack of a closed analytical expression for the solution u of (2.1) in this setting. The right panel in Fig. 4 exhibits our computational orders of convergence. They show a good agreement with the expected log-linear rate from

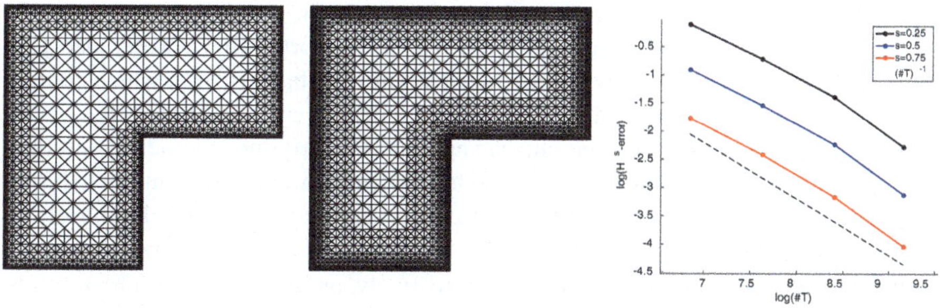

Fig. 4 GREEDY with the surrogate estimator from Remark 2.4 (practical estimator). Left and center: graded bisection meshes with 9504 and 15,118 elements, respectively. Right: errors in the $\widetilde{H}^s(\Omega)$-norm for $s \in \{0.25, 0.5, 0.75\}$ and $f = 1$. Computational rates for the L-shaped domain are consistent with the expected theoretical rate $(\#\mathcal{T})^{-\frac{1}{2}} |\log \#\mathcal{T}|^3$ for solutions $u \in \widetilde{W}^{s+1-\varepsilon}_{1+\varepsilon}(\Omega)$ (Theorem 2.8)

Theorem 2.8, even though Ω does not satisfy the sufficient condition (exterior ball condition) leading to (2.9).

2.6 Nonconforming FEM Based on Dunford–Taylor Representation

A nonconforming finite element method for the approximation of (2.1) has been proposed in [20]. The method is based on the following representation formula for the $\widetilde{H}^s(\Omega)$-inner product:

$$(u, v)_s = \frac{2\sin(s\pi)}{\pi} \int_0^\infty t^{-1-2s} \langle u + w(u,t), v \rangle \, dt \quad \forall u, v \in \widetilde{H}^s(\Omega). \tag{2.67}$$

Above, $w = w(u,t) \in H^1(\mathbb{R}^d)$ is the solution to the *local problem* $w - t^2 \Delta w = -u$ in $\mathbb{R}^d$.

By using this representation, [20] proposes a three-step numerical method.

- *Sinc quadrature:* the change of variables $t = e^{-y/2}$ leads to

$$(u, v)_s = \frac{\sin(s\pi)}{\pi} \int_{-\infty}^{\infty} e^{sy} \langle u + w(u, t(y)), v \rangle \, dy.$$

Therefore, given $N > 0$, one can consider the sinc quadrature approximation

$$(u, v)_s \simeq \frac{\sin(s\pi)}{N\pi} \sum_{\ell=-N}^{N} e^{sy_\ell} \langle u + w(u, t(y_\ell)), v \rangle.$$

- *Domain truncation:* in principle, for any $t > 0$, the function $w(u, t)$ can be supported in the whole space $\mathbb{R}^d$ even if u is supported in Ω. Reference [20] proposes to truncate the local problems to a family of balls $B^M(t)$, which contain Ω and whose radius depends on either t and a certain parameter $M > 0$.
- *Finite element approximation:* one can apply a standard finite element discretization of the local problems on $B^M(t)$. This discretization requires meshes that fit Ω and $B^M(t) \setminus \Omega$ exactly; furthermore, in order to add the contribution of the local problems, the mesh on Ω needs to remain fixed. We denote by $\mathbb{V}_h$ and $\mathbb{V}_h^M$ the discrete spaces on Ω and $B^M(t)$, respectively, and given $\psi \in L^2(\mathbb{R}^d)$, $t > 0$, and $M > 0$, we set $w_h^M = w_h^M(\psi, t)$ to be the unique function in $\mathbb{V}_h^M$ such that

$$\int_{B^M(t)} w_h^M v_h + t^2 \nabla w_h^M \cdot \nabla v_h \, dx = - \int_{B^M(t)} \psi v_h \, dx \quad \forall v_h \in \mathbb{V}_h^M.$$

By combining the three steps outline above, we arrive at the discrete bilinear form

$$a_{\mathcal{T}}^{N,M}(u_h, v_h) := \frac{\sin(s\pi)}{N\pi} \sum_{\ell=-N}^{N} e^{sy_\ell} \langle u_h + w_h^M(u_h, t(y_\ell)), v \rangle \quad \forall u_h, v_h \in \mathbb{V}_h.$$

Strang's lemma implies that the $\widetilde{H}^s(\Omega)$-error between u and u_h is bounded by an approximation error and the sum of the consistency errors from the three steps outlined above. Thus, we have [20, Theorem 7.7]

$$\|u - u_h\|_{\widetilde{H}^s(\Omega)} \le C \left(e^{-c\sqrt{N}} + e^{-cM} + h^{\beta - s} |\log h| \right) \|u\|_{\widetilde{H}^\beta(\Omega)},$$

where $\beta \in (s, 3/2)$. The regularity estimate Theorem 2.3 indicates that for $f \in B_{2,1}^{-s+1/2}(\Omega)$ we have $\beta = s + 1/2 - \varepsilon$. Taking $M = \mathcal{O}(|\log h|)$ and $N = \mathcal{O}(|\log h|^2)$ gives an error bound with order $h^{1/2}|\log h|$. This is comparable with the rate obtained in Theorem 2.6 for quasi-uniform meshes. To the best of the authors' knowledge, implementation of this nonconforming approach over graded meshes, while feasible in theory, has not been addressed yet.

Finally, we comment that formula (2.67) has also served as the starting point for a spectral Galerkin method [98] for problems posed in $\mathbb{R}^d$.

2.7 Other Numerical Approaches

In recent years, there has been substantial progress in the implementation of discretization schemes for problems involving the integral fractional Laplacian (1.1) and other nonlocal operators on bounded domains. Here, we briefly comment on some approaches that are not of finite element type. We refer to [45, 81] for further discussion on these and related approaches.

Finite Difference Methods Reference [74] proposes a method that combines finite differences with numerical quadrature, obtains a discrete convolution operator, and studies the convergence of such a method. However, the finite difference algorithm is only implemented in $d = 1$ dimension, and the convergence analysis requires solutions to be of class C^4 up to the domain boundary. More recent finite difference implementations are able to deal with higher dimensional problems (cf. [53, 54, 87], for example). An interesting two-scale finite difference method on finite element type meshes has been proposed in [72]. The convergence analysis in that work is based on weighted Hölder regularity estimates like (2.10) and the use of suitable discrete barriers, thereby avoiding unrealistic solution regularity assumptions.

Fourier Methods The Fourier representation of the fractional Laplacian offers some opportunities for the discretization of such an operator. In particular, for periodic functions, such a representation can be exploited to develop spectral approximation schemes. Interesting applications of this approach include phase-field modeling [8] and image processing [10].

In contrast, if one truncates the domain on which the Fourier transformation is performed, one instead obtains the fractional Laplacian of a function that was periodically extended outside the truncation domain; such an approach, in conjunction with the use of sinc basis functions, has been exploited in [11] to approximate the Dirichlet problem for the integral fractional Laplacian.

3 Quasilinear Problems

Let $\Omega \subset \mathbb{R}^d$ be an open, bounded set with Lipschitz boundary and functions $f\colon \Omega \to \mathbb{R}$ and $g\colon \Omega^c \to \mathbb{R}$ be given. In this section, we discuss some problems that can be succinctly expressed as follows: find $u\colon \mathbb{R}^d \to \mathbb{R}$ that coincides with g on Ω^c and such that

$$a_u(u, v) = \langle f, v\rangle \tag{3.1}$$

for every v sufficiently smooth and vanishing on Ω^c. Here, the form $a_u(\cdot,\cdot)$ is bilinear albeit it depends on the solution one is seeking,

$$a_u(w, v) := \iint_{Q_\Omega} \widetilde{G}\left(\frac{u(x) - u(y)}{|x - y|^t}\right) \frac{(w(x) - w(y))(v(x) - v(y))}{|x - y|^{d+2\sigma}} dxdy. \tag{3.2}$$

Naturally, to determine the problem, one needs to specify the values of $\sigma \in (0, 1)$, $t \in \{\sigma, 1\}$, and the nonlinearity $\widetilde{G}$ above. In the case $\widetilde{G}$ is a constant function, problem (3.1) reduces to (2.3). Here, we shall be concerned with two examples:

- *Fractional mean curvature.* If we set $\sigma = s + 1/2$ for some $s \in (0, 1/2)$, $t = 1$, and

$$\widetilde{G}(\rho) = C_{d,s} \int_0^1 (1 + \rho^2 r^2)^{-(d+1+2s)/2} dr,$$

then problem (3.1) corresponds to finding a function u that coincides with g on Ω^c and whose subgraph possesses certain *fractional mean curvature* equal to f on the cylinder $\Omega \times \mathbb{R}$. In case $f \equiv 0$, the subgraph of u is an *s-minimal set.* The normalization constant $C_{d,s} := \frac{1-2s}{|D_1(0)|}$ guarantees that, in the limit $s \to \frac{1}{2}^-$, one recovers the classical mean curvature operator [24, Lemmas 5.9 and 5.11]. Because in this work we are not concerned with such a limit, we shall omit the constant in the following.

- *Fractional p-Laplacian.* Setting $\sigma = t = s$ for some $s \in (0, 1)$ and $\widetilde{G}(\rho) = \frac{C_{d,s,p}}{2} |\rho|^{p-2}$ for some $p \in (1, \infty)$, we recover the Dirichlet problem for the (p, s)-Laplace operator,

$$(-\Delta)_p^s v(x) := C_{d,s,p} \int_{\mathbb{R}^d} \frac{|v(x) - v(y)|^{p-2}(v(x) - v(y))}{|x - y|^{d+sp}} dy.$$

The integral above needs to be understood in the principal value sense in case $sp \geq 1$, and we choose the normalizing constant $C_{d,s,p}$ as

$$C_{d,s,p} = \frac{s(1-s)p\, \Gamma(\frac{ps+d}{2})\, 2^{2s-2}}{\pi^{\frac{d-1}{2}} \Gamma(\frac{(p-2)s+3}{2}) \Gamma(2-s)}. \tag{3.3}$$

We remark that this choice is somewhat arbitrary. Nevertheless, in case $p = 2$ it gives rise to the integral fractional Laplacian (1.1); moreover, for every smooth function $v \in C_c^\infty(\mathbb{R}^d)$, we recover the asymptotic behaviors [31, 44, 85]

$$\lim_{s \to 1^-} (-\Delta)_p^s v = -\nabla \cdot (|\nabla v|^{p-2} \nabla v), \qquad \lim_{s \to 0^+} (-\Delta)_p^s v = |v|^{p-2} v.$$

This section is organized as follows. We first review some theory regarding the fractional mean curvature, including the variational problem we aim to solve, and some properties of nonlocal minimal graphs. Afterward, we focus on theoretical aspects related to the fractional p-Laplacian, including the regularity of solutions to the Dirichlet problem for such an operator on Lipschitz domains. Then, we propose a common discretization technique for both operators and study the convergence of the discrete method. The section concludes with some numerical experiments.

3.1 Minimal Graphs

In this section, we describe the notion of fractional perimeter and some results regarding fractional minimal sets. Our focus shall be on *minimal graphs* on $\mathbb{R}^d$, which arise as minimal sets on cylinders in $\mathbb{R}^{d+1}$ when the exterior data is a subgraph.

3.1.1 Fractional Perimeter and Minimal Sets

Next, we briefly review the notions of s-perimeter and s-minimal sets.

Definition 3.1 (s-Perimeter) Given a domain $B \subset \mathbb{R}^{d+1}$ and $s \in (0, 1/2)$, the s-perimeter of a set $A \subset \mathbb{R}^{d+1}$ in B is defined as

$$P_s(A; B) := \frac{1}{2} \iint_{Q_B} \frac{|\chi_A(x) - \chi_A(y)|}{|x - y|^{d+2s}} \, dydx, \tag{3.4}$$

where $Q_B = (\mathbb{R}^{d+1} \times \mathbb{R}^{d+1}) \setminus (B^c \times B^c)$ and $B^c = \mathbb{R}^{d+1} \setminus B$.

Formally, definition (3.4) coincides with

$$P_s(A, B) = \frac{1}{2} \left(|\chi_A|_{W_1^{2s}(\mathbb{R}^{d+1})} - |\chi_A|_{W_1^{2s}(B^c)} \right),$$

and thus we recover the expression (1.3).

The sets that minimize the s-fractional perimeter among those that coincide with them outside B are deemed as *s-minimal sets* in B. The notion of s-minimality involves the behavior of sets in the whole space $\mathbb{R}^{d+1}$, in contrast with the classical (local) minimal sets, that are characterized by their behavior in $\overline{B}$.

Definition 3.2 (s-Minimal Set) A set A is s-minimal in an open set $B \subset \mathbb{R}^{d+1}$ if $P_s(A, B)$ is finite and $P_s(A, B) \leq P_s(A', B)$ among all measurable sets $A' \subset \mathbb{R}^{d+1}$ such that $A' \setminus B = A \setminus B$.

Given an open set B and a fixed set A_0, the Dirichlet or Plateau problem for nonlocal minimal surfaces aims to find an s-minimal set A such that $A \setminus B = A_0 \setminus B$. For a bounded Lipschitz domain B, the existence of solutions to the Plateau problem is established in [38].

3.1.2 Formulation of Minimal Graphs

We focus on s-minimal sets on a cylinder $B = \Omega \times \mathbb{R}$, where Ω is a bounded and sufficiently smooth domain in $\mathbb{R}^d$, with exterior data being a subgraph,

$$A_0 = \left\{ (x', x_{d+1}) \colon x_{d+1} < g(x'), \; x' \in \mathbb{R}^d \setminus \Omega \right\},$$

for some given function $g : \mathbb{R}^d \setminus \Omega \to \mathbb{R}$. We highlight some key features of this problem and refer to [24] for a detailed discussion.

- Definition 3.2 is not appropriate for our problem because every set A that is a subgraph on $\mathbb{R}^{d+1} \setminus B$ has infinite s-perimeter in B. To remedy this issue, one needs to look for *locally* s-minimal sets [82], namely, sets that are s-minimal in every bounded open subset compactly supported in B.
- In our setting, there exists a unique locally s-minimal set, which turns out to be the subgraph of a certain function u (see [47, 83]). We can therefore restrict our minimization problem to the set of subgraphs of functions that coincide with g on Ω^c.
- In case g is a bounded function, we can replace the infinite cylinder $B = \Omega \times \mathbb{R}$ by a truncated cylinder $B_M = \Omega \times (-M, M)$, where $M > 0$ depends on g [83, Proposition 2.5].
- Let A be the subgraph of a certain function v that coincides with g on Ω^c and B_M as above. The fractional s perimeter of A in B_M can be expressed as

$$P_s(A, B_M) = I_s[v] + C(M, d, s, \Omega, g),$$

where I_s is a suitable nonlocal energy functional, see (3.5) below [83, Proposition 4.2.8], [24, Proposition 2.3].

With the considerations above, we can express the Plateau problem for nonlocal minimal graphs as follows [24]: find a function $u : \mathbb{R}^d \to \mathbb{R}$, with the constraint $u = g$ in Ω^c, such that it minimizes the strictly convex energy

$$I_s[u] := \iint_{Q_\Omega} F_s\left(\frac{u(x) - u(y)}{|x - y|}\right) \frac{1}{|x - y|^{d+2s-1}}\, dxdy, \tag{3.5}$$

where F_s is defined as

$$F_s(\rho) := \int_0^\rho \frac{\rho - r}{\left(1 + r^2\right)^{(d+1+2s)/2}} dr.$$

Let $s \in (0, 1/2)$ and $g \in L^\infty(\Omega^c)$ be given. We consider the space

$$\mathbb{V}^g := \{v \colon \mathbb{R}^d \to \mathbb{R} : \ v\big|_\Omega \in W_1^{2s}(\Omega),\ v = g \text{ in } \Omega^c\},$$

equipped with the norm

$$\|v\|_{\mathbb{V}^g} := \|v\|_{L^1(\Omega)} + |v|_{\mathbb{V}^g},$$

where

$$|v|_{\mathbb{V}^g} := \iint_{Q_\Omega} \frac{|v(x) - v(y)|}{|x - y|^{d+2s}} dxdy,$$

and we recall $Q_\Omega = (\mathbb{R}^d \times \mathbb{R}^d) \setminus (\Omega^c \times \Omega^c)$. The seminorm in $\mathbb{V}^g$ does not take into account interactions over $\Omega^c \times \Omega^c$ because these are fixed for the class of functions we consider; therefore, we do not need to assume g to be a function in the class $W_1^{2s}(\Omega^c)$. In particular, g may not decay at infinity. In case g is the zero function, the space $\mathbb{V}^g$ coincides with the standard zero-extension Sobolev space $\widetilde{W}_1^{2s}(\Omega)$; for consistency of notation, we denote such a space by $\mathbb{V}^0$.

Next, given $u \in \mathbb{V}^g$, we consider the bilinear form $a_u : \mathbb{V}^g \times \mathbb{V}^0 \to \mathbb{R}$,

$$a_u(w, v) := \iint_{Q_\Omega} \widetilde{G}\left(\frac{u(x) - u(y)}{|x - y|}\right) \frac{(w(x) - w(y))(v(x) - v(y))}{|x - y|^{d+1+2s}} dxdy, \tag{3.6}$$

with $\widetilde{G}(\rho) = \rho^{-1} F_s'(\rho)$ given by

$$\widetilde{G}(\rho) := \int_0^1 (1 + \rho^2 r^2)^{-(d+1+2s)/2} dr. \tag{3.7}$$

Clearly, the weight $\widetilde{G}$ satisfies $0 < \widetilde{G}(\rho) \leq 1$ for all $\rho \in \mathbb{R}$. We emphasize the behavior $\widetilde{G}(\rho) \to 0$ as $|\rho| \to \infty$, which implies that the weight in (3.2) degenerates whenever the difference quotient $\frac{|u(x)-u(y)|}{|x-y|}$ blows up. We also point out to the exponent $d + 1 + 2s$ in (3.6) that indicates that, whenever the weight $\widetilde{G}(\rho)$ does not degenerate, the form a_u behaves as the one arising in a fractional diffusion problem of order $s + 1/2$ in $\mathbb{R}^d$.

The weak formulation of the fractional minimal graph problem can be obtained in a standard fashion, namely by taking the first variation of $I_s[u]$ in (3.5). As described in [24], such a formulation reads as follows: find $u \in \mathbb{V}^g$ satisfying

$$a_u(u, v) = 0 \quad \forall v \in \mathbb{V}^0. \tag{3.8}$$

3.1.3 Regularity and Stickiness

An outstanding feature of the fractional Plateau problem is the emergence of *stickiness phenomena.* In the setting of this chapter, this means that the minimizer may be discontinuous across $\partial\Omega$. As shown by Dipierro, Savin, and Valdinoci [50], stickiness is indeed the generic behavior of nonlocal minimal graphs in case $\Omega \subset \mathbb{R}$.

Let us illustrate this phenomenon with a quantitative example proposed in [48, Theorem 1.2] and studied numerically in [26]. We solve (3.8) on a fixed mesh for $\Omega = (-1, 1) \subset \mathbb{R}$ and $g(x) = M \operatorname{sign}(x)$, where $M > 0$. Reference [48] proves that, for every $s \in (0, 1/2)$, there exists $M > 0$ such that the corresponding solution u^M is discontinuous across $\partial\Omega$ and that there exists an optimal constant c_0 such that

$$\sup_{x\in\Omega} u^M(x) < c_0 M^{\frac{1+2s}{2+2s}}, \qquad \inf_{x\in\Omega} u^M(x) > -c_0 M^{\frac{1+2s}{2+2s}}. \tag{3.9}$$

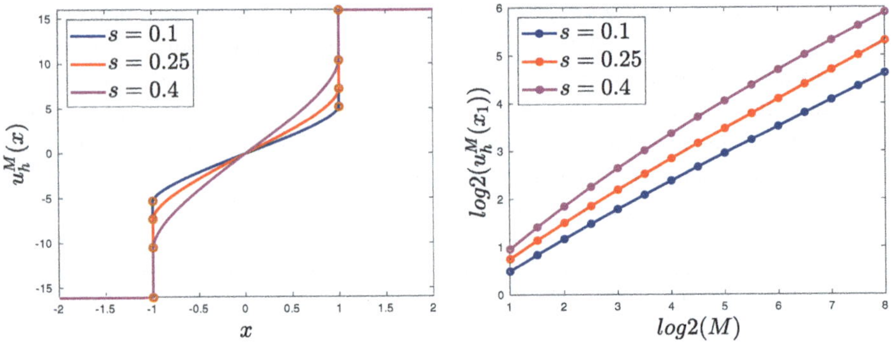

Fig. 5 Stickiness in 1D. The left panel displays the finite element solutions u_h^M for $M = 16$ and $s \in \{0, 1, 0.25, 0.4\}$. The right panel shows the value of $u_h^M(x_1)$ as a function of M for $s \in \{0.1, 0.25, 0.4\}$, which is expected to behave according to (3.9)

The left panel in Fig. 5 shows the computed solutions with $M = 16$ and $s = 0.1, 0.25, 0.4$. Because in all cases the discrete solutions u_h are monotonically increasing in Ω, we regard the value of $u_h^M(x_1)$ as an approximation of $\sup_{x \in \Omega} u^M(x)$, where x_1 is the free node closest to 1. The right panel in Fig. 5 shows how $u_h^M(x_1)$ varies with respect to M for different values of s.

For $s = 0.1$ and $s = 0.25$, the slopes of the curves are slightly larger than the theoretical rate $M^{\frac{1+2s}{2+2s}}$ whenever M is small. However, as M increases, we see a good agreement with theory. Comparing results for $M = 2^{7.5}$ and $M = 2^8$, we observe approximate rates 0.553 for $s = 0.1$ and 0.602 for $s = 0.25$, where the expected rates are $6/11 \approx 0.545$ and $3/5 = 0.600$, respectively. However, the situation is different for $s = 0.4$: the plotted curve does not correspond to a flat line, and the last two nodes plotted, with $M = 2^{7.5}$ and $M = 2^8$, show a relative slope of about 0.57, which is off the expected rate $9/14 \approx 0.643$. This issue is related to poor boundary resolution and can be corrected by refining the mesh accordingly (cf. [26, Table 1]).

Even though stickiness is the typical behavior of fractional minimal graphs, an interesting phenomenon arises when such graphs happen to be continuous at some point on the boundary of the domain. In the case $\Omega \subset \mathbb{R}^2$, reference [49] proves that, at any boundary points at which stickiness does not happen, the tangent planes of the traces from the interior must coincide with those of the exterior datum. Such a hard geometric constraint is in sharp contrast with the case of classical minimal graphs. We illustrate this behavior with the computational results [26] of an experiment first proposed in [49]. We consider $\Omega = (0, 1) \times (-1, 1)$ and the Dirichlet datum

$$g(x, y) = \gamma \left(\chi_{(-1,-a)\times(0,1)}(x, y) - \chi_{(-1,-a)\times(-1,0)}(x, y) \right), \tag{3.10}$$

where $a \in [0, 1]$ and $\gamma > 0$ are parameters to be chosen. Figure 6 (left panel) displays a numerical solution u_h corresponding to $s = 0.25$ and $a = 1/8$. The right

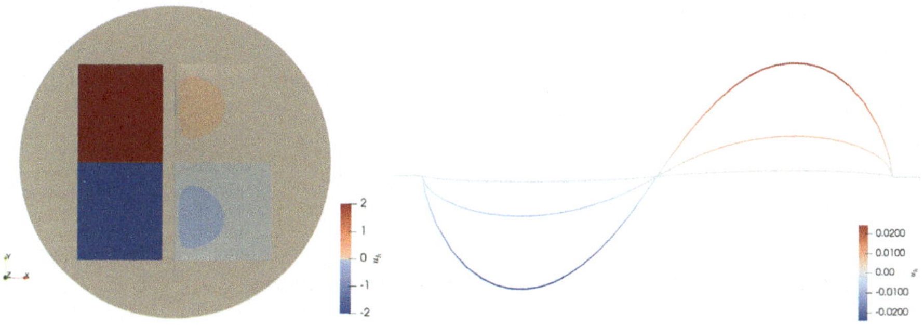

Fig. 6 Plot of the finite element solutions to a fractional graph Plateau problem with g given as in (3.10) and $a = 1/8$, $s = 0.25$. Left panel: top view of the solution. Right panel: slices at $x = 2^{-3}, 2^{-6}$, and 2^{-9}. The fractional minimal graph flattens as $x \to 0^+$, in agreement with the fact that for such a minimizer being continuous at some point $x \in \partial\Omega$ implies having continuous tangential derivatives at such a point

panel in Fig. 6 exhibits slices of the computed solution at $x = 2^{-3}, 2^{-6}$, and 2^{-9}. The flattening of the curves as $x \to 0^+$ is apparent.

In spite of their rich boundary behavior, minimal graphs are smooth in the interior of the domain. The following theorem is stated in [37, Theorem 1.1], where an estimate for the gradient of the minimal function is derived. Once such an estimate is obtained, the claim follows by the arguments from [15] and [65].

Theorem 3.1 (Interior Smoothness of Graph Nonlocal Minimal Sets) *Let $A \subset \mathbb{R}^{d+1}$ be a locally s-minimal set in the cylinder $B = \Omega \times \mathbb{R}$, given by the subgraph of a measurable function u that is bounded in an open set $\Lambda \supset \Omega$. Then, $u \in C^\infty(\Omega)$.*

3.2 *Fractional (p, s)-Laplacians*

In this section, we introduce the fractional (p, s)-Laplacian problem and discuss the regularity of solutions. Given $s \in (0, 1)$, $p \in (1, \infty)$, and a bounded domain $\Omega \subset \mathbb{R}^d$, we introduce the Sobolev space of functions vanishing on Ω^c,

$$\widetilde{W}_p^s(\Omega) := \left\{ v \in W_p^s(\mathbb{R}^d) \colon \operatorname{supp} v \subset \overline{\Omega} \right\}$$

with the norm

$$\|v\|_{\widetilde{W}_p^s(\Omega)} := |v|_{W_p^s(\mathbb{R}^d)} = \left(\frac{C_{d,s,p}}{2} \iint_{Q_\Omega} \frac{|v(x) - v(y)|^p}{|x - y|^{d+sp}} \, dx \, dy \right)^{1/p},$$

where $C_{d,s,p}$ is given by (3.3).

3.2.1 Problem Formulation

For a given $f : \Omega \to \mathbb{R}$, we consider the Dirichlet problem for the fractional (p,s)-Laplacian

$$\begin{cases} (-\Delta)^s_p u = f \ \text{ in } \Omega, \\ u = 0 \ \text{ in } \Omega^c. \end{cases} \tag{3.11}$$

The solution u to (3.11) minimizes the strictly convex energy

$$I_{s,p}[v] := \frac{1}{p}\|v\|^p_{\widetilde{W}^s_p(\Omega)} - \int_\Omega fv = \frac{C_{d,s,p}}{2p}\iint_{Q_\Omega} \frac{|v(x)-v(y)|^p}{|x-y|^{d+sp}}dxdy - \int_\Omega fv \tag{3.12}$$

among functions in $\widetilde{W}^s_p(\Omega)$. From now on, we shall assume that $f \in W^{-s}_{p'}(\Omega) := [\widetilde{W}^s_p(\Omega)]'$, so that we can guarantee the well-posedness of this minimization problem in $\widetilde{W}^s_p(\Omega)$. Taking the first variation $\frac{\delta I_{s,p}[u]}{\delta u}(v)$ of (3.12) in the direction $v \in \widetilde{W}^s_p(\Omega)$ and setting it to zero, such a minimization problem can also be written as the weak formulation: find $u \in \widetilde{W}^s_p(\Omega)$ such that

$$a_u(u,v) = \langle f, v\rangle \quad \forall v \in \widetilde{W}^s_p(\Omega), \tag{3.13}$$

where $\langle\cdot,\cdot\rangle$ denotes the duality pairing between $W^{-s}_{p'}(\Omega)$ and $\widetilde{W}^s_p(\Omega)$ and

$$a_u(w,v) := \frac{C_{d,s,p}}{2}\iint_{Q_\Omega} \frac{|u(x)-u(y)|^{p-2}(w(x)-w(y))(v(x)-v(y))}{|x-y|^{d+sp}}\,dx\,dy. \tag{3.14}$$

Setting $v = u$ in (3.13) leads to the stability estimate

$$\|u\|_{\widetilde{W}^s_p(\Omega)} \lesssim \|f\|^{\frac{1}{p-1}}_{W^{-s}_{p'}(\Omega)}. \tag{3.15}$$

Next, let us recall the following auxiliary identities that follow by [69, Lemmas 5.1–5.4]: for all $a, b \in \mathbb{R}$, we have

$$\left||a|^{p-2}a - |b|^{p-2}b\right| \le \begin{cases} C|a-b|^{p-1} \text{ if } 1 < p \le 2, \\ C|a-b|(|a|+|b|)^{p-2} \text{ if } 2 \le p < \infty, \end{cases}$$

and

$$\left(|a|^{p-2}a - |b|^{p-2}b\right)(a-b) \ge \begin{cases} \alpha|a-b|^2(|a|+|b|)^{p-2} \text{ if } 1 < p \le 2, \\ \alpha|a-b|^p \text{ if } 2 \le p < \infty. \end{cases} \tag{3.16}$$

The constants C and α above only depend on p. The inequalities (3.16) immediately give rise to the following monotonicity estimates for the fractional (p, s)-Laplacian.

Lemma 3.1 (Monotonicity) *If $2 \le p < \infty$, there exists $\alpha > 0$ such that*

$$\begin{aligned}\langle(-\Delta)^s_p u - (-\Delta)^s_p v, u - v\rangle &= a_u(u, u - v) - a_v(v, u - v) \\ &\ge \alpha\|u - v\|^p_{\widetilde{W}^s_p(\Omega)} \quad \forall u, v \in \widetilde{W}^s_p(\Omega).\end{aligned} \tag{3.17}$$

In the case $1 < p < 2$, we have

$$\begin{aligned}\langle(-\Delta)^s_p u - (-\Delta)^s_p v, u - v\rangle &= a_u(u, u - v) - a_v(v, u - v) \\ &\ge \alpha\|u - v|^2_{\widetilde{W}^s_p(\Omega)} \Big(\|u\|_{\widetilde{W}^s_p(\Omega)} \\ &\quad +\|v\|_{\widetilde{W}^s_p(\Omega)}\Big)^{p-2} \quad \forall u, v \in \widetilde{W}^s_p(\Omega).\end{aligned} \tag{3.18}$$

For $p \ge 2$, the operator $(-\Delta)^s_p$ is p-coercive but for $1 < p < 2$, the operator is 2-coercive on bounded sets of $\widetilde{W}^s_p(\Omega)$. Namely, if $u, v \in \widetilde{W}^s_p(\Omega)$ satisfy $\|u\|_{\widetilde{W}^s_p(\Omega)}, \|v\|_{\widetilde{W}^s_p(\Omega)} \le C$, then (3.18) gives

$$C^{2-p}\langle(-\Delta)^s_p u - (-\Delta)^s_p v, u - v\rangle \gtrsim \|u - v\|^2_{\widetilde{W}^s_p(\Omega)}.$$

We can now assess the continuity properties of the solution operator $f \mapsto u_f$ of (3.13). If we assume $f \in W^{-s}_{p'}(\Omega)$ with $\|f\|_{W^{-s}_{p'}(\Omega)} = K$, then $\|u\|_{\widetilde{W}^s_p(\Omega)} \le K^{\frac{1}{p-1}}$ by (3.15). Therefore, the operator $f \mapsto u_f$ defined on

$$\overline{B}_K := \{f \in W^{-s}_{p'}(\Omega) : \|f\|_{W^{-s}_{p'}(\Omega)} \le K\} \mapsto \widetilde{W}^s_p(\Omega)$$

happens to be Lipschitz continuous on $\overline{B}_K$ provided $1 < p < 2$, namely

$$\|u_f - u_g\|_{\widetilde{W}^s_p(\Omega)} \le K^{\frac{2-p}{p-1}}\|f - g\|_{W^{-s}_{p'}(\Omega)} \quad \forall f, g \in \overline{B}_K.$$

In contrast, for $p \ge 2$, the solution map is Hölder continuous on $W^{-s}_{p'}(\Omega)$ in view of (3.17),

$$\|u_f - u_g\|_{\widetilde{W}^s_p(\Omega)} \le \alpha^{-\frac{1}{p-1}}\|f - g\|^{\frac{1}{p-1}}_{W^{-s}_{p'}(\Omega)}.$$

3.2.2 Besov Regularity

Using the coercivity of the energy functional $I_{s,p}$, it is possible to proceed in the same way as in Sect. 2.1.2 to obtain Besov regularity estimates provided the bounded domain Ω has a Lipschitz boundary. As a consequence of Lemma 3.1, the following inequalities hold for the solution u of (3.13):

$$\|v-u\|^p_{\widetilde{W}^s_p(\Omega)} \lesssim I_{s,p}[v] - I_{s,p}[u] \quad \text{if } p \geq 2, \tag{3.19}$$

$$\left(\|u\|_{\widetilde{W}^s_p(\Omega)} + \|v\|_{\widetilde{W}^s_p(\Omega)}\right)^{p-2} \|v-u\|^2_{\widetilde{W}^s_p(\Omega)} \lesssim I_{s,p}[v] - I_{s,p}[u] \quad \text{if } 1 < p < 2. \tag{3.20}$$

In our proof of Besov regularity theory, these inequalities play the same role as (2.14) does in the linear case. We write $I_{s,p}[v] = \mathcal{F}_p(v) - \mathcal{F}_1(v)$, where

$$\mathcal{F}_1(v) := \int_\Omega f v, \quad \mathcal{F}_p(v) := \frac{1}{p}\|v\|^p_{\widetilde{W}^s_p(\Omega)} \quad \forall\, v \in \widetilde{W}^s_p(\Omega),$$

and consider the localized translation operator T_h defined in (2.22). Then, for $\sigma \in (0, 1]$ and $t \in (-1, 1+\sigma)$, we have the following modifications to (2.23) and (2.24) [23]:

$$\sup_{h\in D} \frac{|\mathcal{F}_1(T_h v) - \mathcal{F}_1(v)|}{|h|^\sigma} \lesssim \|f\|_{B^t_{p',1}(D_{2\rho}(x_0)\cap\Omega)} |v|_{B^{\sigma-t}_{p,\infty}(D_{2\rho}(x_0))} \qquad \forall v \in \dot{B}^{\sigma-t}_{p,\infty}(D_{2\rho}(x_0));$$

$$\sup_{h\in D} \frac{|\mathcal{F}_p(T_h v) - \mathcal{F}_p(v)|}{|h|^\sigma} \lesssim \iint_{Q_{D_{2\rho}(x_0)}} \frac{|v(x)-v(y)|^p}{|x-y|^{d+sp}} dydx \qquad \forall v \in \widetilde{W}^s_p(\Omega).$$

Above, $D = \mathcal{C}_\rho(\mathbf{n}(x_0), \theta)$ is a cone of admissible directions (cf. (2.21)), so that we have $T_h v \in \widetilde{W}^s_p(\Omega)$ for all $v \in \widetilde{W}^s_p(\Omega)$ and $h \in D$.

Arguing similarly to the proof of Theorem 2.3, in [23] we obtain the following result.

Theorem 3.2 (Shift Property for the Fractional (p, s)**-Laplacian)** *Let $p \in (1, \infty)$, $r \in (0, 1]$, Ω be a bounded Lipschitz domain, and u be the solution to the fractional (p, s)-Laplace equation* (3.13).

If $p \geq 2$ and $f \in B^{-s+\frac{r}{p'}}_{p',1}(\Omega)$, then we have $u \in \dot{B}^{s+\frac{r}{p}}_{p,\infty}(\Omega)$ with

$$\|u\|_{\dot{B}^{s+\frac{r}{p}}_{p,\infty}(\Omega)} \le C(\Omega,d,p)\|f\|^{\frac{1}{p-1}}_{B^{-s+\frac{r}{p'}}_{p',1}(\Omega)}.$$

If $p < 2$ and $f \in B^{-s+\frac{r}{2}}_{p',1}(\Omega)$, then we have $u \in \dot{B}^{s+\frac{r}{2}}_{p,\infty}(\Omega)$ with

$$\|u\|_{\dot{B}^{s+\frac{r}{2}}_{p,\infty}(\Omega)} \le C(\Omega,d,p)\|f\|^{\frac{2-p}{p-1}}_{W^{-s}_{p'}(\Omega)}\|f\|_{B^{-s+\frac{r}{2}}_{p',1}(\Omega)}.$$

By combining Theorem 3.2 with the embedding $\dot{B}^{\sigma}_{p,\infty}(\Omega) \subset \widetilde{W}^{\sigma-\varepsilon}_p(\Omega)$

$$\|v\|_{\widetilde{W}^{\sigma-\varepsilon}_p(\Omega)} \lesssim \varepsilon^{-1/p}\|v\|_{\dot{B}^{\sigma}_{p,\infty}(\Omega)},$$

valid for $\sigma \in (0,2)$ and $\varepsilon > 0$ such that $\sigma - \varepsilon$ is not an integer,[1] we obtain the Sobolev estimates for the fractional (p,s)-Laplacian for $r \in (0,1]$

$$\|u\|_{\widetilde{W}^{s+\frac{r}{p}-\varepsilon}_p(\Omega)} \le \frac{C(\Omega,d,p)}{\varepsilon^{1/p}}\|f\|^{\frac{1}{p-1}}_{B^{-s+\frac{r}{p'}}_{p',1}(\Omega)} \quad \text{if } p > 2,\ f \in B^{-s+\frac{r}{p'}}_{p',1}(\Omega), \tag{3.21}$$

$$\|u\|_{\widetilde{W}^{s+\frac{r}{2}-\varepsilon}_p(\Omega)} \le \frac{C(\Omega,d,p)}{\varepsilon^{1/p}}\|f\|^{\frac{2-p}{p-1}}_{W^{-s}_{p'}(\Omega)}\|f\|_{B^{-s+\frac{r}{2}}_{p',1}(\Omega)} \quad \text{if } 1 < p < 2,\ f \in B^{-s+\frac{r}{2}}_{p',1}(\Omega). \tag{3.22}$$

These estimates are valid up to the boundary of the domain and, similarly to the linear case, indicate that the upper bound on regularity is due to boundary behavior. In the superquadratic case $p \ge 2$, reference [32] derives the following higher order interior regularity provided $f \in W^{s}_{p'}(\Omega)$:

$$\text{if } s \le \frac{p-1}{p+1}, \quad \text{then } u \in \bigcap_{\varepsilon>0} W^{s\frac{p+1}{p-1}-\varepsilon}_{p,loc}(\Omega),$$

$$\text{if } s > \frac{p-1}{p+1}, \quad \text{then } u \in W^1_{p,loc}(\Omega) \text{ and } \nabla u \in \bigcap_{\varepsilon>0} W^{s\frac{p+1}{p}-\frac{p-1}{p}-\varepsilon}_{p,loc}(\Omega).$$

[1] This estimate follows by the same steps as (2.38), with the caveat that $\dot{B}^1_{p,p}(\Omega) = \widetilde{W}^1_p(\Omega)$ only if $p = 2$ [6, §7.67].

3.2.3 Hölder Regularity

In recent years, there has been a significant progress in the study of the Hölder regularity of solutions to (3.11). We refer to [89] for a thorough discussion and references on the topic, and here we briefly mention some results regarding boundary regularity. If the data f is pointwise bounded and the domain Ω satisfies regularity assumptions similar to Sect. 2.1.1 but stronger than Sect. 3.2.2, a remarkable result follows.

Theorem 3.3 (Global Hölder Regularity for Fractional (p, s)-Laplacians) *If $\Omega \subset \mathbb{R}^d$ satisfies the exterior ball condition, then there exists $\alpha \in (0, s]$ depending only on d, p, and s such that all solutions to the fractional (p, s)-Laplace problem (3.13) satisfy*

$$\|u\|_{C^\alpha(\overline{\Omega})} \leq C(\Omega, d, p, s)\|f\|_{L^\infty(\Omega)}^{1/(p-1)}.$$

In addition, for $p \geq 2$, the above estimate holds for $\alpha = s$.

The theorem above follows from [76, Theorem 1.1] and [33, Theorem 1.4]. In the superquadratic case, the boundary behavior can be refined [77]: the function $x \mapsto u(x)/d(x, \partial\Omega)^s$ is of class C^α up to the boundary of the domain for some $\alpha \in (0, s]$.

Theorem 3.3 can be regarded as an extension of (2.9) to the nonlinear problem with $p \neq 2$. In fact, for $f \in L^\infty(\Omega)$, the boundary behavior of solutions is the same as in the linear problem independently of the value of $p \in (1, \infty)$: [76, Lemma 3.1] shows that $u(x) = x_+^s$ satisfies

$$(-\Delta)_p^s u(x) = 0$$

for $x > 0$. To illustrate the boundary behavior

$$|u(x)| \lesssim \text{dist}(x, \partial\Omega)^s, \tag{3.23}$$

proved in [76, Theorem 4.4] for bounded domains Ω satisfying the exterior ball condition, in Fig. 7, we present some numerical experiments for $\Omega = (-0.5, 0.5) \subset \mathbb{R}$ and $f = 1$ and different choices of p. We plot the computed solution u_h near the boundary and observe a stable boundary behavior for different p, in agreement with (3.23).

Unlike the linear problem with $p = 2$, even if the right-hand side f is smooth, in general one cannot expect the solution u to be smooth in the interior of Ω. In the superquadratic case $p \geq 2$, [33, Theorem 1.4] obtains the following interior Hölder regularity:

$$u \in \bigcap_{\varepsilon>0} C_{loc}^{\theta-\varepsilon}(\Omega) \text{ with } \theta = \min\left\{\frac{1}{p-1}\left(sp - \frac{d}{q}\right), 1\right\},$$

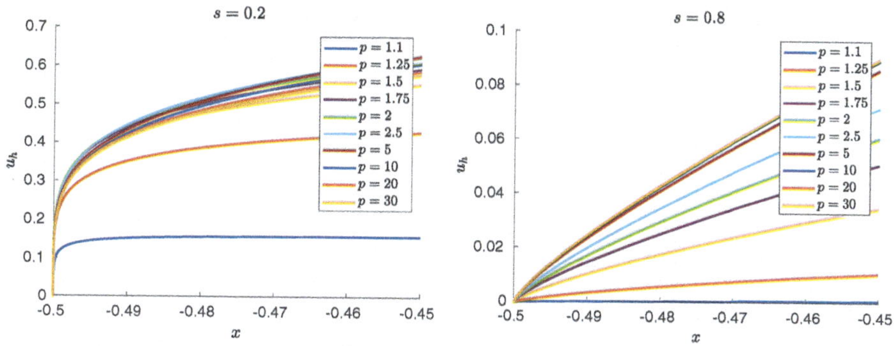

Fig. 7 Boundary behavior of numerical solutions in $\Omega = (-0.5, 0.5)$ for $s = 0.2, 0.8$ and $f = 1$

provided that $f \in L^q_{\text{loc}}(\Omega)$ for $q \geq 1, q > d/(sp)$. As illustrated by [33, Examples 1.5 and 1.6], this regularity is almost sharp when $s \leq (p-1)/p + d/(pq)$.

Finally, we point out that in case f is a Borel measure with finite total mass, reference [80] proves the existence of the so-called SOLA (Solutions Obtained as Limits of Approximations) provided $p > 2 - s/d$. Additionally, in case f belongs to the Lorentz space $L^{\frac{d}{sp},\frac{1}{p-1}}_{\text{loc}}(\Omega)$ with $sp < d$, [80, Corollary 1.2] proves the sharper regularity $u \in C(\Omega)$.

3.3 *Finite Element Discretization and Implementation*

We discuss the finite element approximation of fractional minimal graphs (3.8)–(3.6) (or prescribed fractional mean curvature problems, which correspond to a non-zero right-hand side) and of fractional p-Laplacians (3.13)–(3.14).

An important point to address in the implementation of nonlocal operators with infinite interaction range is the treatment of boundary conditions, which are prescribed on the unbounded set Ω^c. For homogeneous conditions, one can perform tricks as discussed in Sect. 2.2; in general, we resort to computational domains containing Ω and a suitable truncation of the exterior data. Given $H > 0$, we let Ω_H be a bounded open domain with $\Omega \subset \Omega_H$ and $d(x, \overline{\Omega}) \simeq H$ for all $x \in \partial\Omega_H$. We fix a cut-off function $\eta_H \in C^\infty(\Omega^c)$ satisfying

$$0 \leq \eta_H \leq 1, \quad \operatorname{supp}(\eta_H) \subset \overline{\Omega}_{H+1} \setminus \Omega, \quad \eta_H(x) = 1 \quad \text{in} \quad \Omega_H \setminus \Omega.$$

We replace g by $g_H := g\eta_H$ as Dirichlet condition in problem (3.8). Here we shall not discuss the effect of this boundary truncation but refer to [26] for a detailed study; naturally, one requires $H \to \infty$ as $h \to 0$ in order to guarantee convergence. From now on, we assume there exists a bounded set Λ such that $\operatorname{supp}(g) \subset \Lambda \subset \Omega_H$, where Ω_H is the computational domain.

To impose the condition $u = g$ in Ω^c at the discrete level, we introduce an exterior interpolation operator

$$\Pi_h^c g := \sum_{\mathtt{x}_i \in \mathcal{N}_h^c} (\Pi_h^{\mathtt{x}_i} g)(\mathtt{x}_i)\, \varphi_i,$$

where $\Pi_h^{\mathtt{x}_i} g$ is the L^2-projection of $g\big|_{S_i \cap \Omega^c}$ onto $\mathcal{P}_1(S_i \cap \Omega^c)$. Thus, $\Pi_h^c g(\mathtt{x}_i)$ coincides with the standard Clément interpolation of g on $\mathtt{x}_i$ for all nodes $\mathtt{x}_i$ such that $S_i \subset \mathbb{R}^d \setminus \overline{\Omega}$. On the other hand, for nodes $\mathtt{x}_i \in \partial\Omega$, Π_h^c only averages over the elements in S_i that lie in Ω^c.

In the same fashion as in Sect. 2.2, we use discrete spaces consisting in piecewise linear, continuous functions over $\mathcal{T}$,

$$\mathbb{V}_h := \{v \in C(\Omega_H) \colon v|_T \in \mathcal{P}_1 \ \forall T \in \mathcal{T}\}.$$

It is clear that these spaces depend on the computational domain size parameter $H > 0$; for easiness of notation and because we assume H is fixed (and sufficiently large so that $\Lambda \subset \Omega_H$), we shall omit such a dependence. To account for the exterior data, we define the discrete counterpart of $\mathbb{V}^g$,

$$\mathbb{V}_h^g := \{v \in \mathbb{V}_h \colon \ v|_{\Omega_H \setminus \Omega} = \Pi_h^c g\}.$$

Additionally, we denote by $\mathbb{V}_h^0$ the corresponding space in case $g \equiv 0$. The discrete counterpart to (3.1) reads as follows: find $u_h \in \mathbb{V}_h^g$ such that

$$a_{u_h}(u_h, v_h) = \langle f, v_h\rangle \quad \text{for all } v_h \in \mathbb{V}_h^0. \tag{3.24}$$

Algorithm 3.1 Damped Newton algorithm

Select an arbitrary initial $u_h^0 \in \mathbb{V}_h^g$ and let $k = 0$. Choose a small number Tol > 0.
while $\|\{a_{u_h^k}(u_h^k, \varphi_i)\}_{i=1}^m\|_{l^2} >$ Tol **do**
  Find $w_h^k \in \mathbb{V}_h^0$ such that

$$\frac{\delta a_{u_h^k}(u_h^k, v_h)}{\delta u_h^k}(w_h^k) = -a_{u_h^k}(u_h^k, v_h), \qquad \forall v_h \in \mathbb{V}_h^0. \tag{3.25}$$

  Determine the minimum $n \in \mathbb{N}$ such that $u_h^{k,n} := u_h^k + 2^{-n} w_h^k$ satisfies

$$\|\{a_{u_h^k}(u_h^{k,n}, \varphi_i)\}_{i=1}^m\|_{l^2} \le (1 - 2^{-n-1})\|\{a_{u_h^k}(u_h^k, \varphi_i)\}_{i=1}^m\|_{l^2}$$

  Let $u_h^{k+1} = u_h^{k,n}$ and $k = k + 1$.
end while

To solve this nonlinear discrete problem, we use the damped Newton scheme described in Algorithm 3.1. We exploit that discrete functions $v_h \in \mathbb{V}_h$ are Lipschitz and rewrite

$$a_{u_h}(u_h, v_h) = \iint_{Q_\Omega} G\Big(\frac{u_h(x) - u_h(y)}{|x-y|}\Big) \frac{v_h(x) - v_h(y)}{|x-y|^{d+2\sigma-1}} dxdy,$$

where $G(r) := r\widetilde{G}(r)$ and $\widetilde{G}(r)$ is given in (3.2) with $t = 1$. We now compute the first variation $\frac{\delta a_{u_h}(u_h, v_h)}{\delta u_h}(w_h)$ of $a_{u_h}(u_h, v_h)$ with respect to u_h in the direction $w_h \in \mathbb{V}_h^0$, or equivalently the second variation of $I_{s,p}[u_h]$ in the directions $v_h, w_h \in \mathbb{V}_h^0$, to get

$$\begin{aligned}\frac{\delta a_{u_h}(u_h, v_h)}{\delta u_h}(w_h) = \iint_{Q_\Omega} &\psi\left(\frac{u_h(x) - u_h(y)}{|x-y|}\right) \\ &\times \frac{(w_h(x) - w_h(y))(v_h(x) - v_h(y))}{|x-y|^{d+2\sigma}} dxdy,\end{aligned}$$

where $\psi(r) := G'(r)$. For the nonlocal minimal graph problem, we have

$$\psi(r) = (1 + r^2)^{-(d+1+2s)/2}, \quad \sigma = s + \frac{1}{2}.$$

In contrast, for the fractional (p, s)-Laplace equation, we let

$$\psi(r) = \frac{(p-1)C_{d,s,p}}{2} |r|^{p-2}, \quad \sigma = 1 + \frac{(s-1)p}{2},$$

but realize that $\psi(0)$ is not well defined whenever $p \in (1, 2)$. To overcome this issue in such a case, we introduce a small parameter $\varepsilon > 0$ and regularize the energy in (3.12), namely

$$\widetilde{I}_{s,p}[v] := \frac{C_{d,s,p}}{2p} \iint_{Q_\Omega} \left(\frac{|v(x) - v(y)|^2}{|x-y|^2} + \varepsilon^2\right)^{p/2} \frac{dxdy}{|x-y|^{d+(s-1)p}} - \int_\Omega fv.$$

Moreover, the discrete minimizer $u_h \in \mathbb{V}_h^g$ of $\widetilde{I}_{s,p}$ satisfies

$$\widetilde{a}_{u_h}(u_h, v_h) = \langle f, v_h \rangle \quad \text{for all } v_h \in \mathbb{V}_h^0, \tag{3.26}$$

with

$$\widetilde{a}_{u_h}(u_h, v_h) = \iint_{Q_\Omega} G_\varepsilon\Big(\frac{u_h(x) - u_h(y)}{|x-y|}\Big) \frac{v_h(x) - v_h(y))}{|x-y|^{d+1+(s-1)p}} dxdy,$$

$$G_\varepsilon(r) = \frac{C_{d,s,p}}{2}\left(r^2+\varepsilon^2\right)^{\frac{p}{2}-1} r.$$

We then apply Algorithm 3.1 to find the discrete solution to the regularized problem (3.26). Since

$$\frac{\delta \widetilde{a}_{u_h}(u_h, v_h)}{\delta u_h}(w_h) = \iint_{Q_\Omega} \psi\left(\frac{u_h(x)-u_h(y)}{|x-y|}\right) \times \frac{(w_h(x)-w_h(y))(v_h(x)-v_h(y))}{|x-y|^{d+2+(s-1)p}} dxdy,$$

with $\psi(r) = G_\varepsilon'(r)$ given by

$$\psi_\varepsilon(r) = \frac{C_{d,s,p}}{2}\left((p-1)r^2+\varepsilon^2\right)\left(r^2+\varepsilon^2\right)^{\frac{p}{2}-2}, \quad p \in (1,2),$$

we deduce that $\frac{\delta \widetilde{a}_{u_h}(u_h, v_h)}{\delta u_h}(w_h)$ is well defined for all $u_h \in \mathbb{V}_h^g$, $v_h, w_h \in \mathbb{V}_h^0$.

At each step of Algorithm 3.1, problem (3.25) boils down to solving a linear system $\mathbf{K}^k \mathbf{W}^k = \mathbf{F}^k$. The matrix $\mathbf{K}^k = (\mathbf{K}_{ij}^k)$, given by

$$\mathbf{K}_{ij}^k = \iint_{Q_\Omega} \psi\left(\frac{u_h^k(x)-u_h^k(y)}{|x-y|}\right) \frac{(\varphi_i(x)-\varphi_i(y))(\varphi_j(x)-\varphi_j(y))}{|x-y|^{d+2\sigma}} dxdy,$$

is the stiffness matrix for a weighted linear problem of order $\sigma < 1$: in the nonlocal minimal graph problem, we have $\sigma = s + 1/2$, while $\sigma = 1-(1-s)p/2$ for the fractional (p,s)-Laplacian. We compute this matrix and the right-hand side vector $\mathbf{F}_i^k = -a_{u_h^k}(u_h^k, \varphi_i)$ in a similar way as in the linear problem discussed in Sect. 2.2.

3.4 Convergence

We now discuss the convergence of the finite element solutions to (3.24) towards the solution of the continuous problem (3.1) as the mesh size h tends to zero. We shall follow different strategies for the fractional mean curvature problem and the Dirichlet problem for the fractional p-Laplacian.

3.4.1 Minimal Graphs

To prove the convergence of the finite element scheme, the approach in [24] consists of proving that the discrete energy is consistent and using a compactness argument.

Theorem 3.4 (Convergence for the Nonlocal Minimal Graph Problem) *Let $s \in (0, 1/2)$, Ω be a bounded Lipschitz domain, and g be uniformly bounded and satisfy $supp(g) \subset \Lambda$ for some bounded set Λ. Let u and u_h be the solutions to* (3.8) *and* (3.24), *respectively. Then, it holds that*

$$\lim_{h\to 0} I_s[u_h] = I_s[u] \quad \text{and} \quad \lim_{h\to 0} \|u - u_h\|_{W_1^{2r}(\Omega)} = 0 \quad \forall r \in [0, s).$$

The theorem above has the important feature of guaranteeing convergence without any regularity assumption on the solution. However, it does not offer any convergence rates. We now show estimates for a geometric notion of error that mimics the one analyzed in [64] for the classical Plateau problem (see also [14, 43]). In the local setting, such a notion of error is given by

$$\begin{aligned} e^2(u, u_h) &:= \int_\Omega \left|\widehat{\nu}(\nabla u) - \widehat{\nu}(\nabla u_h)\right|^2 \frac{Q(\nabla u) + Q(\nabla u_h)}{2}\, dx, \\ &= \int_\Omega \left(\widehat{\nu}(\nabla u) - \widehat{\nu}(\nabla u_h)\right) \cdot \left(\nabla(u - u_h), 0\right) dx, \end{aligned}$$

where $Q(\boldsymbol{a}) = \sqrt{1 + |\boldsymbol{a}|^2}$, $\widehat{\nu}(\boldsymbol{a}) = \frac{(\boldsymbol{a}, -1)}{Q(\boldsymbol{a})}$. Because $\widehat{\nu}(\nabla u)$ is the normal unit vector to the graph of u, the quantity $e(u, u_h)$ is a weighted L^2-discrepancy between the normal vectors. For the nonlocal minimal graph problem, we introduced in [24] the fractional counterpart e_s of e

$$e_s(u, u_h) := \left(\widetilde{C}_{d,s} \iint_{Q\Omega} \left(G_s\left(d_u(x, y)\right) - G_s\left(d_{u_h}(x, y)\right)\right) \frac{d_{u-u_h}(x, y)}{|x - y|^{d-1+2s}} dx dy\right)^{1/2},$$

where $G_s(\rho) = F_s'(\rho)$, the constant $\widetilde{C}_{d,s} = \frac{1-2s}{\alpha_d}$, α_d is the volume of the d-dimensional unit ball, and d_v is the difference quotient of the function v,

$$d_v(x, y) := \frac{v(x) - v(y)}{|x - y|}.$$

In [24], we showed this novel quantity $e_s(u, u_h)$ to be connected with a notion of nonlocal normal vector and established its asymptotic behavior as $s \to 1/2^-$.

Theorem 3.5 (Asymptotics of e_s) *For all $u, v \in H_0^1(\Lambda)$, we have*

$$\lim_{s\to \frac{1}{2}^-} e_s(u, v) = e(u, v).$$

A simple Galerkin orthogonality-type argument allows us to derive an error estimate for $e_s(u, u_h)$ without additional regularity (cf. [24, Theorem 5.1]).

Theorem 3.6 (Geometric Error) *Under the same hypothesis as in Theorem 3.4, it holds that*

$$e_s(u,u_h) \leq C(d,s) \inf_{v_h \in \mathbb{V}_h^g} \left(\iint_{Q\Omega} \frac{|(u-v_h)(x)-(u-v_h)(y)|}{|x-y|^{d+2s}} dxdy \right)^{1/2}. \tag{3.27}$$

Therefore, to obtain convergence rates with respect to $e_s(u, u_h)$, it suffices to prove interpolation estimates for the nonlocal minimizer in a sort of W_1^{2s}-norm. Although minimal graphs are expected to be discontinuous across the boundary, we still expect that $u \in BV(\Lambda)$ in general. Under this further regularity assumption, the error estimate (3.27) leads to

$$e_s(u,u_h) \leq C(d,s)\, h^{1/2-s} |u|_{BV(\Lambda)}^{1/2}.$$

3.4.2 Fractional (p, s)-Laplacian

To prove the convergence of the finite element solution u_h in (3.24) to the continuous solution u in (3.13), one needs to control $\|u - u_h\|_{\widetilde{W}_p^s(\Omega)}$ in terms of the best approximation error $\inf_{v_h \in \mathbb{V}_h} \|u - v_h\|_{\widetilde{W}_p^s(\Omega)}$. Following the idea of [40] for the classical p-Laplacian, in [23], we obtain such a result for its fractional counterpart.

Theorem 3.7 (Error Bounds for the Fractional (p, s)-Laplacian) *Let u and u_h be the respective solutions of the continuous and discrete Dirichlet problems for the fractional (p, s)-Laplacian* (3.13) *and* (3.24). *If $p \in (1, 2]$, then there holds*

$$\|u-u_h\|_{\widetilde{W}_p^s(\Omega)} \leq C \inf_{v_h \in \mathbb{V}_h} \|u-v_h\|_{\widetilde{W}_p^s(\Omega)}^{\frac{p}{2}}.$$

On the other hand, if $p \in [2, \infty)$, then the following alternative error bound is valid:

$$\|u-u_h\|_{\widetilde{W}_p^s(\Omega)} \leq C \inf_{v_h \in \mathbb{V}_h} \|u-v_h\|_{\widetilde{W}_p^s(\Omega)}^{\frac{2}{p}}.$$

Combining Theorem 3.7 with the local interpolation estimates from Lemma 2.1 and the Sobolev regularity estimates (3.21)–(3.22), the following convergence rates are derived in [23].

Theorem 3.8 (Convergence Rates) *Let $p > 1$, Ω be a bounded Lipschitz domain, and let u and u_h be the solutions of the continuous and discrete fractional (p, s)-Laplacian problems* (3.13) *and* (3.24), *respectively. Let $q = \max\{1/p', 1/2\}$ and assume that $f \in B_{p',1}^{-s+rq}(\Omega)$ for some $r \in (0, 1]$. If $p \in (1, 2]$, then there holds*

$$\|u-u_h\|_{\widetilde{W}_p^s(\Omega)} \leq C h^{\frac{pr}{4}} |\log h|^{\frac{1}{2}} \|f\|_{W_{p'}^{-s}(\Omega)}^{\frac{2-p}{p-1}} \|f\|_{B_{p',1}^{-s+\frac{r}{2}}(\Omega)}.$$

On the other hand, if $p \in [2, \infty)$, then the following alternative error bound is valid:

$$\|u - u_h\|_{\widetilde{W}^s_p(\Omega)} \leq C h^{\frac{2r}{p^2}} |\log h|^{\frac{2}{p^2}} \|f\|^{\frac{1}{p-1}}_{B^{-s+\frac{r}{p'}}_{p',1}(\Omega)} .$$

3.5 Numerical Experiments

We conclude by presenting some numerical experiments for the prescribed fractional mean curvature problem and the Dirichlet problem for the fractional (p, s)-Laplacian. Regarding nonlocal minimal graphs, we recall the two examples we provided in Sect. 3.1.3, concretely in Figs. 5 and 6. We refer to [24, 26] for several additional experiments that illustrate qualitative features of minimal graphs and explore computational aspects of the problem, such as conditioning and the effect of data truncation.

Example 3.1 (Effect of Boundary Curvature on Nonlocal Minimal Graphs) We present examples of graphs with prescribed nonlocal mean curvature in three two-dimensional domains Ω with qualitatively different $\partial\Omega$ and examine the impact on stickiness. We fix data $g = 0$ and $f = -1$ in (3.1) and solve (3.6)–(3.7). We first consider the annulus $\Omega = B(0, 1) \setminus B(0, 1/4)$ and $s = 0.25$. The top row in Fig. 8 depicts a top view of the discrete solution u_h and a radial slice of it. We observe that the discrete solution is about three times stickier in the inner boundary than in the outer one. The middle and bottom rows in Fig. 8 display different views of the solution in the square $\Omega = (-1, 1)^2$ for $s = 0.01$. Near the boundary of the domain Ω, we observe a steep slope in the middle of the edges; however, stickiness is not observed at the convex corners of Ω. We finally investigate stickiness at the boundary of the L-shaped domain $\Omega = (-1, 1)^2 \setminus (0, 1) \times (-1, 0)$ with $s = 0.25$. We observe in Fig. 9 that stickiness is most pronounced at the reentrant corner but is again absent at the convex corners of Ω.

The previous example indicates that there is a connection between the jump of solutions across the boundary $\partial\Omega$ and the curvature of $\partial\Omega$. Stickiness is stronger at the concave portions of the boundary than at the convex ones. Moreover, we find no numerical evidence of stickiness at convex corners. We refer to [26, §7.6.3] for a heuristic explanation supporting these observations.

Next, we perform some experiments involving the fractional (p, s)-Laplacian $(-\Delta)^s_p$.

Example 3.2 (Dependence of $(-\Delta)^s_p$ with Respect to s and p) Let $\Omega = (-0.5, 0.5) \subset \mathbb{R}$, and set $f = 1$ in (3.13), with the form given by (3.14). We use uniform grids with mesh size $h = 2^{-11}$. In Fig. 10, we exhibit numerical solutions for different choices of s and p. We observe that the boundary behavior depends on s but is independent of p.

We also measure the convergence rates in the $\widetilde{W}^s_p(\Omega)$ norm on uniform meshes and meshes graded according to (2.59) with $d = 1$ and grading parameter $\mu = p(2 - s)$. We now justify the choice of μ by the expected boundary behavior

$$u(x) \approx d(x, \partial\Omega)^s,$$

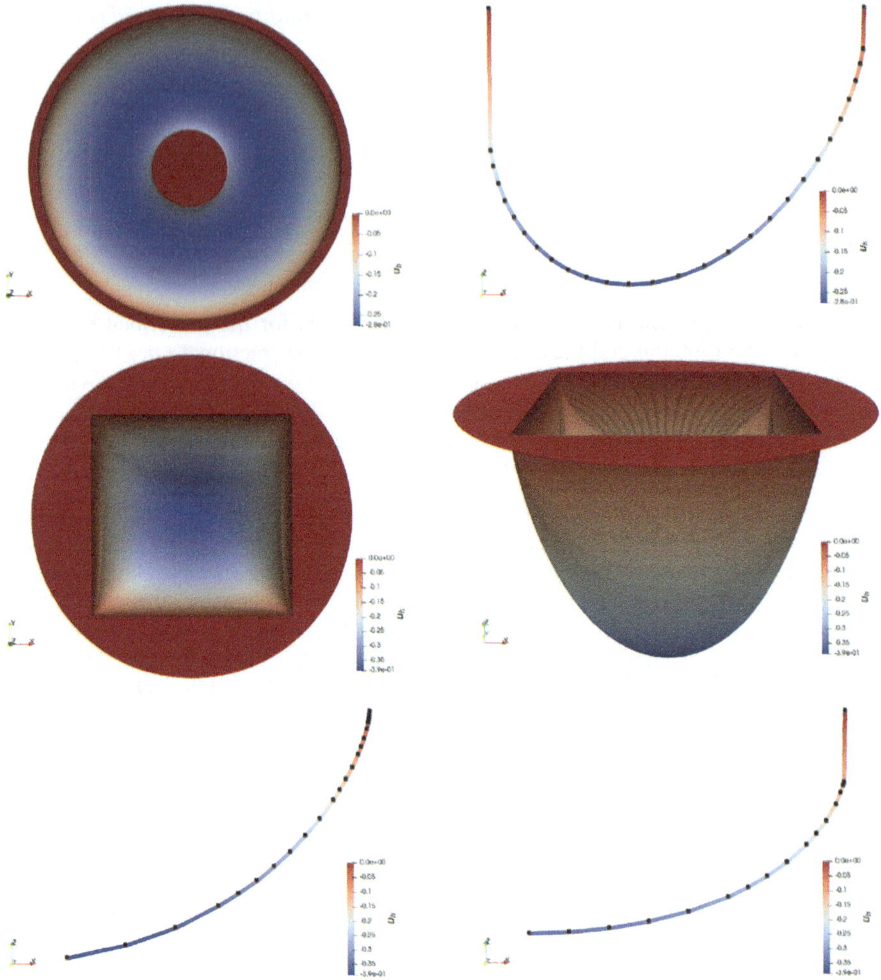

Fig. 8 Functions with prescribed fractional mean curvature $f = -1$ in Ω that vanish in Ω^c. The domain Ω is either an annulus (top row) or a square (middle and bottom rows). The plot in the top-right panel corresponds to a radial slice ($y = 0,\ 0.25 \leq x \leq 1$) of the annulus, while the ones in the bottom left and bottom right show parts of slices along the diagonal ($0 \leq y = x \leq 1$) and perpendicular to an edge of the square ($y = 0.5, 0 \leq x \leq 1$), respectively. The discrete solutions exhibit a much steeper gradient near the concave portions of the boundary than near the convex ones, and they seem to be continuous in the corners of the square

and corresponding second derivative of u, obtained heuristically and similarly to (2.12),

$$|u''(x)| \approx d(x, \partial\Omega)^{s-2}.$$

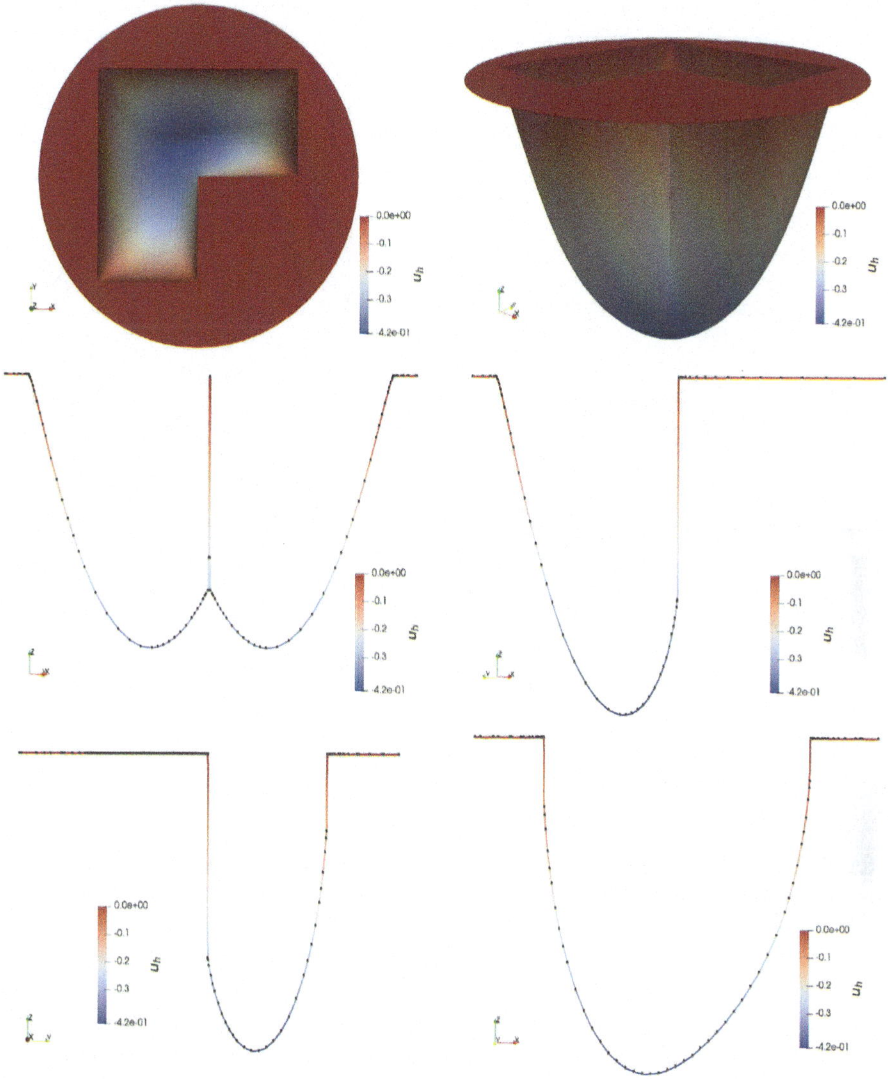

Fig. 9 Stickiness on the L-shaped domain $\Omega = (-1, 1)^2 \setminus (0, 1) \times (-1, 0)$ with prescribed fractional mean curvature $f = -1$ in Ω and zero Dirichlet condition in Ω^c. The plots in the center row correspond to slices along $y = x$ and $y = -x$, respectively, while the ones in the bottom are slices along $x = 0$ or $y = 0.5$, respectively. We observe evidence of stickiness at the reentrant corner, while such a phenomenon is absent at the convex corners

Defining weighted Sobolev spaces $W^2_{p,\kappa}(\Omega)$ in the spirit of (2.11) (modifying the integrability index accordingly from 2 to p), we would then expect to have

$$u \in W^2_{p,\alpha}(\Omega) \quad \text{if } \alpha > 2 - s - \frac{1}{p}.$$

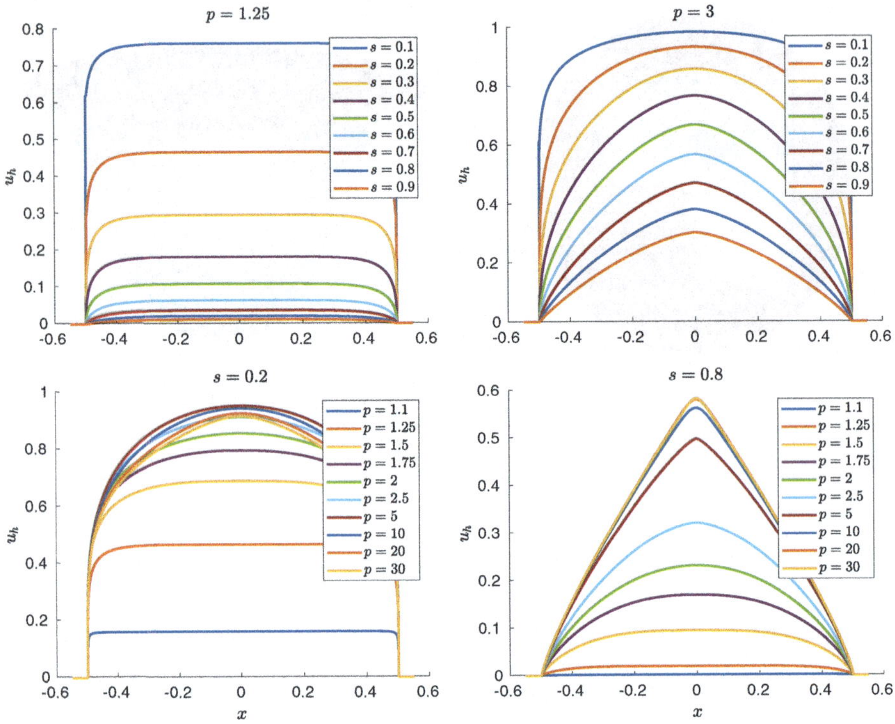

Fig. 10 Numerical solutions of fractional (p, s)-Laplacians in $\Omega = (-0.5, 0.5)$ for $f = 1$. The boundary behavior of solutions depends on s but not on p

Revisiting Lemma 2.1 with this regularity and constructing meshes according to (2.59), we immediately find that the optimal (in terms of h) interpolation estimate

$$\|u - \Pi u\|_{\widetilde{W}^s_p(\Omega)} \leq Ch^{2-s}|u|_{W^2_{p,\alpha}(\Omega)}$$

holds provided $\mu \geq p(2-s)$. By (2.60) with $d = 1$, we know already that $\dim \mathbb{V}_h \approx h^{-1}$ regardless of the value of μ, and we therefore expect the interpolation error to be of order $2-s$ in such a case. In contrast, if $d > 1$, mesh shape-regularity limits the optimal range of μ in (2.59) and one cannot expect to recover optimal interpolation error estimates unless some additional conditions on p and s are satisfied.

Table 7 records convergence rates on uniform and graded meshes for $p = 1.25$, $p = 3$, and a range of values of s. On uniform meshes, the energy-norm error decays with order approximately $h^{1/p}$. This rate is consistent with the interpolation error corresponding to the regularity of solutions derived in Theorem 3.2 for $p \geq 2$ but suggests that our result for $p < 2$ is non-optimal in the sense that the best regularity one can expect given the boundary behavior (2.12) is $W^{s+1/p-\varepsilon}_p(\Omega)$. Moreover, the rates in Table 7 are better than the convergence rate predicted by Theorem 3.8 for a general $f \in B^{-s+q}_{p',1}(\Omega)$, $q = \max\{1/p', 1/2\}$. In contrast, on graded meshes, we observe

Table 7 Convergence rates α for $\|u - u_h\|_{\widetilde{W}_p^s(\Omega)}$ for $p = 1.25, 3$ and different s on uniform mesh and graded mesh with $\mu = p(2 - s)$

	Value of s	0.1	0.2	0.3	0.4	0.5	0.6	0.7	0.8	0.9
Uniform $\mathcal{T}_h$	$p = 1.25$	0.796	0.777	0.780	0.788	0.797	0.807	0.821	0.85	0.898
	$p = 3$	0.337	0.334	0.332	0.332	0.333	0.333	0.334	0.335	0.339
Graded $\mathcal{T}_h$	$p = 1.25$	1.848	1.747	1.634	1.523	1.417	1.313	1.216	1.128	1.053
	$p = 3$	1.880	1.790	1.673	1.579	1.475	1.306	1.141	1.011	0.906

Table 8 Convergence rates α for $\|u - u_h\|_{\widetilde{W}_p^s(\Omega)}$ for $p = 1.25, 3$ and different s on graded mesh with $\mu = p(2 - s)$ and mesh grading satisfying (3.28). We observe an improvement with respect to Table 7 for $p = 3$ and $s > 0.5$

	Value of s	0.1	0.2	0.3	0.4	0.5	0.6	0.7	0.8	0.9
Graded $\mathcal{T}_h$	$p = 1.25$	1.846	1.735	1.623	1.514	1.411	1.308	1.213	1.125	1.052
	$p = 3$	1.895	1.762	1.653	1.549	1.446	1.345	1.250	1.160	1.085

convergence with order h^{2-s} as expected unless for $p = 3$ and s close to 1. Let us provide a possible explanation for this issue. In the local case ($s = 1$), the solution corresponding to $f = 1$ is

$$u(x) = C\left(1 - |2x|^{\frac{p}{p-1}}\right)_+,$$

which is locally of class W_p^2 only if $\frac{p}{p-1} + \frac{1}{p} \geq 2$. Thus, for the local problem, if $p \geq (3 + \sqrt{5})/2 \approx 2.62$, we do not expect to recover optimal convergence rates in the interior unless some adaptive refinement near the origin is performed. In the fractional-order case, we expect a similar behavior to occur when s is close to 1, and this may justify the reduced rates for $p = 3$ and $s > 0.5$ in Table 7.

Furthermore, for this specific example, we modify the mesh grading (2.59) as follows:

$$h_T \leq C(\sigma) \begin{cases} h^{\mu} \text{ if } T \cap \{\partial\Omega, 0\} \neq \emptyset, \\ h \operatorname{dist}(T, \{\partial\Omega, 0\})^{(\mu-1)/\mu} \text{ if } T \cap \{\partial\Omega, 0\} = \emptyset. \end{cases} \tag{3.28}$$

Above, $\mu \geq 1$ is a parameter and a simple calculation yields the mesh cardinality

$$\#\mathcal{T} \approx \dim \mathbb{V}_h \approx h^{-1}$$

because of $d = 1$. Let us comment that the grading near the origin $x = 0$ seems to be an overkill because the expected singularity there is weaker than the one near the boundary $\partial\Omega$. In Table 8, we report the convergence rates for $p = 1.25, 3$ on meshes satisfying (3.28) with $\mu = p(2 - s)$. The rates for $p = 3$ and s close to 1 are now in good agreement with the predicted rate $2 - s$, supporting our claim about the singularity of solutions near $x = 0$ in this example.

Our last experiment involves the fractional (p, s)-Laplacian on an L-shaped domain in two-dimensional space.

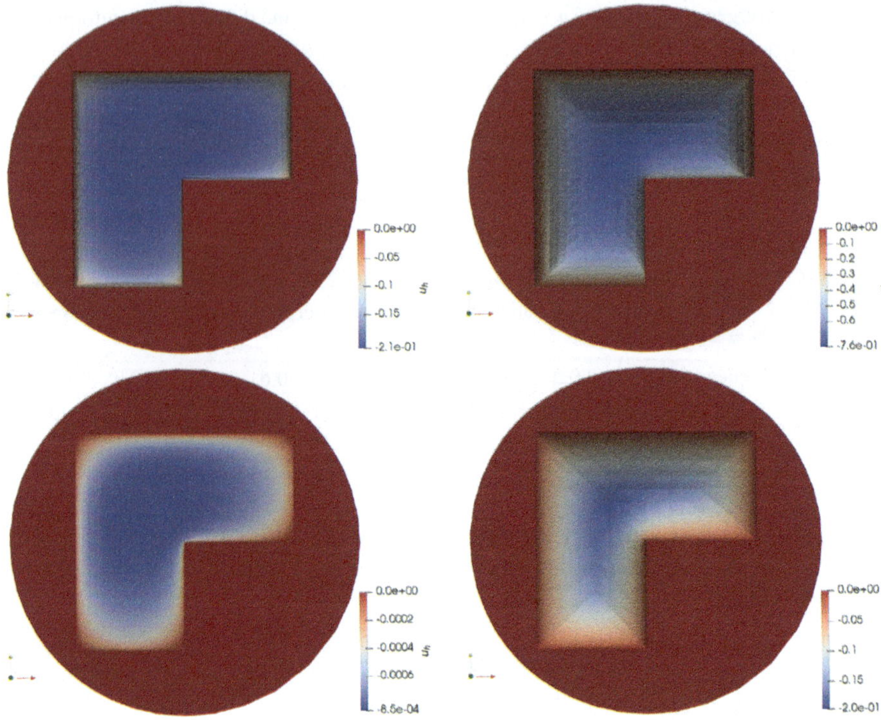

Fig. 11 Numerical solutions of fractional (p, s)-Laplacians in $\Omega = (-0.5, 0.5)^2 \setminus [(0, 0.5) \times (-0.5, 0)] \subset \mathbb{R}^2$ for $f = -1$. Top left: $p = 1.25, s = 0.2$; top right: $p = 3, s = 0.2$; bottom left: $p = 1.25, s = 0.8$; bottom right: $p = 3, s = 0.8$. We observe a stronger boundary behavior as $s \searrow 0$, while the solution approaches $d(\cdot, \partial\Omega)^s$ as $p \nearrow \infty$

Example 3.3 ((p, s)-Laplacian on L-Shaped Domain) Let $\Omega = (-0.5, 0.5)^2 \setminus [(0, 0.5) \times (-0.5, 0)] \subset \mathbb{R}^2$ and $f = -1$. Figure 11 depicts numerical solutions for $s = 0.2, 0.8$ and $p = 1.25$, $p = 3$. For a fixed p, we observe a much stronger boundary behavior for $s = 0.2$ than for $s = 0.8$, while for a fixed s the solutions converge to the distance to boundary to the power s as $p \to \infty$.

References

1. N. Abatangelo and X. Ros-Oton. Obstacle problems for integro-differential operators: higher regularity of free boundaries. *Adv. Math.*, 360:106931, 2020.
2. H. Abels and G. Grubb. Fractional-order operators on nonsmooth domains. *Journal of the London Mathematical Society*, 107(4):1297–1350, 2023.
3. G. Acosta, F. Bersetche, and J.P. Borthagaray. A short FE implementation for a 2D homogeneous Dirichlet problem of a fractional Laplacian. *Comput. Math. Appl.*, 74(4):784–816, 2017.

4. G. Acosta and J.P. Borthagaray. A fractional Laplace equation: regularity of solutions and finite element approximations. *SIAM J. Numer. Anal.*, 55(2):472–495, 2017.
5. G. Acosta, J.P. Borthagaray, O. Bruno, and M. Maas. Regularity theory and high order numerical methods for the (1D)-fractional Laplacian. *Math. Comp.*, 87(312):1821–1857, 2018.
6. R.A. Adams and J.J.F. Fournier. *Sobolev spaces*. Elsevier, 2003.
7. M. Ainsworth and C. Glusa. Aspects of an adaptive finite element method for the fractional Laplacian: a priori and a posteriori error estimates, efficient implementation and multigrid solver. *Comput. Methods Appl. Mech. Engrg.*, 327:4–35, 2017.
8. M. Ainsworth and Z. Mao. Analysis and approximation of a fractional Cahn–Hilliard equation. *SIAM J. Numer. Anal.*, 55(4):1689–1718, 2017.
9. M. Ainsworth, W. McLean, and T. Tran. The conditioning of boundary element equations on locally refined meshes and preconditioning by diagonal scaling. *SIAM J. Numer. Anal.*, 36(6):1901–1932, 1999.
10. H. Antil and S. Bartels. Spectral approximation of fractional PDEs in image processing and phase field modeling. *Comput. Methods Appl. Math.*, 17(4):661–678, 2017.
11. H. Antil, P. Dondl, and L. Striet. Approximation of integral fractional Laplacian and fractional PDEs via sinc-basis. *SIAM J. Sci. Comput.*, 43(4):A2897–A2922, 2021.
12. C. Atkinson and C.W. Jones. Similarity solutions in some non-linear diffusion problems and in boundary-layer flow of a pseudo-plastic fluid. *Quart. J. Mech. Appl. Math.*, 27(2):193–211, 1974.
13. I. Babuška, R.B. Kellogg, and J. Pitkäranta. Direct and inverse error estimates for finite elements with mesh refinements. *Numer. Math.*, 33(4):447–471, 1979.
14. E. Bänsch, P. Morin, and R.H. Nochetto. Surface diffusion of graphs: variational formulation, error analysis, and simulation. *SIAM J. Numer. Anal.*, 42(2):773–799, 2004.
15. B. Barrios, A. Figalli, and E. Valdinoci. Bootstrap regularity for integro-differential operators, and its application to nonlocal minimal surfaces. *Ann. Sc. Norm. Super. Pisa Cl. Sci. (5)*, 13(3):609–639, 2014.
16. M. Bauer, M. Bebendorf, and B. Feist. Kernel-independent adaptive construction of $\mathcal{H}^2$-matrix approximations. *Numer. Math.*, 150:1–32, 2022.
17. U. Biccari, M. Warma, and E. Zuazua. Local elliptic regularity for the Dirichlet fractional Laplacian. *Adv. Nonlinear Stud.*, 17(2):387–409, 2017.
18. P. Binev, W. Dahmen, and R. DeVore. Adaptive finite element methods with convergence rates. *Numer. Math.*, 97(2):219–268, 2004.
19. A. Bonito, J.P. Borthagaray, R.H. Nochetto, E. Otárola, and A.J. Salgado. Numerical methods for fractional diffusion. *Comput. Vis. Sci.*, 19(5):19–46, Mar 2018.
20. A. Bonito, W. Lei, and J.E. Pasciak. Numerical approximation of the integral fractional Laplacian. *Numer. Math.*, 142(2):235–278, 2019.
21. J. P. Borthagaray and P. Ciarlet Jr. On the convergence in H^1-norm for the fractional Laplacian. *SIAM J. Numer. Anal.*, 57(4):1723–1743, 2019.
22. J.P. Borthagaray, D. Leykekhman, and R.H. Nochetto. Local energy estimates for the fractional Laplacian. *SIAM J. Numer. Anal.*, 59(4):1918–1947, 2021.
23. J.P. Borthagaray, W. Li, and R.H. Nochetto. Quasi-linear fractional-order operators in Lipschitz domains. arXiv preprint arXiv:2305.17818, 2023.
24. J.P. Borthagaray, W. Li, and R.H. Nochetto. Finite element discretizations for nonlocal minimal graphs: Convergence. *Nonlinear Anal.*, 189:111566, 31, 2019.
25. J.P. Borthagaray, W. Li, and R.H. Nochetto. Linear and nonlinear fractional elliptic problems. In *75 Years of Mathematics of Computation*, volume 754 of *Contemp. Math.*, pages 69–92. Amer. Math. Soc., Providence, RI, 2020.
26. J.P. Borthagaray, W. Li, and R.H. Nochetto. Finite element algorithms for nonlocal minimal graphs. *Mathematics in Engineering*, 4(2):1–29, 2021.
27. J.P. Borthagaray and R.H. Nochetto. Besov regularity for the Dirichlet integral fractional Laplacian in Lipschitz domains. *Journal of Functional Analysis*, 284(6):109829, 2023.

28. J.P. Borthagaray and R.H. Nochetto. Constructive approximation on graded meshes for the integral fractional Laplacian. Constructive Approximation, 57:463–487, 2023.
29. J.P. Borthagaray, R.H. Nochetto, and A.J. Salgado. Weighted Sobolev regularity and rate of approximation of the obstacle problem for the integral fractional Laplacian. *Math. Models Methods Appl. Sci.*, 29(14):2679–2717, 2019.
30. J.P. Borthagaray, R.H. Nochetto, S. Wu, and J. Xu. Robust BPX preconditioner for fractional Laplacians on bounded Lipschitz domains. arXiv preprint arXiv:2103.12891, 2021.
31. J. Bourgain, H. Brezis, and P. Mironescu. Another look at Sobolev spaces. In *Optimal Control and Partial Differential Equations*, pages 439–455, 2001.
32. L. Brasco and E. Lindgren. Higher Sobolev regularity for the fractional p-Laplace equation in the superquadratic case. *Adv. Math.*, 304:300–354, 2017.
33. L. Brasco, E. Lindgren, and A. Schikorra. Higher Hölder regularity for the fractional p-Laplacian in the superquadratic case. *Adv. Math.*, 338:782–846, 2018.
34. S.C. Brenner and L.R. Scott. *The mathematical theory of finite element methods*, volume 15 of *Texts in Applied Mathematics*. Springer, New York, third edition, 2008.
35. C. Bucur and E. Valdinoci. *Nonlocal diffusion and applications*, volume 20. Springer, 2016.
36. O. Burkovska and M. Gunzburger. Regularity analyses and approximation of nonlocal variational equality and inequality problems. *J. Math. Anal. Appl.*, 478(2):1027–1048, 2019.
37. X. Cabré and M. Cozzi. A gradient estimate for nonlocal minimal graphs. *Duke Math. J.*, 168(5):775–848, 2019.
38. L. Caffarelli, J.-M. Roquejoffre, and O. Savin. Nonlocal minimal surfaces. *Comm. Pure Appl. Math.*, 63(9):1111–1144, 2010.
39. A. Chernov, T. von Petersdorff, and C. Schwab. Exponential convergence of hp quadrature for integral operators with Gevrey kernels. *ESAIM Math. Model. Numer. Anal.*, 45(3):387–422, 2011.
40. S.-S. Chow. Finite element error estimates for non-linear elliptic equations of monotone type. *Numer. Math.*, 54(4):373–393, 1989.
41. P. Ciarlet, Jr. Analysis of the Scott-Zhang interpolation in the fractional order Sobolev spaces. *J. Numer. Math.*, 21(3):173–180, 2013.
42. M. Cozzi. Interior regularity of solutions of non-local equations in Sobolev and Nikol'skii spaces. *Ann. Mat. Pura Appl. (4)*, 196(2):555–578, 2017.
43. K. Deckelnick, G. Dziuk, and C.M. Elliott. Computation of geometric partial differential equations and mean curvature flow. *Acta Numer.*, 14:139–232, 2005.
44. F. del Teso, D. Gómez-Castro, and J.L. Vázquez. Three representations of the fractional p-Laplacian: semigroup, extension and Balakrishnan formulas. *Fract. Calc. Appl. Anal.*, 24(4):966–1002, 2021.
45. M. D'Elia, Q. Du, C. Glusa, M. Gunzburger, X. Tian, and Z. Zhou. Numerical methods for nonlocal and fractional models. *Acta Numerica*, 29:1–124, 2020.
46. J.I. Diaz and F. De Thelin. On a nonlinear parabolic problem arising in some models related to turbulent flows. *SIAM J. Math. Anal.*, 25(4):1085–1111, 1994.
47. S. Dipierro, O. Savin, and E. Valdinoci. Graph properties for nonlocal minimal surfaces. *Calc. Var. Partial Differential Equations*, 55(4):86, 2016.
48. S. Dipierro, O. Savin, and E. Valdinoci. Boundary behavior of nonlocal minimal surfaces. *J. Funct. Anal.*, 272(5):1791–1851, 2017.
49. S. Dipierro, O. Savin, and E. Valdinoci. Boundary properties of fractional objects: flexibility of linear equations and rigidity of minimal graphs. *J. Reine Angew. Math.*, 769:121–164, 2020.
50. S. Dipierro, O. Savin, and E. Valdinoci. Nonlocal minimal graphs in the plane are generically sticky. *Comm. Math. Phys.*, 376(3):2005–2063, 2020.
51. Z. Ditzian. On the Marchaud-type inequality. *Proc. Amer. Math. Soc.*, 103(1):198–202, 1988.
52. Q. Du. *Nonlocal Modeling, Analysis, and Computation: Nonlocal Modeling, Analysis, and Computation*. SIAM, 2019.

53. S. Duo, H.W. van Wyk, and Y. Zhang. A novel and accurate finite difference method for the fractional Laplacian and the fractional Poisson problem. *J. Comput. Phys.*, 355:233–252, 2018.
54. S. Duo and Y. Zhang. Accurate numerical methods for two and three dimensional integral fractional Laplacian with applications. *Comput. Methods Appl. Mech. Engrg.*, 355:639–662, 2019.
55. B. Dyda and M. Kassmann. On weighted Poincaré inequalities. *Ann. Acad. Sci. Fenn. Math*, 38(2):721–726, 2013.
56. V. Ervin. Regularity of the solution to fractional diffusion, advection, reaction equations in weighted Sobolev spaces. *J. Differential Equations*, 278:294–325, 2021.
57. V. Ervin, N. Heuer, and J. Roop. Regularity of the solution to 1-d fractional order diffusion equations. *Math. Comp.*, 87(313):2273–2294, 2018.
58. G. I. Èskin. *Boundary value problems for elliptic pseudodifferential equations*, volume 52. Amer Mathematical Society, 1981.
59. B. Faermann. Localization of the Aronszajn-Slobodeckij norm and application to adaptive boundary element methods. I. The two-dimensional case. *IMA J. Numer. Anal.*, 20(2):203–234, 2000.
60. B. Faermann. Localization of the Aronszajn-Slobodeckij norm and application to adaptive boundary element methods. II. The three-dimensional case. *Numer. Math.*, 92(3):467–499, 2002.
61. M. Faustmann, M. Karkulik, and J.M. Melenk. Local convergence of the FEM for the integral fractional Laplacian. *SIAM Journal on Numerical Analysis*, 60(3):1055–1082, 2022.
62. M. Faustmann, C. Marcati, J.M. Melenk, and C. Schwab. Weighted analytic regularity for the integral fractional Laplacian in polygons. *SIAM Journal on Mathematical Analysis*, 54(6): 6323–6357, 2022.
63. M. Faustmann, J.M. Melenk, and M. Parvizi. On the stability of Scott-Zhang type operators and application to multilevel preconditioning in fractional diffusion. *ESAIM Math. Model. Numer. Anal.*, 55(2), 2021.
64. F. Fierro and A. Veeser. On the a posteriori error analysis for equations of prescribed mean curvature. *Math. Comp.*, 72(244):1611–1634, 2003.
65. A. Figalli and E. Valdinoci. Regularity and Bernstein-type results for nonlocal minimal surfaces. *J. Reine Angew. Math.*, 2017(729):263–273, 2017.
66. S. Funken, D. Praetorius, and P. Wissgott. Efficient implementation of adaptive P1-FEM in Matlab. *Comput. Methods Appl. Math.*, 11(4):460–490, 2011.
67. H. Gimperlein, E. Stephan, and J. Stocek. Corner singularities for the fractional Laplacian and finite element approximation. Preprint available at https://mat1.uibk.ac.at/heiko/corners.pdf, 2019.
68. H. Gimperlein, J. Stocek, and C. Urzúa-Torres. Optimal operator preconditioning for pseudodifferential boundary problems. *Numer. Math.*, 148:1–41, 2021.
69. R. Glowinski and A. Marrocco. Sur l'approximation, par éléments finis d'ordre un, et la résolution, par pénalisation-dualité, d'une classe de problèmes de Dirichlet non linéaires. *Rev. Française Automat. Informat. Recherche Opérationnelle Sér. Rouge Anal. Numér.*, 9(R-2):41–76, 1975.
70. P. Grisvard. *Elliptic problems in nonsmooth domains*, volume 24 of *Monographs and Studies in Mathematics*. Pitman (Advanced Publishing Program), Boston, MA, 1985.
71. G. Grubb. Fractional Laplacians on domains, a development of Hörmander's theory of μ-transmission pseudodifferential operators. *Adv. Math.*, 268:478–528, 2015.
72. R. Han and S. Wu. A monotone discretization for integral fractional Laplacian on bounded Lipschitz domains: Pointwise error estimates under Hölder regularity. *SIAM Journal on Numerical Analysis*, 60(6): 3052–3057, 2022.
73. Z. Hao and Z. Zhang. Optimal regularity and error estimates of a spectral Galerkin method for fractional advection-diffusion-reaction equations. *SIAM J. Numer. Anal.*, 58(1):211–233, 2020.

74. Y. Huang and A. Oberman. Numerical methods for the fractional Laplacian: A finite difference-quadrature approach. *SIAM J. Numer. Anal.*, 52(6):3056–3084, 2014.
75. N. E. Humphries, H. Weimerskirch, N. Queiroz, E. Southall, and D. Sims. Foraging success of biological Lévy flights recorded in situ. *Proceedings of the National Academy of Sciences*, 109(19):7169–7174, 2012.
76. A. Iannizzotto, S. Mosconi, and M. Squassina. Global Hölder regularity for the fractional p-Laplacian. *Rev. Mat. Iberoam.*, 32(4):1353–1392, 2016.
77. A. Iannizzotto, S. Mosconi, and M. Squassina. Fine boundary regularity for the degenerate fractional p-Laplacian. *J. Funct. Anal.*, 279(8):108659, 54, 2020.
78. C.-Y. Kao, Y. Lou, and W. Shen. Random dispersal vs. nonlocal dispersal. *Discrete Contin. Dyn. Syst.*, 26(2):551–596, 2010.
79. M. Karkulik and J.M. Melenk. $\mathcal{H}$-matrix approximability of inverses of discretizations of the fractional Laplacian. *Adv. Comput. Math.*, 45(5-6):2893–2919, 2019.
80. T. Kuusi, G. Mingione, and Y. Sire. Nonlocal equations with measure data. *Comm. Math. Phys.*, 337(3):1317–1368, 2015.
81. A. Lischke, G. Pang, M. Gulian, F. Song, C. Glusa, X. Zheng, Z. Mao, W. Cai, M.M. Meerschaert, M. Ainsworth, and G.E. Karniadakis. What is the fractional Laplacian? A comparative review with new results. *J. Comput. Phys.*, 404:109009, 2020.
82. L. Lombardini. Approximation of sets of finite fractional perimeter by smooth sets and comparison of local and global s-minimal surfaces. *Interfaces Free Bound.*, 20(2):261–296, 2018.
83. L. Lombardini. *Minimization Problems Involving Nonlocal Functionals: Nonlocal Minimal Surfaces and a Free Boundary Problem.* PhD thesis, Universita degli Studi di Milano and Universite de Picardie Jules Verne, 2018.
84. Z. Mao, S. Chen, and J. Shen. Efficient and accurate spectral method using generalized Jacobi functions for solving Riesz fractional differential equations. *Appl. Numer. Math.*, 106:165–181, 2016.
85. V. Maz'ya and T. Shaposhnikova. On the Bourgain, Brezis, and Mironescu theorem concerning limiting embeddings of fractional Sobolev spaces. *Journal of Functional Analysis*, 195(2):230–238, 2002.
86. W. McLean. *Strongly elliptic systems and boundary integral equations*. Cambridge university press, 2000.
87. V. Minden and L. Ying. A simple solver for the fractional Laplacian in multiple dimensions. *SIAM J. Sci. Comput.*, 42(2):A878–A900, 2020.
88. L. Modica and S. Mortola. Un esempio di Γ^--convergenza. *Boll. Un. Mat. Ital. B (5)*, 14(1):285–299, 1977.
89. S. Mosconi and M. Squassina. Recent progresses in the theory of nonlinear nonlocal problems. In *Bruno Pini Mathematical Analysis Seminar 2016*, volume 7 of *Bruno Pini Math. Anal. Semin.*, pages 147–164. Univ. Bologna, Alma Mater Stud., Bologna, 2016.
90. J.A. Nitsche and A.H. Schatz. Interior estimates for Ritz-Galerkin methods. *Math. Comp.*, 28:937–958, 1974.
91. R.H. Nochetto, K.G. Siebert, and A. Veeser. Theory of adaptive finite element methods: an introduction. In *Multiscale, nonlinear and adaptive approximation*, pages 409–542. Springer, Berlin, 2009.
92. R.H. Nochetto and A. Veeser. Primer of adaptive finite element methods. In *Multiscale and adaptivity: modeling, numerics and applications*, volume 2040 of *Lecture Notes in Math.*, pages 125–225. Springer, Heidelberg, 2012.
93. G. Ramos-Fernández, J. L. Mateos, O. Miramontes, G. Cocho, H. Larralde, and B. Ayala-Orozco. Lévy walk patterns in the foraging movements of spider monkeys (Ateles geoffroyi). *Behavioral Ecology and Sociobiology*, 55(3):223–230, 2004.
94. X. Ros-Oton and J. Serra. The Dirichlet problem for the fractional Laplacian: regularity up to the boundary. *J. Math. Pures Appl.*, 101(3):275–302, 2014.
95. S.A. Sauter and C. Schwab. *Boundary element methods*, volume 39 of *Springer Series in Computational Mathematics*. Springer-Verlag, Berlin, 2011.

96. G. Savaré. Regularity results for elliptic equations in Lipschitz domains. *J. Funct. Anal.*, 152(1):176–201, 1998.
97. O. Savin and E. Valdinoci. Γ-convergence for nonlocal phase transitions. *Ann. Inst. H. Poincaré Anal. Non Linéaire*, 29(4):479–500, 2012.
98. C. Sheng, J. Shen, T. Tang, L.-L. Wang, and H. Yuan. Fast Fourier-like mapped Chebyshev spectral-Galerkin methods for PDEs with integral fractional Laplacian in unbounded domains. *SIAM J. Numer. Anal.*, 58(5):2435–2464, 2020.
99. D. Sims, E. Southall, N. Humphries, G. Hays, C. Bradshaw, J. Pitchford, A. James, M. Ahmed, A. Brierley, M. Hindell, D. Morritt, M. Musyl, D. Righton, E. Shepard, V. Wearmouth, R. Wilson, M. Witt, and J. Metcalfe. Scaling laws of marine predator search behaviour. *Nature*, 451(7182):1098–1102, 2008.
100. R. Stevenson. The completion of locally refined simplicial partitions created by bisection. *Math. Comp.*, 77(261):227–241, 2008.
101. X. Tian and Q. Du. Asymptotically compatible schemes and applications to robust discretization of nonlocal models. *SIAM J. Numer. Anal.*, 52(4):1641–1665, 2014.
102. J.L. Vázquez. The mathematical theories of diffusion: nonlinear and fractional diffusion. In *Nonlocal and nonlinear diffusions and interactions: new methods and directions*, pages 205–278. Springer, 2017.
103. M. I. Višik and G. I. Èskin. Convolution equations in a bounded region. *Uspehi Mat. Nauk*, 20(3 (123)):89–152, 1965. English translation in *Russian Math. Surveys*, 20:86–151, 1965.
104. G. Viswanathan, S. Buldyrev, S. Havlin, M. Da Luz, E. Raposo, and H. Stanley. Optimizing the success of random searches. *Nature*, 401(6756):911–914, 1999.
105. L.-B. Wahlbin. Local behavior in finite element methods. In *Handbook of numerical analysis, Vol. II*, pages 353–522. North-Holland, Amsterdam, 1991.
106. X. Zhao, X. Hu, W. Cai, and G.E. Karniadakis. Adaptive finite element method for fractional differential equations using hierarchical matrices. *Comput. Methods Appl. Mech. Engrg.*, 325:56–76, 2017.

Regularity Estimates and Open Problems in Kinetic Equations

Luis Silvestre

Abstract We survey some new results regarding a priori regularity estimates for the Boltzmann and Landau equations conditional to the boundedness of the associated macroscopic quantities. We also discuss some open problems in the area. In particular, we describe some ideas related to the global well-posedness problem for the space homogeneous Landau equation for Coulomb potentials.

Keywords Kinetic equations · A non-cutoff Boltzmann equation · Landau equation · Long-range interactions · Conditional regularity

1 Introduction

In this note, we discuss some recent developments and open problems concerning regularity estimates for kinetic equations. The material is based on the minicourse given by the author in the Barret Memorial lectures in May 2021. The lectures are focused on the conditional regularity estimates for the Boltzmann equation. In these notes, we also discuss some additional topics. We describe the conditional regularity estimates for the Boltzmann and Landau equations and the main techniques leading to that result. We explain how some techniques that were initially explored in the context of elliptic nonlocal equations apply to the context of the Boltzmann equation. We also discuss open problems in the space homogeneous setting and present some ideas and conjectures related to those problems.

The program on conditional regularity consists of studying solutions to the inhomogeneous Boltzmann or Landau equation under the assumption that their

Luis Silvestre is supported by NSF grants 2054888 and 1764285.

L. Silvestre (✉)
Mathematics Department, University of Chicago, Chicago, IL, USA
e-mail: luis@math.uchicago.edu

T. Mengesha, A. J. Salgado (eds.), *A^3N^2M: Approximation, Applications, and Analysis of Nonlocal, Nonlinear Models*, The IMA Volumes in Mathematics and its Applications 165, https://doi.org/10.1007/978-3-031-34089-5_3

associated macroscopic hydrodynamic quantities stay pointwise bounded. This assumption rules out the formation of an implosion singularity that would be visible macroscopically. The conditional regularity estimates tell us, essentially, that no other type of singularity other than macroscopic implosions is possible for the non-cutoff Boltzmann or Landau equations.

There are various tools that play a role in obtaining the conditional regularity result. The first step, which is about L^∞ bounds, works for the case of hard or moderately soft potentials. In the very soft potential case (which corresponds to $\gamma + 2s < 0$, where γ and s are certain parameters in the collision kernel B defined below), the lower order reaction term in the collision operator is too singular and difficult to control with the diffusion term. Currently, there are no upper bounds in the very soft potential case even for space homogeneous solutions. The most extreme case, and also arguably the most interesting, is the Landau–Coulomb equation. We discuss the well-known open problem of existence of global-in-time solutions for the Landau–Coulomb equation in Sect. 7. In this section we present informally some ideas that lead to new points of view on the problem. We include some conjectures.

We describe the regularity results including minimal technical details about their proofs. A more detailed description of the methods involved in obtaining the conditional regularity result for the Boltzmann equations can be found, still in survey form, in [60].

2 Description of the Equations

Kinetic equations describe the evolution in time of a function $f(t, x, v)$ representing the density of particles in a dilute gas with respect to their position and velocity. The kinetic representation of a fluid lies at an intermediate scale between the large-dimensional dynamical system following the position and velocity of each and every particle and the macroscopic hydrodynamic description given by the Euler and Navier–Stokes equations.

A kinetic equation typically has the following form:

$$f_t + v \cdot \nabla_x f + F(t, x) \cdot \nabla_v f = Q(f, f).$$

The first two terms on the left-hand side form the pure transport equation. In the absence of any force, each particle moves at a straight line with its corresponding velocity v. A model in which there is no interaction between particles would simply be $f_t + v \cdot \nabla_x f = 0$. The third term, $F(t, x) \cdot \nabla_v f$, accounts for the macroscopic force. The force $F(t, x)$ may be the result of external forces or a mean field potential force computed in terms of the distribution f itself. The right-hand side, $Q(f, f)$, accounts for local interactions between particles, typically in the form of collisions.

In these notes, we concentrate on models that are limited to local interactions. We take $F \equiv 0$ and reduce the equation to

$$f_t + v \cdot \nabla_x f = Q(f, f). \tag{2.1}$$

The right-hand side $Q(f, f)$ is a quadratic nonlocal operator acting on $f(t, x, \cdot)$, for each value of (t, x). The Boltzmann collision operator has the following form:

$$Q(f, f)(v) = \int_{\mathbb{R}^d} \int_{S^{d-1}} (f'_* f' - f_* f) B(|v - v_*|, \cos\theta) \, d\sigma \, dv_*. \tag{2.2}$$

Let us make some clarifications to understand the expression for $Q(f, f)$. It is a double integral, with respect to two parameters, $v_* \in \mathbb{R}^d$ and a spherical variable $\sigma \in S^{d-1}$. The notations f, f_*, f', and f'_* denote the values of the function f evaluated at the same value of (t, x), but with different velocities v, v_*, v', and v'_*, respectively. The last two values depend on the integration parameters through the formulas

$$v' = \frac{v + v_*}{2} + \frac{|v - v_*|}{2}\sigma,$$
$$v'_* = \frac{v + v_*}{2} - \frac{|v - v_*|}{2}\sigma.$$

The first term in the integral, $f'_* f'$, accounts for particles with pre-collisional velocities v'_* and v' that may be colliding at the point (t, x) and turning their velocities to v and v_*. The loss term $-f_* f$ accounts for particles of velocities v_* and v that are colliding and changing their velocities. The rate by which these collisions occur is measured by the nonnegative kernel B. There are different choices for this kernel depending on modelling assumptions.

The angle θ measures the deviation between the pre- and post-collisional velocities. It is precisely given by the formula

$$\cos(\theta) = \frac{(v' - v'_*) \cdot (v - v_*)}{|v - v_*|^2}.$$

We write the collision kernel as $B(|v - v_*|, \cos\theta)$ to emphasize that it only depends on these two values. In particular, it is an even and 2π-periodic function of θ. A standard reference for the more classical mathematical theory related to these models, as well as some comments about their derivation, can be found in [94].

Since the Boltzmann collision operator acts on the velocity variable, for each fixed value of t and x, it is often convenient to study how it applies to functions $f = f(v)$ in terms of v only.

The Boltzmann collision operator (2.2), for any kernel $B \geq 0$ and any function $f \geq 0$, has remarkable cancellation properties that we list here

- $\int Q(f, f) \, dv = 0$.
- $\int v Q(f, f) \, dv = 0$.
- $\int |v|^2 Q(f, f) \, dv = 0$.
- $\int Q(f, f) \log f \, dv \leq 0$.

Each of these items corresponds to a conserved or monotone physical quantity. Any solution f of the Boltzmann equation (2.1) formally satisfies the following conservation laws:

- Conservation of mass: the integral $\iint f(t,x,v)\,\mathrm{d}v\,\mathrm{d}x$ is constant in time.
- Conservation of momentum: the integral $\iint v\,f(t,x,v)\,\mathrm{d}v\,\mathrm{d}x$ is constant in time.
- Conservation of energy: the integral $\iint |v|^2\,f(t,x,v)\,\mathrm{d}v\,\mathrm{d}x$ is constant in time.
- Entropy dissipation: the integral $\iint f(t,x,v)\log f(t,x,v)\,\mathrm{d}v\,\mathrm{d}x$ is monotone decreasing in time.

It can be verified directly that the only functions $f:\mathbb{R}^d\to[0,\infty)$ so that $Q(f,f)=0$ are the Gaussians. In other words, $Q(f,f)=0$ if and only if $f(v)=a\exp(-b|v-u|^2)$ for some $a,b\geq 0$ and $u\in\mathbb{R}^d$. In the kinetic context, these stationary solutions of (2.1) are called *Maxwellians*.

2.1 Common Collision Kernels and Variants

The exact form of the collision kernel B depends on modelling choices. Particles that bounce against each other like billiard balls lead to a different collision kernel than if we modeled particles that behave like sponges. There are many different possible microscopic interactions between particles, leading to a variety of choices for the collision kernel B. Rigorously justifying the derivation of the Boltzmann equation from microscopic particle interactions is a delicate mathematical problem, with several fundamental open questions. In many cases, one can compute informal derivations based on heuristics. In this section, we merely list the most common collision kernels B that are found in the literature, without any attempt to justify them.

The hard spheres model is the one that results from particles that bounce like billiard balls. Its collision kernel B takes a particularly simple form

$$B(r,\cos\theta)=r.$$

Another natural model is to consider point particles that repel each other by a potential of the form $1/r^q$, with $q\geq 1$. In that case, the kernel B is not explicit, but it satisfies the following bounds:

$$B(r,\cos\theta)\approx r^\gamma|\sin(\theta/2)|^{-d+1-2s}. \tag{2.3}$$

Here, we write $a\approx b$ to denote the fact that there is a constant C so that $a/C\leq b\leq Ca$. The parameters γ and s depend on the power q by the formulas

$$\gamma=\frac{q-2d+2}{q},\qquad 2s=\frac{2}{q}.$$

These kernels have a singularity at $\theta = 0$. Remarkably, for any values of v and v_*,

$$\int_{S^{d-1}} B(|v - v_*|, \cos(\theta))\, d\sigma = \infty.$$

The expression (2.2) still makes sense for any smooth function f when $s \in (0, 1)$. This is because the cancellation in the factor $(f'_* f' - f_* f)$ compensates the singularity in B. It is the same way we normally make sense of general integro-differential operators with singular kernels. Indeed, as we will see in Sect. 4, the Boltzmann collision operator consists of a diffusion term of fractional order $2s$ plus a lower order term.

When $s = 1$, the kernel B is too singular to make sense of the integral in (2.2). Taking $B(r, \cos(\theta)) = c_d(1-s)r^\gamma \sin(\theta/2)^{-d+1-2s}$, for some dimensional constant c_d, the expression (2.2) converges to the following expression as $s \to 1$:

$$Q(f, f)(v) = \bar{a}_{ij}(v)\partial_{ij} f(v) + \bar{c}(v) f(v). \tag{2.4}$$

Here, the coefficients $\bar{a}_{ij}$ and $\bar{c}$ depend on f through the formulas

$$\bar{a}_{ij} = \int_{\mathbb{R}^d} (|w|^2\delta_{ij} - w_i w_j)|w|^\gamma f(v-w)\, dw, \qquad \bar{c} = \partial_{ij}\bar{a}_{ij} = cf * |\cdot|^\gamma. \tag{2.5}$$

The case of Coulombic potentials in 3D corresponds to $d = 3$, $q = 1$, $s = 1$, and $\gamma = -3$. In that case, the convolution in (2.2) also becomes too singular. An appropriate asymptotic limit (2.5) as $\gamma \to -3$ leads to the Landau–Coulomb operator with

$$\bar{a}_{ij} = \partial_{ij}(-\Delta)^{-2} f = \frac{1}{8\pi} \int_{\mathbb{R}^3} (|w|^2\delta_{ij} - w_i w_j)|w|^{-3} f(v - w)\, dw, \qquad \bar{c} = f. \tag{2.6}$$

There are alternative equivalent expressions to (2.4) that are useful depending on the type of computation that we want to do. The following integral expression is equivalent to (2.4) and is useful, for example, to prove the dissipation of entropy.

$$Q(f, f) = \partial_i \int_{\mathbb{R}^d} (|w|^2\delta_{ij} - w_i w_j)|w|^\gamma (f(v - w)\partial_j f(v) - \partial_j f(v - w) f(v))\, dw.$$

The Landau operator is the limit of the Boltzmann collision operator as its angular singularity approaches its limit point $s = 1$. One could argue that because it is the most singular case, then the Landau operator should be *harder* to study than the Boltzmann collision operator. However, the Landau operator involves only classical differentiation instead of an integro-differential expression. At the end, depending on what we want to prove, sometimes the Landau equation is harder, and sometimes it is simpler, than the Boltzmann equation.

It is useful to analyze the Landau operator as a way to understand the nature of the Boltzmann collision operator as well. The first term is a diffusion operator whose coefficients depend on the solution f. It is a quasilinear second-order operator. The second term is of lower order. Note that the Landau operator is **not** the Laplacian plus a lower order term (i.e., it is not semi-linear), in the same way that the Boltzmann collision operator is **not** the fractional Laplacian plus a lower order term.

The Boltzmann collision operator (2.2) does not make sense in one dimension ($d = 1$). The constraints on the pre- and post-collisional velocities force $\{v', v'_*\} = \{v, v_*\}$. There is a common one-dimensional toy model proposed by Kac in which there is conservation of energy but not of momentum. The Kac collision operator takes the following form:

$$Q(f, f)(v) = \int_{\mathbb{R}} \int_{-\pi}^{\pi} (f'_* f' - f_* f) B(|v - v_*|, |\theta|) \, d\theta \, dv_*.$$

Here, we write

$$\begin{aligned} v' &= v \cos\theta - v_* \sin\theta, \\ v'_* &= v \sin\theta + v_* \cos\theta. \end{aligned}$$

In the original Kac model, $B(r, |\theta|) = 1/2\pi$. It makes sense to consider non-cutoff kernels with the bounds (2.3). It is natural to expect that the regularity estimates that we prove for the Boltzmann equation in Theorem 3.1 should apply to the Kac model as well. However, so far it has not been explicitly analyzed as far as we are aware.

2.2 *Mathematical Problems in Kinetic Equations*

There are many mathematical problems related to kinetic equations. An important area of research is on the rigorous derivation of the models. We can study the derivation of every variant of the Boltzmann equation and the Landau equation from particle models. We can also study the derivation of the Euler and Navier–Stokes equation from the Boltzmann equation.

Another, rather different, direction of research is on the well-posedness of the equations. Ultimately, the key to prove a well-posedness result lies on the a priori estimates that we are able to obtain for the solution. Using merely the conservation of mass and the entropy dissipation, one can construct a very weak notion of global solution. They are the *renormalized solutions* for the cutoff case and the *renormalized solutions with defect measure* for the non-cutoff case. The uniqueness of solutions within this class is currently unknown.

From an optimistic perspective, one may expect a solution f to be C^∞ smooth, strictly positive everywhere and with a strong decay as $|v| \to \infty$. Proving the existence of such a solution would require estimates that ensure these properties. The main focus of this note is on these a priori estimates.

There are several kinds of estimates that have attracted people's attention through the years. The following is a rough attempt to classify them:

(1) **Moment estimates**. They refer to upper bounds on quantities of the form

$$\iint f(t,x,v)\omega(v)\,\mathrm{d}v\,\mathrm{d}x.$$

Typically, $\omega(v) = \langle v\rangle^q$. A moment estimate for some large value of q should be understood as a type of decay estimate for $|v| \to \infty$.

(2) **Pointwise upper bounds**. They refer to an inequality of the form $f(t,x,v) \lesssim \omega(v)$. Typically, $\omega(v) = \langle v\rangle^{-q}$. One can easily argue that pointwise upper bounds are more desirable than moment estimates. Note that a bound of the form $f(t,x,v) \lesssim \langle v\rangle^{-d-q}$ implies a moment estimate for the weight $\langle v\rangle^p$ for any $p < q$.

(3) **Regularity estimates**. They may be upper bounds on some weighted Sobolev norm and on some Hölder norm or an upper bound on the derivatives of f.

(4) **Lower bounds**. They quantify how far f is from vacuum. They have the form $f \gtrsim \omega(v)$. Because the Maxwellians are stationary solutions, the best possible lower bound one can hope for would have a Gaussian decay as $v \to \infty$.

(5) **Convergence rates to equilibrium**. As time $t \to +\infty$, it is natural to expect that the solution f will converge to the equilibrium Maxwellian distribution. In some cases, it is possible to quantify this rate of convergence.

We use the Japanese bracket convention $\langle v\rangle := \sqrt{1+|v|^2}$.

Estimates of every kind described above are known in some circumstances. Below, we describe some of the mathematical difficulties and common simplifications.

2.2.1 Cutoff vs Non-cutoff

It may seem convenient to be able to make sense of the formula (2.2) for $Q(f,f)(v)$, even when f is not necessarily smooth. For that, it would be desirable that the collision kernel B is integrable with respect to the variable σ. The condition is known as Grad's cutoff assumption. It says that for any $v, v_* \in \mathbb{R}^d$,

$$\int_{S^{d-1}} B(|v-v_*|,\cos\theta)\,\mathrm{d}\sigma < +\infty. \tag{2.7}$$

The collision kernel B corresponding to the hard spheres model satisfies this condition. The kernels B that arise from inverse power law models (2.3) do not satisfy (2.7). The Boltzmann equation with a kernel of the form (2.3) would be referred to as the non-cutoff Boltzmann equation.

In the cutoff case, when (2.7) holds, it is possible to split the Boltzmann operator $Q(f, f)$ into its *gain* and *loss* terms. That is, $Q = Q_+ - Q_-$, where

$$Q_+ = \int_{\mathbb{R}^d} \int_{S^{d-1}} f'_* f' \, B(|v - v_*|, \cos\theta) \, d\sigma \, dv_*,$$

$$Q_- = \int_{\mathbb{R}^d} \int_{S^{d-1}} f_* f \, B(|v - v_*|, \cos\theta) \, d\sigma \, dv_*.$$

Both terms Q_+ and Q_- are nonnegative. It is common to use this decomposition when studying upper and lower bounds for the cutoff Boltzmann equation.

In the non-cutoff case, it makes no sense to split Q between a gain and a loss term. Indeed, the terms Q_+ and Q_- defined above would be both equal to $+\infty$ in general.

Interestingly, the non-cutoff Boltzmann equation exhibits a regularization effect that does not hold in the cutoff case. The Boltzmann collision operator (2.2) is in some sense a quasilinear integro-differential elliptic operator of order $2s$. This regularization effect is the reason why this note focuses on the non-cutoff case.

2.3 *Homogeneous vs Inhomogeneous*

When we consider solutions to (2.1) that are constant in x, the equation is significantly simpler. The second term vanishes and we have

$$f_t = Q(f, f).$$

Thinking that $Q(f, f)$ is a quasilinear elliptic operator, what we have is a nonlinear parabolic equation that satisfies

- **Conservation of mass:** $\int_{\mathbb{R}^d} f(t, v) \, dv$.
- **Conservation of momentum:** $\int_{\mathbb{R}^d} v f(t, v) \, dv$.
- **Conservation of energy:** $\int_{\mathbb{R}^d} |v|^2 f(t, v) \, dv$.
- **Monotone decreasing entropy:** $\int_{\mathbb{R}^d} f(t, v) \log f(t, v) \, dv$.

From a macroscopic perspective, the space homogeneous equation is rather plain. It corresponds to a fluid whose macroscopically observable quantities are homogeneous and stationary. From a mathematical perspective, the homogeneous problem is significantly simpler than the inhomogeneous. The well-posedness is well understood in the homogeneous non-cutoff Boltzmann equation when $\gamma + 2s \geq 0$. Remarkably, the existence of global in time classical solutions when $\gamma + 2s < 0$ remains open.

3 On the Existence of Solutions

For any evolution equation that comes from mathematical physics, we are interested in whether the problem is well-posed or not. In the case of the Boltzmann equation, we consider the Cauchy problem for a given initial value $f_0(x, v) \geq 0$, and we would want to determine if there exists a solution f to (2.1) so that $f(0, x, v) = f_0(x, v)$.

We know that a generalized notion of solution exists globally. It was established first in the cutoff case [34] and then also for non-cutoff [4]. It is unknown, for any kind of physically relevant collision kernel, whether these solutions may develop singularities in finite time starting from smooth data. It is unclear if the solution will stay unique past a potential singularity. Establishing the existence of global smooth solutions depends crucially on obtaining a priori regularity estimates.

Results about the existence of classical solutions can be classified in three rough sub-categories. They are the following:

(1) **Global well-posedness near equilibrium.** If the initial value f_0 is sufficiently close to a Maxwellian with respect to some appropriate norm, then it is often possible to prove that there exists a smooth solution with f_0 as its initial data.
(2) **Short-time existence results.** For generic values of f_0 in some functional space, a short-time existence result says that there exists a solution f in some interval of time $[0, T]$ for T sufficiently small (depending on f_0).
(3) **Unconditional global well-posedness.** For generic values of f_0 in some functional space, we would want to prove that there exists a solution f that exists forever. This is the ultimate goal, but it might also be too ambitious.

The analysis of solutions for short time is a basic type of well-posedness. However, it is a more complex issue than one might initially expect. The first result of this kind for the inhomogeneous non-cutoff Boltzmann equation appeared in [7] (see also [9]). It applies for $0 < s < 1/2$ and $\gamma + 2s < 1$. It also requires the initial data to belong to a Sobolev space involving at least four derivatives in x and v, with a weight that forces f_0 to have Gaussian decay as $v \to \infty$. In particular, the result in [7] cannot be applied for initial data that has polynomial, or even simply exponential decay, for large velocities.

The first result that allows initial data with algebraic decay rate appeared in [82]. It was later improved in [53]. It applies in the case of soft potentials ($\gamma \leq 0$) and requires the initial data in H^4 with a polynomial weight.

As far as we are aware, there are no further short-time existence results for the inhomogeneous Boltzmann equation without cutoff explicitly covered in the literature. The case of initial data with algebraic decay rate (as in $f_0 \lesssim \langle v \rangle^{-q}$), with hard potentials ($\gamma > 0$), is not handled by any documented result. It is remarkable that even though the Boltzmann equation is one of the most studied problems in mathematical physics, such a basic question still remains unanswered. Note also that the issue of uniqueness of solutions depends exclusively on our short-time well-posedness theory.

There are a number of papers dealing with solutions near equilibrium, for practically the full range of parameters γ and s. See [2, 3, 10, 12, 36, 43, 55, 91, 98]. From the proofs in any of these papers, in theory, one should be able to extract a proof of short-time existence. Its conditions on the initial data would be milder than for the global existence result. However, in most of these papers, the short-time existence results are not explicitly stated, and it is difficult to determine what can be done in each case.

For any inhomogeneous model like (2.1), with a nontrivial and physically meaningful collision kernel Q, the global existence of classical solutions for generic initial data f_0 remains an outstanding open problem.

A short-time existence result is a first step in the Cauchy theory. In order to make the solution global, we would need sufficiently strong a priori estimates. If the solution stays bounded in a sufficiently strong norm, one would be able to reapply the short-time existence result arbitrarily many times and extend the solution forever. The key for global well-posedness would be to combine a short-time existence result with good enough a priori estimates. It is not necessary to work with weak solutions. A priori estimates for classical solution would suffice, since these are the type of solutions provided by short-time existence results. We review a priori estimates for the inhomogeneous Boltzmann equation in these notes. As we will see, they are not strong enough to ensure the unconditional global well-posedness of the equation yet.

While global well-posedness results near equilibrium are interesting in themselves, they say hardly anything about the solutions for arbitrary initial data. Moreover, any appropriate short-time existence result combined with the convergence to equilibrium [28], and Theorem 3.1, implies the existence of global solutions when the initial data is sufficiently close to a nonzero Maxwellian (see [91]). It is also possible to study the global well-posedness of the equations when the initial data is near zero (see [76] and [26]). It is a completely different setting, driven by dispersion rather than by diffusion.

3.1 Conditional Regularity Estimates

We now describe our recent results on a priori regularity estimates. Let us start with some discussion of what we would want to prove and what we can realistically achieve.

The hydrodynamic equations, like Euler and Navier–Stokes, describe the evolution of a fluid at a coarser scale than kinetic equations. The velocity $u(t, x)$, density $\rho(t, x)$, and temperature $\theta(t, x)$ in the compressible Euler equation make sense at a macroscopic scale. Microscopically, they correspond to some local average over particles of the fluid around each point (t, x). The compressible Euler equation can be obtained formally as an asymptotic limit of solutions to the Boltzmann equation. The compressible Euler equation develops singularities in finite time. Should we expect the Boltzmann equation to develop singularities as well? Or is there any reason to believe that the Boltzmann equation will act as a regularization of the usual hydrodynamic equations?

There are two very different kinds of singularities that one observes in the compressible Euler equation: shocks and implosions. A *shock* is the same type of singularity that emerges from the flow of Burgers equation. The values of the function stay bounded, and a discontinuity is created when two characteristic curves collide. It is as if the equation was pushing the solution to take two contradictory values at the same point, resulting in a discontinuity. In the kinetic setting, on the contrary, there can be no such thing as a shock singularity. The function $f(t, x, v)$ allows a distribution of different velocities coexisting at each point (t, x). An *implosion* singularity has a very different nature. The flows concentrate at one point, where the energy concentrates and its density explodes. While some implosion profiles for the Euler equations appeared a long time ago, it was only recently that smooth implosion profiles and their stability were properly understood [77, 78]. There is no obvious reason to rule out the existence of implosion singularities for the Boltzmann equation. It seems conceivable that there may exist a solution to the Boltzmann equation (2.1) whose mass, momentum, and temperature density behave similarly as in the implosion singularity of the compressible Euler equation. Rigorously constructing, or ruling out, such a solution seems like a very difficult problem right now.

Beyond the possibility of implosion singularities, we may still wonder if there exist any other kind of kinetic singularities for the Boltzmann equation that are not related to anything that we can see macroscopically in the Euler equation. Our conditional regularity results described below rule it out. Essentially, they say that the solution to the non-cutoff Boltzmann (and Landau) equation will be C^∞ smooth for as long as there is no implosion singularity in its associated macroscopically observable quantities.

Let us state our conditions. They say that the mass density is bounded from below and above, the energy density is bounded above, and the entropy density is bounded above. The following three inequalities will be assumed to hold at every point (t, x):

$$m_0 \leq \int_{\mathbb{R}^d} f(t, x, v)\, dv \leq M_0, \tag{3.1}$$

$$\int_{\mathbb{R}^d} f(t, x, v)|v|^2\, dv \leq E_0, \tag{3.2}$$

$$\int_{\mathbb{R}^d} f(t, x, v) \log f(t, x, v)\, dv \leq H_0. \tag{3.3}$$

We stress that we will not prove the hydrodynamic inequalities (3.1), (3.2), and (3.3). We take them as an assumption. From the discussion above, one would be inclined to believe that there may exist some initial data that evolves into an implosion singularity for which all these quantities blow up at a point. However, such a solution has not been constructed yet for any physically reasonable collision operator Q. While the assumptions (3.1), (3.2), and (3.3) will not be proved in general, it would not be difficult to verify that for any reasonable initial data they hold for some short period of time or globally if the initial data is sufficiently close to a Maxwellian (see, for example, [91]).

Here is our conditional regularity theorem proved in [63].

Theorem 3.1 *Let f be a (classical) solution to the inhomogeneous Boltzmann equation* (2.1)*, periodic in space, where the collision operator Q has the form* (2.2) *and B is the standard non-cutoff collision kernel of the form* (2.3) *with $\gamma + 2s \in [0, 2]$.*

Assume that there are constants $m_0 > 0$, M_0, E_0, and H_0 such that the hydrodynamic bounds (3.1), (3.2), *and* (3.3) *hold. Moreover, if $\gamma \leq 0$, we also assume that the initial data f_0 decays rapidly at infinity: $f_0 \leq C_q \langle v\rangle^{-q}$ for all $q > 0$.*

Then, for $\tau > 0$, all derivatives $D^\alpha f$, for any multi-index α, are bounded in $[\tau, \infty) \times \mathbb{R}^d \times \mathbb{R}^d$ and decay as $|v| \to \infty$. More precisely, there are constants $C_{\tau,\alpha,q}$ so that

$$\langle v\rangle^q |D^\alpha f(t, x, v)| \leq C_{\tau,\alpha,q} \qquad \textit{for all } t > \tau, x \in \mathbb{R}^d, v \in \mathbb{R}^d.$$

The proof of theorem 3.1 contains several ingredients that originated in the study of nonlocal parabolic equations. We describe some of the key ingredients in this manuscript. A more detailed description of the proof, still in survey form, is given in [60].

There is a parallel result for the inhomogeneous Landau equation that is proved by Chris Henderson and Stan Snelson in [51]. We state it here.

Theorem 3.2 *Let f be a (possibly weak) solution to the inhomogeneous Landau equation* (2.1) *where the collision operator Q has the form* (2.4) *and* (2.5) *with $\gamma \in (-2, 0)$.*

Assume that there are constants $m_0 > 0$, M_0, E_0, and H_0 such that the hydrodynamic bounds (3.1), (3.2), *and* (3.3) *hold. There is a $\mu > 0$, depending on m_0, M_0, E_0, H_0, and γ, so that if*

$$f_0(x, v) \leq Ce^{-\mu|v|^2},$$

then, for any $\tau > 0$, $\mu' \in (0, \mu)$, all derivatives $D^\alpha f$, for any multi-index α, are bounded in $[\tau, \infty) \times \mathbb{R}^d \times \mathbb{R}^d$ and decay as $|v| \to \infty$. More precisely, there are constants $C_{\tau,\mu',\alpha,q}$ so that

$$e^{\mu'|v|^2} |D^\alpha f(t, x, v)| \leq C_{\tau,\mu',\alpha,q} \qquad \textit{for all } t > \tau, x \in \mathbb{R}^d, v \in \mathbb{R}^d.$$

The proof of Theorem 3.2 contains similar steps as the proof of Theorem 3.1, except that instead of regularity techniques for nonlocal equations, it involves more classical second-order diffusion. It uses the upper bounds from [23] instead of the ones from [58, 89] and the Harnack inequality from [40] instead of [61].

The regularity estimates from Theorems 3.1 and 3.2 stay uniform as $t \to \infty$, or for as long as (3.1), (3.2) and (3.3) hold. This is important. For example, the celebrated convergence to equilibrium result by Desvillettes and Villani in [28] applies to solutions provided that they are uniformly C^∞ as $t \to \infty$. However, if we only seek to construct a global smooth solution, it will suffice to have regularity estimates that are allowed to deteriorate for large time. In that case, it is shown in [52, 54] that the upper bounds on mass density and energy density suffice. We summarize both results in the following statement.

Theorem 3.3 *Let f be a solution to* (2.1)*, periodic in space, where the collision operator Q is either the non-cutoff Boltzmann collision operator* (2.2) *with B as in* (2.3) *with $s \in (0, 1)$ and $\gamma + 2s \in [0, 2]$ or the Landau operator* (2.4) *and* (2.5) *with $\gamma \in (-2, 0)$.*

Assume that there are upper bounds for the mass and energy density. That means that there is a constant N_0 so that for all t, x,

$$\int f(t, x, v)(1 + |v|^2) \, dv \leq N_0.$$

Assume that f_0 is continuous and periodic in x. It does not need to be strictly positive. There might be values of x so that $f_0(x, \cdot)$ vanishes.

In the case that $\gamma \leq 0$, for the Boltzmann equation, we also assume that $f_0(x, v) \leq C_q \langle v \rangle^{-q}$ for all $q > 0$.

In the case of the Landau equation, we also assume $f_0(x, v) \leq Ce^{-\mu|v|^2}$.

Then, for all derivatives $D^\alpha f$, for any multi-index α, there exist upper bounds of the form

$$|D^\alpha f(t, x, v)| \leq \begin{cases} C_{\alpha,q,f_0}(t)\langle v \rangle^{-q} & \text{for any } q \geq 0, \text{ for Boltzmann,} \\ C_{\alpha,\mu',f_0}(t)e^{-\mu'|v|^2} & \text{for any } \mu' \in [0, \mu), \text{ for Landau.} \end{cases}$$

Unlike the upper bounds in Theorems 3.1 and 3.2, the upper bounds in Theorem 3.3 deteriorate as $t \to \infty$. They suffice to construct a global smooth solution, but they would not let us apply the result in [28] and deduce their relaxation to equilibrium.

Theorem 3.3 builds upon Theorems 3.1 and 3.2 by propagating a lower bound that is stronger than the first inequality in (3.1). Indeed, for any nonzero continuous initial data f_0, there will be a clear ball $B = B_r(x_0, v_0)$ and $\delta > 0$ so that $f_0(x, v) \geq \delta$ whenever $(x, v) \in B$. The upper bounds on mass and energy densities turn out to suffice to propagate and expand this lower bound to the full space by a barrier-like argument. Thus, the lower bound in (3.1) holds for positive time. The upper bound on the entropy (3.3) is not proved directly. Instead, the authors observe that the proofs of Theorems 3.1 and 3.2 go through without (3.3) as soon as we have a barrier from below. The right-hand side of the regularity estimates in Theorem 3.3 depends on the initial data in terms of this clear ball.

4 Integro-Differential Diffusion Inside the Boltzmann Collision Operator

The analysis of equations involving integro-differential diffusion has been an intensely researched area since the beginning of the twenty-first century. A linear *elliptic* integro-differential operator has the general form

$$Lf(v) = \int (f(v') - f(v))K(v, v')\, \mathrm{d}v', \tag{4.1}$$

for some nonnegative kernel K. The typical linear parabolic integro-differential equation would be $f_t = Lf$, for a nonlocal operator L as above.

The operator L in (4.1) is the nonlocal analog of more classical second-order elliptic operators. In the case the $K(v, v') = |v - v'|^{-d-2s}$, it is a multiple of the fractional Laplacian: $Lf(v) = -c(-\Delta)^s f(v)$. The operator L would be considered elliptic of order $2s$ if its kernel is comparable with the kernel of the fractional Laplacian $|v - v'|^{-d-2s}$. However, this concept is more subtle than it may seem initially. There are several different ellipticity conditions in the literature depending on different interpretations of the word *comparable*.

There are two types of second-order elliptic operators: divergence form operators $Lf = \partial_i a_{ij}(v)\partial_j f$ and nondivergence form operators $Lf = a_{ij}(v)\partial_{ij} f$. In both cases, the uniform ellipticity condition consists in requiring the coefficients $a_{ij}(v)$ to be strictly positive definite at every point v, with some specific upper and lower bounds $\lambda I \leq \{a_{ij}\} \leq \Lambda I$. In the nonlocal setting (4.1), the divergence or nondivergence structure of the operator will be reflected in different symmetry conditions on the kernel K. A possible uniform ellipticity condition would be to require $\lambda|v - v'|^{-d-2s} \leq K(v, v') \leq \Lambda|v - v'|^{-d-2s}$. It is a common condition in the literature of nonlocal equations. However, the richness of the class of integral kernels gives us a lot of flexibility, and more general notions of ellipticity exist.

Most of the central regularity results for elliptic and parabolic equations have a nonlocal counterpart. In particular, there are nonlocal versions of De Giorgi–Nash–Moser theorem [14, 18, 20, 22, 37, 65, 68, 71], Krylov–Safonov theorem [15–19, 21, 25, 66, 67, 74, 84, 87, 88, 92], and Schauder estimates [35, 57, 64, 79, 86].

In order to apply methods that originate in the study of elliptic equations in divergence form (like in the theorem of De Giorgi, Nash, and Moser), we would need the kernel K to satisfy a symmetry condition that allows us to apply variational techniques. The usual elliptic operators in divergence form $f \mapsto \partial_i a_{ij} \partial_j f$ are self-adjoint. Analogously, the integro-differential operator L above is self-adjoint when $K(v, v') = K(v', v)$.

Conversely, in order to apply methods that originate in the study of elliptic equations in non-divergence form (like in the theorem of Krylov and Safonov), we need different symmetry assumptions on the kernel K. The key property of non-divergence operators like $f \mapsto a_{ij}\partial_{ij} f$ is that for any smooth function f the values of the operator make sense pointwise. This is always true for integro-differential operators when $s < 1/2$. If $s \geq 1/2$, we would need the symmetry condition $K(v, v+w) = K(v, v-w)$. The type of techniques that one can apply to integro-differential equations vary according to the symmetry conditions on the kernel.

The Boltzmann collision operator (2.2) involves an integral of differences. However, it is not immediately clear how to relate it with results about nonlocal operators of the form (4.1). With that in mind, let us analyze the expression for $Q(f, f)$ in (2.2) and derive an equivalent formulation.

We add and subtract a term $f'_* f$ inside the integrand.

$$\begin{aligned} Q(f, f)(v) &= \int_{\mathbb{R}^d} \int_{S^{d-1}} (f'_* f' - f'_* f + f'_* f - f_* f)\, B \,\mathrm{d}\sigma\, \mathrm{d}v_*, \\ &= \int_{\mathbb{R}^d} \int_{S^{d-1}} (f' - f)\, [f'_* B]\, \mathrm{d}\sigma\, \mathrm{d}v_* \\ &\quad + f(v) \int_{\mathbb{R}^d} \int_{S^{d-1}} (f'_* - f_*) B \,\mathrm{d}\sigma\, \mathrm{d}v_*. \end{aligned}$$

For the first term, we use a change of variables known as *Carleman coordinates*. For the second term, we refer to the *cancellation lemma*.

The cancellation lemma appeared first in [1] (see also [89]). It reduces the integral in the second term to a convolution.

Lemma 4.1 (Cancellation Lemma) *There is a nonnegative function $b : \mathbb{R}^d \to [0, \infty)$, depending only on the collision kernel B on* (2.2), *so that*

$$\int_{\mathbb{R}^d} \int_{S^{d-1}} (f'_* - f_*) B \,\mathrm{d}\sigma\, \mathrm{d}v_* = [f * b](v).$$

In particular, if B has the standard non-cutoff form (2.3), *then $b(v) \approx |v|^\gamma$.*

For the first term, we want to change the variables of integration from $v_* \in \mathbb{R}^d$ and $\sigma \in S^{d-1}$ to $v' \in \mathbb{R}^d$ and $w \in \langle v' - v \rangle^\perp$, so that $v'_* = v + w$ and $v_* = v' + w$. With these new variables, it can be checked (see [89]) that

$$\mathrm{d}\sigma\, \mathrm{d}v_* = \frac{2^{2d-1}}{|v' - v||v - v_*|^{d-2}}\, \mathrm{d}w\, \mathrm{d}v'.$$

Thus, the first term becomes

$$\int_{\mathbb{R}^d} \int_{S^{d-1}} (f' - f)\, [f'_* B]\, \mathrm{d}\sigma\, \mathrm{d}v_* = \int_{\mathbb{R}^d} (f' - f) K_f(v, v')\, \mathrm{d}v',$$

where

$$K_f(v, v') = \frac{2^{2d-1}}{|v' - v|} \int_{w \perp v' - v} |v - v_*|^{-d+2} f(v'_*) B(|v - v_*|, \cos\theta)\, \mathrm{d}w. \qquad (4.2)$$

For a standard non-cutoff kernel B satisfying (2.3), we obtain

$$K_f(v, v') \approx |v' - v|^{-d-2s} \int_{w \perp v'-v} |w|^{\gamma+2s+1} f(v+w) \, dw. \tag{4.3}$$

When this last integral factor is bounded from below and above, we recover the most classical notion of classical ellipticity $K_f(v,' v) = |v - v'|^{-d-2s}$. However, the integral is on a hyperplane. If f is a function in L^1, with finite second moment, it is not possible to bound this integral pointwise either from below or from above. For that reason, we must explore more general notions of ellipticity.

Let us suppose that all we know about the function $f : \mathbb{R}^d \to [0, \infty)$ is that it has finite mass (integral), energy (second moment), and entropy. Moreover, its mass is bounded from below. We state these assumptions precisely below.

$$0 < m_0 \le \int f(v) \, dv \le M_0,$$

$$\int f(v)|v|^2 \le E_0,$$

$$\int f(v) \log f(v) \, dv \le H_0.$$

Based on these four constants only, we can deduce the following nondegeneracy conditions on the kernel K_f (see [60, 61, 63, 89]).

Symmetry in the *nondivergence* way: This is automatic from the formula (4.2).

$$K(v, v + w) = K(v, v - w).$$

Lower bound on a cone of directions: There is a constant $\lambda > 0$ depending on m_0, M_0, E_0, and H_0 such that

$$K(v, v') \ge \lambda |v' - v|^{-d-2s},$$

when v' belongs to a cone of directions emanating from v. This cone is *thick* in the sense that the measure of its intersection with the unit sphere centered at v has a lower bound depending also on m_0, M_0, E_0 and H_0.

Upper bound on average: There is a constant Λ depending only on M_0 and E_0 such that

$$\int_{B_r(v)} K_f(v, v')|v - v'|^2 \, dv' \le \Lambda r^{2-2s}.$$

Cancellation: While the exact *divergence form* symmetry does not hold. We do have some symmetry in the form of cancellation estimates

$$\left| \int_{B_r(v)} K_f(v, v') - K_f(v', v) \, dv' \right| \leq \Lambda r^{-2s}.$$

If $s \geq 1/2$, we also have

$$\left| \int_{B_r(v)} (v - v')(K_f(v, v') - K_f(v', v)) \, dv' \right| \leq \Lambda r^{1-2s}.$$

The first of these inequalities is a rewriting of the classical cancellation lemma from [1] in terms of the kernel K_f.

These are the estimates that we get for K_f based on minimal physically meaningful assumptions on f. Interestingly, much of the theory of elliptic and parabolic integro-differential equations can be recovered for equations involving kernels that satisfy these assumptions only. In that sense, the lower bound on the cone of nondegeneracy and the upper bound on average, described above, are some mild form of integro-differential uniform ellipticity of order $2s$.

The values of λ and Λ, and the thickness of the cone of degeneracy, depend on m_0, M_0, E_0, and H_0 and also on $|v|$. Their values degenerate in certain precise way as $|v| \to \infty$.

A key property of divergence form elliptic operators that is central in the proof of the theorem of De Giorgi, Nash, and Moser is their coercivity and boundedness with respect to the H^1 norm. In the case of integro-differential operators, we work with fractional order Sobolev spaces. To have a nonlocal analog of this important regularity theorem, we want the integro-differential operator L to be bounded and coercive with respect to the H^s norm. The following results tell us that the cone condition and the upper bound on average described above for the Boltzmann kernel K_f are enough.

Proposition 4.2 *Assume that a nonnegative kernel $K : \mathbb{R}^d \times \mathbb{R}^d \to \mathbb{R}$ satisfies the following two conditions:*

- *There is a constant Λ such that*

$$\int_{B_r(v)} K(v, v')|v - v'|^2 \, dv' \leq \Lambda r^{2-2s}.$$

- *The cancellation condition holds*

$$\left| \int_{B_r(v)} K(v, v') - K(v', v) \, dv' \right| \leq \Lambda r^{-2s}.$$

If $s \geq 1/2$, we also have

$$\left| \int_{B_r(v)} (v - v')(K(v, v') - K(v', v)) \, dv' \right| \leq \Lambda r^{1-2s}.$$

Then the operator L defined in (4.1) *is a well-defined bounded linear operator from* H^s *to* H^{-s}.

As we discussed, the kernel K_f of the Boltzmann equation satisfies the assumption of Proposition 4.2. Yet, Proposition 4.2 is a general statement of integro-differential operators. It applies to any kernel K, regardless of whether it was obtained from the Boltzmann collision operator or from any other origin. A proof of Proposition 4.2 is given in [61].

A form of coercivity follows from the cone of nondegeneracy described above. It is a consequence of the following general result for integro-differential operators proved in [24]

Theorem 4.3 *Let* $K : B_2 \times B_2 \to \mathbb{R}$ *be a nonnegative kernel. Assume that there exist two constants* $\lambda > 0$ *and* $\mu > 0$ *so that for any* $v \in B_2$ *and any ball* $B \subset B_2$ *that contains* v, *we have*

$$|\{v' \in B : K(v, v') \geq \lambda |v - v'|^{-d-2s}\}| \geq \mu |B|.$$

Then, the following coercivity holds:

$$\begin{aligned}\iint_{B_2 \times B_2} & |f(v') - f(v)|^2 K(v, v') \, dv' \, dv \\ & \geq c\lambda \iint_{B_1 \times B_1} |f(v') - f(v)|^2 |v - v'|^{-d-2s} \, dv' \, dv = \tilde{c}\lambda \|f\|_{\dot{H}^s}.\end{aligned}$$

Here, the constants c *and* $\tilde{c}$ *depend on* μ, s, *and dimension only.*

The cone of nondegeneracy condition satisfied by the Boltzmann kernel K_f is naturally stronger than the assumption of Theorem 4.3. The coercivity of the Boltzmann collision operator was proved several times in the literature using various techniques. It is interesting to realize that it follows as a consequence of a general statement of integro-differential quadratic forms.

5 Regularity Estimates for the Kinetic Fokker–Planck Equation

The collision operator $Q(f, f)$ in (2.1) acts as a diffusion in the velocity variable. It may be a nonlocal diffusion in the case of the Boltzmann equation or a more classical second-order diffusion in the case of Landau equation. In any case, we must address the fact that the equation has a regularization effect with respect to all variables t, x, and v, even though the diffusion is with respect to velocity only. It is a hypoelliptic effect that comes from the interaction between the diffusion in velocity and the kinetic transport.

In 1934, Kolmogorov studied the following equation:

$$f_t + v \cdot \nabla_x f = \Delta_v f.$$

It is a simpler linear model than (2.1), where the collision operator Q is replaced by the usual Laplacian. Kolmogorov observed that for any initial data $f_0 \in L^1 + L^\infty$, the equation has a smooth solution. It follows immediately after explicitly computing its fundamental solution.

$$K(t,x,v) = \begin{cases} c_d t^{-2d} \exp\left(-\frac{|v|^2}{4t} - \frac{3|x-tv/2|^2}{t^3}\right) & \text{for } t > 0, \\ 0 & \text{for } t \le 0. \end{cases}$$

We compute the solution f, for any initial data f_0, by a modified convolution of f_0 with K. From this formula, we observe immediately that the solution is C^∞ smooth in all variables.

For fractional diffusion, a similar analysis applies, but this time the fundamental solution is not explicit. If we consider, for $s \in (0,1)$,

$$f_t + v \cdot \nabla_x f + (-\Delta_v)^s f = 0,$$

there is a fundamental solution K_s whose Fourier transform with respect to x and v is given by

$$\hat{K}_s(t,\psi,\xi) = c \exp\left(-\int_0^t |\xi + \tau\psi|^{2s} \, d\tau\right).$$

It is easy to see that K_s is smooth in all variables, so the fractional Kolmogorov equation enjoys analogous regularization properties as the classical one. It is also possible to see that the kernels K_s are bounded, nonnegative, and integrable. They have some polynomial decay as x and v go to infinity. However, the exact asymptotics have not been precisely described yet.

The regularity analysis for nonlinear equations depends on regularity results for *linear* equations with variable (possibly rough) coefficients. In this section, we describe the three most fundamental regularity results, which bring to the kinetic setting the theorems of De Giorgi–Nash–Moser and Schauder.

5.1 De Giorgi Meets Kinetic Equations

The theorem of De Giorgi, Nash, and Moser is arguably the most fundamental regularity result for nonlinear elliptic PDE. It is usually stated in terms of *linear* equations. But since it does not have any regularity assumption on the coefficients, it is readily applicable to solutions of quasilinear elliptic equations whose coefficients depend somehow on the solution.

For kinetic equations with classical second-order diffusion in divergence form, we study equations of the following form.

$$f_t + v \cdot \nabla_x f - \partial_{v_i} a_{ij}(t,x,v) \partial_{v_j} f = b \cdot \nabla_v f + h. \tag{5.1}$$

Here, the function h is a source term. We also included an extra drift term $b \cdot \nabla_v f$ on the right-hand side. The coefficients $a_{ij}(t,x,v)$ are assumed to be uniformly elliptic. It means that there exist constants $\Lambda \geq \lambda > 0$ so that for all (t,x,v) in the domain of the equation, $\lambda \mathrm{I} \leq \{a_{ij}(t,x,v)\} \leq \Lambda \mathrm{I}$. It is important that there is no regularity assumption on the coefficients a_{ij} beyond boundedness and measurability.

The kinetic version of De Giorgi, Nash, and Moser theorem provides Hölder continuity estimates for solutions to (5.1). When the right-hand side h vanishes, there is also some form of Harnack inequality that holds. The following theorem was developed in [83], [75], [96], [97], and [40]. The presentation below matches the result in the last one of these papers.

Theorem 5.1 *Assume $f : [0,1] \times B_1 \times B_1 \to \mathbb{R}$ is a weak solution of (5.1) in $(0,1] \times B_1 \times B_1$ for some bounded functions b and h. Then, there exists an $\alpha > 0$ depending only on dimension and the ellipticity parameters λ and Λ and a constant C depending on dimension, λ, Λ, and $\|b\|_{L^\infty}$, so that*

$$\|f\|_{C^\alpha((1/2,1)\times B_{1/2} \times B_{1/2})} \leq C \left(\|f\|_{L^\infty([0,1]\times B_1 \times B_1)} + \|h\|_{L^\infty([0,1]\times B_1 \times B_1)} \right).$$

Theorem 5.1 can be used to conclude that solutions to the inhomogeneous Landau equation that satisfy (3.1), (3.2), and (3.3) are Hölder continuous. It is a key ingredient in the proof of Theorem 3.2.

In order to study the regularity of the Boltzmann equation, we need a similar result but for integro-differential diffusion. Let us now consider equations of the form

$$f_t + v \cdot \nabla_x f - \int_{\mathbb{R}^d} (f(t,x,v') - f(t,x,v)) K(t,x,v,v') \, \mathrm{d}v' = h. \tag{5.2}$$

Here h is a source term. The kernel K is a nonnegative function. We want to study Eq. (5.2) with an eye on possible applications to the Boltzmann equation. Thus, we will assume that the kernel K satisfies the nondegeneracy conditions described in Sect. 4.

The following theorem is proved in [61].

Theorem 5.2 *Assume $f : [0,1] \times B_1 \times \mathbb{R}^d \to \mathbb{R}$ is a bounded weak solution of (5.2) for $(t,x,v) \in (0,1] \times B_1 \times B_1$ and some bounded function h. We make the following assumptions on the kernel K:*

- ***Upper bound on average.*** *There is a constant Λ such that for all $(t,x,v) \in (0,1] \times B_1 \times B_1$ and $r > 0$,*

$$\int_{B_r(v)} K(t,x,v,v')|v-v'|^2 \, \mathrm{d}v' \leq \Lambda r^{2-2s}.$$

- ***Nondegeneracy condition.*** *There exist two constants $\lambda > 0$ and $\mu > 0$ so that for any $(t,x,v) \in (0,1] \times B_1 \times B_1$ and any ball $B \subset B_2$ that contains v, we have*

$$|\{v' \in B : K(t,x,v,v') \geq \lambda |v-v'|^{-d-2s}\}| \geq \mu |B|.$$

- ***The cancellation condition.*** *For all $(t,x,v) \in (0,1] \times B_1 \times B_1$ and $r \in [0,1]$,*

$$\left| \int_{B_r(v)} K(t,x,v,v') - K(t,x,v',v) \, \mathrm{d}v' \right| \leq \Lambda r^{-2s}.$$

If $s \geq 1/2$, we also have

$$\left| \int_{B_r(v)} (v-v')(K(t,x,v,v') - K(t,x,v',v)) \, \mathrm{d}v' \right| \leq \Lambda r^{1-2s}.$$

Then f is Hölder continuous and it satisfies the following estimate for some $\alpha > 0$:

$$\|f\|_{C^\alpha((1/2,1)\times B_{1/2}\times B_{1/2})} \leq C \left(\|f\|_{L^\infty([0,1]\times B_1 \times \mathbb{R}^d)} + \|h\|_{L^\infty([0,1]\times B_1\times B_1)} \right).$$

The constants α and C depend on dimension, λ, Λ, and μ only.

Note that the assumptions on Theorem 5.2 are exactly the union of the assumptions of Proposition 4.2 and Theorem 4.3. The Boltzmann kernel satisfies a cone of nondegeneracy condition that implies the nondegeneracy hypothesis of Theorem 5.2.

Remarkably, Theorems 5.1 and 5.2 are not specifically for the Landau and Boltzmann equation as in (2.1). They apply to a more general family of kinetic Fokker–Planck equations. In the case of Landau equation, the coefficients a_{ij} will be related to the solution through (2.5). In the case of the Boltzmann equation, the kernel K will be related to f through (4.2). Theorems 5.1 and 5.2 apply regardless of these extra relations.

5.2 *Schauder Meets Kinetic Equations*

The Schauder estimates provide an a priori estimate in higher order Hölder spaces when we know, in addition to uniform ellipticity, that the coefficients of the equation are Hölder continuous.

Unlike the Hölder estimate obtained using De Giorgi's method, the Schauder estimates provide an estimate in a Hölder norm with a precise exponent. When we consider a kinetic Fokker–Planck equation like (5.1), we gain exactly two derivatives in v. In the fractional order case (5.2), we should expect to gain $2s$ derivatives in v. The exact gain of regularity with respect to the variables t and x is less obvious. We need to keep track of the precise balance of scales between these variables. For that, it is important to analyze the group of transformations that leave the class of equations (5.1) or (5.2) invariant.

Let us focus on the fractional order case (5.2), which is the more complex of the two. We observe that if f solves (5.2) for some kernel K satisfying the hypothesis of Theorem 5.2, then the same is true for the scaled function

$$f_r(t, x, v) = f(r^{2s}t, r^{1+2s}x, rv).$$

The kernel K would have to be replaced with $r^{d+2s}K(r^{2s}t, r^{1+2s}x, rv)$, which satisfies the same assumptions as the original K.

Because this is the natural scaling of the equation, we define the kinetic cylinders at the origin Q_r as

$$Q_r := (-r^{2s}, 0] \times B_{r^{1+2s}} \times B_r.$$

The class of equations (5.2) is not translation invariant because the coefficient in the drift term depends on v. It is *Galilean* invariant, in the sense that we can change coordinates to another inertial frame of reference. That is, given (s, y, w), if we define

$$(s, y, w) \circ (t, x, v) = (s + t, y + x + tw, w + v),$$

then $\tilde{f}(t, x, v) = f((s, y, w) \circ (t, x, v))$ would satisfy also an equation like (5.2) with K replaced with $\tilde{K}(t, x, v) = K((s, y, w) \circ (t, x, v))$.

In order to state the Schauder estimates for the Eq. (5.2), we need to use a Hölder norm that is compatible with the scaling and Galilean invariance of the equation. The best way to achieve it is by defining an appropriate notion of distance.

Definition 5.3 For any two points $z_1 = (t_1, x_1, v_1)$ and $z_2 = (t_2, x_2, v_2)$ in $\mathbb{R}^{1+2d}$, we define the following: *kinetic* distance

$$d_\ell(z_1, z_2) := \min_{w \in \mathbb{R}^d} \left\{ \max\left(|t_1 - t_2|^{\frac{1}{2s}}, |x_1 - x_2 - (t_1 - t_2)w|^{\frac{1}{1+2s}}, |v_1 - w|, |v_2 - w| \right) \right\}.$$

Strictly speaking, this is only a distance (it satisfies the triangle inequality) when $s \geq 1/2$. For smaller values of $s \in (0, 1/2)$, d_ℓ^{2s} is a distance. In either case, we use the expression for d_ℓ as in Definition 5.3 to keep consistency of the 1-homogeneity with respect to the variable v.

The distance d_ℓ has the following two properties:

- It is homogeneous with respect to the scaling of the equation, in the sense that

$$d_\ell((r^{2s}t_1, r^{1+2s}x_1, rv_1), (r^{2s}t_2, r^{1+2s}x_2, rv_2)) = rd_\ell((t_1, x_1, v_1), (t_2, x_2, v_2)).$$

- It is invariant by left Galilean translations. That is, for every $z_0, z_1, z_2 \in \mathbb{R}^{2d+1}$,

$$d_\ell(z_0 \circ z_1) = d_\ell(z_0 \circ z_2).$$

The subindex "ℓ" is meant to indicate that the distance is **"l"**eft-invariant, as opposed to right-invariant.

In terms of this kinetic distance, we define the kinetic Hölder norms.

Definition 5.4 Let $\Omega \subset \mathbb{R}^{2d+1}$. For any $\alpha \in (0, \infty)$, we define the $C^\alpha(z_0)$ semi-norm of a function $f : \Omega \to \mathbb{R}$ as the smallest constant C so that for all $z \in \Omega$

$$|f(z) - p(z)| \leq C d_\ell(z, z_0)^\alpha,$$

for some polynomial $p(t, x, v)$ whose kinetic degree is less than α.

Moreover, the C^α semi-norm of f in Ω is the supremum of $\|f\|_{C^\alpha(z_0)}$ for all $z_0 \in \Omega$.

The full norm $\|f\|_{C^\alpha_\ell(D)}$ is defined as $[f]_{C^\alpha_\ell(D)} + [f]_{L^\infty(D)}$.

We left out the definition of kinetic degree of a polynomial. It is adjusted by the scaling defined above so that the variable v has degree one, t has degree $2s$, and x has degree $1 + 2s$.

Now that we have a precise notion of Hölder continuity that takes into account the scaling and Galilean invariance of the equation, we are ready to state the kinetic Schauder estimates. The following theorem is proved in [62].

Theorem 5.5 (The Schauder Estimate) *Let $\alpha \in (0, \min(1, 2s))$ and $\alpha' = 2s\alpha/(1+2s)$. Let $K : Q_1 \times \mathbb{R}^d \to \mathbb{R}$ be a nonnegative kernel satisfying the following assumptions:*

- ***Upper bound on average**. There is a constant Λ such that for all $(t, x, v) \in Q_1$ and $r > 0$,*

$$\int_{B_r(v)} K(t, x, v, v')|v - v'|^2 \, dv' \leq \Lambda r^{2-2s}.$$

- ***Nondegeneracy condition**. There exist two constants $\lambda > 0$ and $\mu > 0$ so that for any $(t, x, v) \in Q_1$ and any ball $B \subset B_2$ that contains v, we have*

$$|\{v' \in B : K(t, x, v, v') \geq \lambda |v - v'|^{-d-2s}\}| \geq \mu |B|.$$

- ***Nondivergence symmetry**. For all $(t, x, v) \in Q_1$ and $w \in \mathbb{R}^d$, $K(t, x, v, v + w) = K(t, x, v, v - w)$.*

- ***Holder continuity***. *For all $(t_1, x_1, v_1), (t_2, x_2, v_2) \in Q_1$ and $r > 0$,*

$$\int_{B_r} |K(t_1, x_1, v_1, v_1 + w) - K(t_2, x_2, v_2, v_2 + w)||w|^2 \, dw$$
$$\leq A_0 r^{2-2s} d_\ell((t_1, x_1, v_1), (t_2, x_2, v_2))^{\alpha'}.$$

Let $h : Q_1 \to \mathbb{R}$ be α'-Hölder continuous. Assume further that $2s + \alpha' \notin \{1, 2\}$. If f satisfies (5.2) *in Q_1, then*

$$[f]_{C^{2s+\alpha'}(Q_{1/2})} \leq C(\|f\|_{C^{\alpha}((-1,0]\times B_1 \times \mathbb{R}^d)} + \|h\|_{C^{\alpha'}(Q_1)}).$$

The constant C only depends on dimension, α, the order $2s$ of the integral diffusion, ellipticity constants μ, λ, Λ, and A_0.

The first two conditions on the kernel are the same as in Theorem 5.2. The symmetry condition replaces the cancellation condition. Schauder theorem is, in this case, a statement for equations in nondivergence form. The last condition on the kernel is the integral version, with respect to the kinetic distance, of the Hölder continuity of the coefficients in the classical case.

The Schauder estimates for kinetic equations with standard second-order diffusion are a particular case of a more general theory developed in the context of hypoelliptic equations (see [85] and the references therein). A simplified version of the proof in [62] also leads to the following result.

Theorem 5.6 (The Schauder Estimate for Kinetic Equations with Second-Order Diffusion) *Let $\alpha \in (0, 1)$. Let $h : Q_1 \to \mathbb{R}$ be α-Hölder continuous.*

Suppose f is a classical solution of

$$f_t + v \cdot \nabla_x f - a_{ij}(t, x, v)\partial_{v_i v_j} f = h \text{ in } Q_1,$$

for some C^{α} function h and uniformly elliptic coefficients a_{ij} that are C^{α} Hölder continuous in Q_1 with respect to the d_ℓ distance defined with $s = 1$. Then

$$[f]_{C^{2+\alpha}(Q_{1/2})} \leq C(\|f\|_{C^{\alpha}(Q_1)} + \|h\|_{C^{\alpha}(Q_1)}).$$

The constant C depends on dimension, ellipticity constants λ, Λ, and the Hölder norm of a_{ij}.

6 Proving the Conditional Regularity Estimates of Theorem 3.1

Let us outline the ingredients in the proof of Theorem 3.1. It has the following steps:

(1) Obtain pointwise upper bounds for the solution f depending on the constants in (3.1), (3.2), and (3.3). These upper bounds should decay faster than $\langle v\rangle^{-q}$, for any exponent $q > 0$.
(2) Use Theorem 5.2 to deduce Hölder estimates for f.
(3) Once we know that f is Hölder, we see through the formula (4.2) that the kernel K_f is Hölder continuous as well. Thus, we use the Schuader estimates of Theorem 5.5 to achieve higher order Hölder estimates for f.
(4) There is a change of variables that makes the ellipticity of K_f uniform for large velocities. In that way, the local regularity estimates given by Theorems 5.2 and 5.5 are transformed into global estimates.
(5) We iterate the application of the Schauder estimates to derivatives of the function f to obtain C^∞ estimates.

While Theorems 5.2 and 5.5 are results about generic kinetic equations with integral diffusion, the upper bounds and the change of variables are two steps that are specific about the Boltzmann equation. We describe these two ingredients below. A more detailed description of the ideas in the proof of Theorem 3.1 can be found in [60].

6.1 The Pointwise Upper Bounds

The following theorem is proved in [58].

Theorem 6.1 *Let f be a (classical) solution to the inhomogeneous Boltzmann equation* (2.1)*, periodic in space, where the collision operator Q has the form* (2.2) *and B is the standard non-cutoff collision kernel of the form* (2.3) *with $\gamma + 2s \in [0, 2]$.*

Assume that there are constants $m_0 > 0$, M_0, E_0, and H_0 such that the hydrodynamic bounds (3.1), (3.2), *and* (3.3) *hold. Then*

- *If $\gamma > 0$, for $t > 0$, we get $f \leq C_q\langle v\rangle^{-q}$ for all $q > 0$. This is a self-generated bound independent of the initial data.*
- *If $\gamma \leq 0$ and we assume in addition $f_0 \leq A_q\langle v\rangle^{-q}$ for all $q > 0$, then we have $f_0 \leq C_q\langle v\rangle^{-q}$.*

The constants C_q depend on dimension, γ, s, m_0, M_0, E_0, H_0, and, in the case $\gamma \leq 0$, also on the constants A_q.

The proof of Theorem 6.1 is based on some type of nonlocal barrier argument. We argue by contradiction. We look at the first point where the function invalidates our pointwise upper bound, and somehow a contradiction is reached by carefully analyzing the equation at that point. Using similar techniques, we also obtain Maxwellian lower bounds in [59].

The first time a barrier argument was used in the context of the Boltzmann equation was in [38] to obtain Maxwellian upper bounds in the case of the space homogeneous cutoff Boltzmann. See also [11, 39] for related results for the space homogeneous non-cutoff Boltzmann.

These pointwise upper bounds tell us that the solution f decays as $|v| \to \infty$ faster than any algebraic rate. We may compare them with the classical moment estimates for the Boltzmann equation. Needless to say, pointwise estimates are a much stronger result than moment estimates. Theorem 6.1 implies in particular that for every $p > 0$ there is a C_p so that for all $(t, x) \in [0, \infty) \times \mathbb{R}^d$,

$$\int_{\mathbb{R}^d} f(t, x, v) \langle v \rangle^p \, dv \leq C_p.$$

Moment estimates like these are well known in the space homogeneous case. However, there is no direct proof in the space inhomogeneous case, other than as a consequence of the pointwise estimates.

6.2 A Change of Variables to Handle Large Velocities

Theorems 5.2 and 5.5 are used to obtain estimates in Hölder norms for the solution f of the Boltzmann equation (2.1). The hydrodynamic assumptions (3.1), (3.2), and (3.3) are used to ensure that the kernel K_f satisfies the appropriate nondegeneracy conditions. However, a careful analysis of the formula (4.2) reveals that while the three assumptions of Theorem 5.2 hold in any bounded set of velocities v, its parameters degenerate as $|v| \to \infty$.

In order to obtain global estimates in Hölder spaces, we must figure out what happens to the kernels K_f as $|v| \to \infty$. The following change of variables resolves this problem altogether.

For any value of $v_0 \notin B_1$, we define the linear transformation T_{v_0} by the formula

$$T_{v_0}(av_0 + w) := \frac{a}{|v_0|} v_0 + w \qquad \text{whenever } w \perp v_0.$$

Note that T_{v_0} maps the unit ball B_1 into an ellipsoid that is flattened by the factor $1/|v_0|$ in the direction of v_0. If $v_0 \in B_1$, we simply take T_{v_0} to be the identity. Given any $z_0 = (t_0, x_0, v_0)$, we further define

$$\mathcal{T}_{z_0} := z_0 \circ \left(|v_0|^{-\gamma-2s} t, |v_0|^{-\gamma-2s} T_{v_0} x, T_{v_0} v \right).$$

Here, $\circ$ is the Galilean group operator in $\mathbb{R}^{1+2d}$. This transformation $\mathcal{T}_{z_0}$ maps Q_1 into certain neighborhood of z_0 in $\mathbb{R}^{2d+1}$.

Let f be a solution to the non-cutoff Boltzmann equation. Consider $\bar{f}(z) = f(\mathcal{T}_{z_0}(z))$. This function $\bar{f}$ solves a modified equation in Q_1

$$\partial_t \bar{f} + v \cdot \nabla_x \bar{f} - \int_{\mathbb{R}^d} (f(v') - f(v)) \bar{K}_f(t, x, v, v') \, dv' = \bar{h},$$

where

$$\bar{h}(z) = c|v_0|^{-\gamma-2s}\bar{f}(z)(f *_v |\cdot|^\gamma)(\mathcal{T}_{z_0}z),$$

and

$$\bar{K}_f(t,x,v,v') = |v_0|^{-\gamma-2s}K(\mathcal{T}_{z_0}z, v_0 + T_{v_0}v').$$

Remarkably, the kernel $\bar{K}_f$ satisfies the assumptions of Theorem 5.2 and 5.5 with parameters that are independent of v_0. Thus, by working out how the Hölder norms are affected by this linear change of variables, we have a recipe to compute the precise asymptotics as $|v| \to \infty$ of all our regularity estimates.

We make a precise statement in the following proposition.

Proposition 6.2 *Let $f : [0,T] \times \mathbb{R}^d \times \mathbb{R}^d \to [0,\infty)$ be a function that satisfies* (3.1), (3.2), *and* (3.3). *Let K_f be the Boltzmann kernel that is obtained by the formula* (4.2) *and $\bar{K}_f : Q_1 \times \mathbb{R}^d \to [0,\infty)$ be the modified kernel as above. Then, there are constants $\lambda > 0$, Λ, and μ, depending only on m_0, M_0, E_0, and H_0, but **not** on v_0, such that for all $(t,x,v) \in Q_1$*

- *For all $r > 0$,*

$$\int_{B_r(v)} K(t,x,v,v')|v-v'|^2 \, dv' \leq \Lambda r^{2-2s}.$$

- *For any ball $B \subset \mathbb{R}^d$ that contains v, we have*

$$|\{v' \in B : K(v,v') \geq \lambda |v-v'|^{-d-2s}\}| \geq \mu r^d.$$

- *For all $r \in [0,1]$,*

$$\left| \int_{B_r(v)} K(t,x,v,v') - K(t,x,v',v) \, dv' \right| \leq \Lambda r^{-2s}.$$

If $s \geq 1/2$, we also have

$$\left| \int_{B_r(v)} (v-v')(K(t,x,v,v') - K(t,x,v',v)) \, dv' \right| \leq \Lambda r^{1-2s}.$$

Proposition 6.2 tells us that the assumptions of Theorem 5.2 hold uniformly after the change of variables. A similar computation holds for the Hölder continuity assumption of Theorem 5.5. Effectively, Proposition 6.2 turns the local estimates of Theorems 5.2 and 5.5 into global regularity estimates, with optimal asymptotics as $|v| \to \infty$.

Proposition 6.2 is proved in [63]. It is inspired by a similar change of variables for the Landau equation that first appeared in [23].

7 The Homogeneous Problem

A simplified form of the Boltzmann or Landau equation (2.1) takes place when we consider solutions that are constant in the space variable x. In that case, the kinetic transport term disappears from the equation, and we are left with

$$f_t = Q(f, f). \tag{7.1}$$

Here, $Q(f, f)$ may be the Boltzmann or Landau operators described in (2.2) and (2.4). The equation takes the more classical form of quasilinear parabolic equations. Moreover, the conserved quantities become stronger since they do not involve integration in x. More precisely, we have the following:

Conservation of mass:

$$\int_{\mathbb{R}^d} f(t, v) \, dv \qquad \text{is constant in time.}$$

Conservation of energy:

$$\int_{\mathbb{R}^d} f(t, v)|v|^2 \, dv \qquad \text{is constant in time.}$$

Conservation of momentum:

$$\int_{\mathbb{R}^d} f(t, v)v \, dv \qquad \text{is constant in time.}$$

Entropy:

$$\int_{\mathbb{R}^d} f(t, v) \log f(t, v) \, dv \qquad \text{is monotone decreasing.}$$

In particular, the hydrodynamic assumptions (3.1), (3.2), and (3.3) are automatically true for any solution [1] of the space homogeneous problem whose initial data have finite mass, energy, and entropy.

Establishing the existence of global smooth solutions of (7.1) does not face the same difficulties as in the space in-homogeneous case. Clearly, Theorems 3.1

[1] We mean for *classical* solutions with sufficient decay as $|v| \to \infty$. Some of these conservation laws might break if we work with an overly weak notion of solution.

and 3.2 can be combined with any short-time well-posedness result to produce a global-in-time solution. Anyhow, the global well-posedness of the space homogeneous problem, at least for some range of the values γ and s, can be proved directly and it was done earlier. An incomplete list of references would be [32, 90, 93] for the homogeneous Landau equation and [5, 6, 8, 27, 29, 31, 33, 50, 56, 80, 81] for the homogeneous Boltzmann equation.

Regularity estimates for the space homogeneous problem face a strong limitation when $\gamma + 2s < 0$. This range is called the *very soft potential* case. The Landau equation corresponds to $s = 1$, and thus very soft potentials are those with $\gamma < -2$. Note that the very soft potential range is excluded from Theorems 3.1 and 3.2. The problem is that, even in the space homogeneous scenario, we do not know how to obtain an a priori estimate for the solution in L^∞ when $\gamma + 2s < 0$. We explore this major open problem in the rest of this section.

In order to understand the difficulties with very soft potentials, it is best to focus on the most extreme case: the Landau–Coulomb equation. It corresponds to $s = 1$ and $\gamma = -3$ in three dimensions. It is a model for plasma dynamics, what makes it particularly interesting as a natural equation from mathematical physics. The operator $Q(f, f)$ takes the following form (2.6), which we recall here:

$$Q(f, f) = \bar{a}_{ij}\partial_{ij} f + f^2, \tag{7.2}$$

where

$$\bar{a}_{ij}(v) = -\partial_{ij}(-\Delta)^{-2} f(v) = \frac{1}{8\pi}\int_{\mathbb{R}^3} (|w|^2\delta_{ij} - w_i w_j)|w|^{-3} f(v - w)\, dw.$$

Some lower bound on the coefficients is computed in terms of the mass, energy, and entropy of f. Let us state it for a general power γ in the next lemma.

Lemma 7.1 *Let $f : \mathbb{R}^d \to [0, \infty)$ be a function with finite mass, energy, and entropy. Assume that*

$$\int_{\mathbb{R}^d} f(v)\, dv \geq m_0,$$

$$\int_{\mathbb{R}^d} f(v)|v|^2\, dv \leq E_0,$$

$$\int_{\mathbb{R}^d} f(v)\log f(v)\, dv \leq H_0.$$

Consider the coefficients $\bar{a}_{ij}$ given by (2.5) *for any $\gamma \in (-d - 2, 0]$. That is,*

$$\bar{a}_{ij}(v) = \int_{\mathbb{R}^d} (|w|^2\delta_{ij} - w_i w_j)|w|^\gamma f(v - w)\, dw.$$

Then, we have the following lower bounds:

$$\bar{a}_{ij}(v)e_ie_j \geq c \begin{cases} \langle v\rangle^{\gamma} & \textit{for any } e \in S^{d-1}, \\ \langle v\rangle^{\gamma+2} & \textit{if } e \cdot v = 0. \end{cases}$$

Here, the constant c depends on γ, dimension, m_0, E_0, and H_0 only.

The proof of Lemma 7.1 can be found in [90] (see the proof of Lemma 3.1 inside that paper). See [23, Lemma 2.1] for the corresponding upper bounds for $\gamma \in [-2, 0]$.

If $\gamma < -2$, there is no pointwise upper bound for $\bar{a}_{ij}$ in terms of the mass, energy and entropy of f only. We may see this fact as a hint that analyzing the very soft potential case $\gamma < -2$ will be delicate. In order to obtain pointwise upper bounds for the coefficients $\bar{a}_{ij}$, we would need to have some extra information about the function f. For example, it would suffice to have an upper bound of its weighted L^p norm

$$\|f\|_{L^p_k(\mathbb{R}^d)} = \left(\int |f(v)|^p \langle v\rangle^{pk} \,\mathrm{d}v\right)^{1/p} \leq K.$$

In that case, if $p > d/(d+2+\gamma)$ and k is sufficiently large (depending on p, γ, and dimension), the upper bounds for the coefficients $\bar{a}_{ij}$ would differ from the lower bounds by a constant factor. A conditional a priori estimate for the solution to (7.1) follows in that case (see [90]).

The bounds for the coefficients $\bar{a}_{ij}$ might suggest a comparison between the homogeneous Landau–Coulomb equation and the nonlinear heat equation $f_t = \Delta f + f^2$. Since the latter blows up in finite time, one may be inclined to believe that the Landau–Coulomb equation also develops singularities. It is an example of how we may arrive to the wrong conclusions when we think of the collision operator $Q(f, f)$ as if it was semilinear.[2] The coefficients $\bar{a}_{ij}$ are not bounded above for $\gamma < -2$. Intuitively, the coefficients $\bar{a}_{ij}$ will become large around the areas where the mass of f is accumulating. This enhanced dissipation seems to prevent blow-up altogether. Nowadays, the consensus among researchers in the area is that everybody expects the homogeneous Landau–Coulomb equation to have global smooth solutions for any reasonable initial data.

There are two positive, unconditional, results about the Landau–Coulomb equation that provide some partial progress in the problem. They are the following:

(1) It is shown in [30] that we can get a bound for $\|f\|_{L^1_t L^3_{-k}}$, for some suitable exponent k, using the entropy dissipation. Here

$$\|f\|_{L^1_t L^3_{-k}} = \int_0^T \|\langle v\rangle^{-k} f(t,v)\|_{L^3(\mathbb{R}^3)} \,\mathrm{d}t.$$

[2] It would be equally misleading to think of the Boltzmann collision operator as the fractional Laplacian plus lower order terms.

(2) In [41], the authors show that the potential set of times where a weak solution blows up has Hausdorff dimension at most $1/2$.

There are also conditional regularity estimates for the homogeneous Landau equation, which say that the solution will not blow up unless certain quantities become infinite. Results of that kind can be found in [44, 45, 90].

7.1 The Krieger–Strain Model

A simplified toy model that captures the properties and key difficulties of the Landau–Coulomb operator was proposed by Krieger and Strain in [72]. The idea is to replace the Landau–Coulomb operator (7.2) with

$$Q(f,f) = (-\Delta)^{-1} f(v) \cdot \Delta f(v) + f(v)^2.$$

Solutions to Eq. (7.1) with this new operator Q still conserve mass and dissipate entropy, but they do not conserve energy. Note also that the Maxwellian functions are no longer stationary solutions. In fact, the only function f so that $Q(f,f) = 0$ is $f \equiv 0$.

We can modify the operator further by introducing a parameter $\alpha \in [0,1]$ and letting

$$Q_\alpha(f,f) = (-\Delta)^{-1} f(v) \cdot \Delta f(v) + \alpha f(v)^2.$$

The idea is that by setting $\alpha < 1$, we are making the dissipation term stronger than the reaction term. In [72], the authors study the Cauchy problem for (7.1) with Q_α and establish the existence of global smooth solutions for radially symmetric, positive, and monotone initial data, and $\alpha < 2/3$. Soon after, in [42], the result is extended to the range $\alpha \in (0, 74/75)$. The milestone $\alpha = 1$ is achieved in [45], also for symmetric, positive, and radial initial data.

Note that all the regularity estimates for the Krieger–Strain equation in the current literature are limited to radially symmetric solutions.

7.2 Toward Global Smooth Solutions

We mentioned above that we believe that the space homogeneous equation (7.1), with the Landau–Coulomb operator (7.2), has global smooth solutions for any reasonably smooth initial data that decays quickly as $|v| \to \infty$. But how could we ever possibly prove it? In this subsection, we outline some potential ideas to attack this problem. This is a purely speculative section. We state two conjectures.

We also state two lemmas, with complete proofs, that support possible directions to study Conjectures 7.3 and 7.5.

If we had an upper bound in L^∞ for the solution f of (7.1), further regularity would follow by a standard application of the Schauder estimates. The key difficulty is to determine, given a (classical) solution $f : [0, T] \times \mathbb{R}^d \to [0, \infty)$ of (7.1), whether $\|f(t, \cdot)\|_{L^\infty}$ may blow up at time T or not.

Given mass, momentum, and energy, the Maxwellian distribution is the function that minimizes the entropy. It is also easy to verify that the entropy dissipation of a nonnegative function f is zero if and only if f is a Maxwellian. Thus, a global solution to (7.1) will naturally converge to the unique Maxwellian function $M(v)$ whose mass, momentum, and energy coincide with those of the initial data f_0. Before the limit $t \to \infty$ can effectively take place, we need to make sure that no singularity emerges from the flow of the equation. If we had a solution $f : [0, T] \times \mathbb{R}^d \to [0, \infty)$ so that $\lim_{t \to T} f(t, x_t) = +\infty$, along some curve x_t we should also have the L^p norm of f blowing up around the points x_t, for any $p > 3/2$. In that case, it is reasonable to expect some separation of scales: to have some local blow-up profile at a small scale, separated from the bulk of the mass. The Landau–Coulomb operator (7.2) is nonlocal only through the formula of the coefficients $\bar{a}_{ij}$. The values of $f(t, y)$, for y far from x_t, would only influence the equation around x_t by enhancing its dissipation. Intuitively, more dissipation reduces the chances of a blow-up. The more negative γ is in (2.5), the more localized the formula for $\bar{a}_{ij}$ is in terms of f. In the very soft potential case, and in particular for Landau–Coulomb, we might expect that the local blow-up profile solves its own Landau equation (2.5), but at accelerated time scale. We will do a more precise blow-up analysis below, which justifies this intuition in part. If the local blow-up profile also solves (7.1), then it should also converge to a Maxwellian. The mass, momentum, and energy of the local blow-up profile are completely uncorrelated with the global function f (and we will see that they may even be infinite). The following proposition shows an interesting inequality that is independent of their values.

Proposition 7.2 *Let $m > 0$, $q \in \mathbb{R}^d$, and $e > 0$ be given. Let $M : \mathbb{R}^d \to [0, \infty)$ be the Maxwellian distribution so that*

$$m = \int_{\mathbb{R}^d} M(v)\, dv, \qquad q = \int_{\mathbb{R}^d} M(v) v\, dv, \qquad e = \int_{\mathbb{R}^d} M(v) |v|^2\, dv.$$

For any nonnegative function $f \in L^1_2(\mathbb{R}^d)$ so that

$$m = \int_{\mathbb{R}^d} f(v)\, dv, \qquad q = \int_{\mathbb{R}^d} f(v) v\, dv, \qquad e = \int_{\mathbb{R}^d} f(v) |v|^2\, dv,$$

we have

$$\frac{\|M\|_{L^\infty(\mathbb{R}^d)}}{\|f\|_{L^\infty(\mathbb{R}^d)}} \le C_d,$$

for some constant C_d depending on dimension only. Moreover, the equality is achieved if and only if f is the characteristic function of some ball.

We can compute the constant C_d explicitly for each dimension. In three dimensions, we have

$$C_3 = \sqrt{\frac{250}{9\pi}}.$$

Proof By translating and scaling the functions f and M if necessary, we assume without loss of generality that $q = 0$, $m = 1$, and $\|f\|_{L^\infty} = 1$. We analyze, under these conditions, what the maximum possible value of $\|M\|_{L^\infty}$ is.

The Maxwellian M is uniquely determined by the values of m, q, and e. In this case, after fixing $q = 0$ and $m = 1$, we observe that $\|M\|_{L^\infty}$ will be maximal if e takes the minimum possible value. Indeed, if M_1 is the Maxwellian with $q = 0$, $m = 1$, and $e = 1$, we scale it to recover a Maxwellian M with any arbitrary energy e by

$$M(v) = e^{-d/2} M_1(e^{-1/2} v).$$

Thus, the problem is reduced to minimize the energy of the function f constrained to $0 \le f \le 1$, having unit mass and zero momentum.

Let B_R be the ball centered at the origin with volume one and $\chi(v)$ be its characteristic function. We claim that the energy of f is always larger or equal to the energy of χ. We compute

$$\int f(v)|v|^2 \,\mathrm{d}v - \int \chi(v)|v|^2 \,\mathrm{d}v = \int_{B_R} (f(v) - \chi(v))|v|^2 \,\mathrm{d}v + \int_{\mathbb{R}^d \setminus B_R} (f(v) - \chi(v))|v|^2 \,\mathrm{d}v.$$

Since f takes values in $[0, 1]$, then $f(v) - \chi(v)$ is nonpositive in B_R and nonnegative in $\mathbb{R}^d \setminus B_R$. In both cases, we obtain an inequality in the same direction by replacing $|v|^2$ with R^2.

$$\ge \int_{B_R} (f(v) - \chi(v))R^2 \,\mathrm{d}v + \int_{\mathbb{R}^d \setminus B_R} (f(v) - \chi(v))R^2 \,\mathrm{d}v$$

$$= R^2 \int_{\mathbb{R}^d} (f(v) - \chi(v)) \,\mathrm{d}v = 0.$$

The last equality is a consequence of f and χ having the same mass.

The inequality will be strict unless $f = \chi$ a.e. □

Proposition 7.2 says that the ratio between the maximum of the initial data f_0 and the maximum of the asymptotic limit of $f(t, \cdot)$ as $t \to \infty$ is bounded by a universal constant depending on dimension only. Our intuition of separation of scales suggests that this ratio is the most the L^∞ norm of a function should be able to grow by the evolution of the equation. We formulate the following bold conjecture.

Conjecture 7.3 *Let $f : [0, T] \times \mathbb{R}^3 \to [0, \infty)$ be a classical solution of* (7.1), *where Q is the Landau–Coulomb operator* (7.2), *and $f(0, v) = f_0(v)$. Then, for all values of $(t, v) \in [0, T] \times \mathbb{R}^3$, we have the inequality*

$$f(t, v) \leq \sqrt{\frac{250}{9\pi}} \max f_0(v).$$

Conjecture 7.3, if true, rules out the finite time blow-up for solutions to (7.1) in the Landau–Coulomb case. Moreover, since the inequality is independent of the initial mass, momentum, and energy, it suggests that the required analysis should be oblivious of these conserved quantities.

While we formulated Conjecture 7.3 for the Landau–Coulomb case, similar intuition applies for solutions to (7.1) where Q is the Boltzmann or Landau operator in the soft potential case. The intuition is less clear for hard potentials, but even in that case there is no apparent counterexample to rule out this inequality.

The analysis of blow-up limits is a common tool in the analysis of regularity of solutions across partial differential equations. Let us describe some attempt to apply that logic to the homogeneous Landau–Coulomb equation. Our objective is to prove that no solution to the Landau–Coulomb equation may blow up in a finite time. Let us consider the simpler scenario of a radially symmetric solution that is monotone decreasing along the radial direction. It is a class of functions that is preserved by the evolution, and they can only possibly blow up at the origin.

We argue by contradiction. Let us suppose that the radially symmetric, monotone solution f blows up at time T. That is, we have a function $f : [0, T] \times \mathbb{R}^3 \to [0, \infty)$, which is smooth for $t < T$ and solves (7.1), where Q is the Landau–Coulomb operator (7.2). We assume that f depends only on t and $|v|$ (it is radially symmetric), $v \cdot \nabla_v f \leq 0$, and therefore $\max f(t, \cdot) = f(t, 0)$. Since it is blowing up at time T, we have $f(t, 0) \to +\infty$ as $t \to T$. Ideally, we would want to extract a blow-up limit profile and find some contradiction.

Since f attains its maximum always at $v = 0$, we must have $D^2 f(t, 0) \leq 0$. Thus, $f_t(t, 0) = Q(f, f)(t, 0) \leq f(t, 0)^2$. We deduce a minimum blow-up rate: if f is blowing up at time $t = T$, then $f(t, 0) \geq (T - t)^{-1}$.

Let $t_k \to T$ be such that

$$f(t_k, 0) = \sup\{f(t, v) : t \leq t_k, v \in \mathbb{R}^d\},$$

and, for some arbitrary $\varepsilon > 0$,

$$\left(\frac{1}{2}+\varepsilon\right) f(t_k,0) \geq f(t_k - f(t_k,0)^{-1},0). \tag{7.3}$$

The reason why we can find a sequence satisfying the second inequality is precisely the blow-up rate $f(t,0) \geq (T-t)^{-1}$.

We consider the rescaled solutions

$$f_k(t,v) = a_k^{-1} f(t_k + a_k^{-1} t, b_k v).$$

We choose $a_k = f(t_k,0)$ and b_k so that $(-\Delta_v)^{-1} f_k(0,0) = 1$. By construction, the functions f_k solve the homogeneous Landau–Coulomb equation in $(-a_k t_k, 0] \times \mathbb{R}^d$. Moreover, $0 \leq f_k \leq 1$ in the same domain, $f_k(0,0) = 1$, and $(-\Delta_v)^{-1} f_k(0,0) = 1$. Also, (7.3) implies that

$$f_k(-1,0) \leq \left(\frac{1}{2}+\varepsilon\right). \tag{7.4}$$

Since we are working with a radially symmetric function f, the coefficients $\bar{a}_{ij}$ are isotropic at $v = 0$. We have $\{\bar{a}_{ij}(t,0)\} = \frac{1}{3}(-\Delta)^{-1} f(t,0)\mathrm{I}$. The same can be said about the rescaled functions f_k. The purpose of the normalization $(-\Delta)^{-1} f_k(0,0) = 1$ is so that we fix the corresponding coefficients $\{\bar{a}_{ij}^k(0,0)\} = \frac{1}{3}\mathrm{I}$.

We would like to pass to the limit as $k \to \infty$ and recover an ancient solution. We have $t_k \to T$ and $a_k \to \infty$, so the domain of the functions will grow to $(-\infty, 0] \times \mathbb{R}^3$ as $k \to \infty$.

The following lemma says that if a radially symmetric solution of the Landau–Coulomb equation is bounded by one and its coefficients are bounded from below and above at one point, then we can obtain a bound for the coefficients in any bounded domain.

Lemma 7.4 *Let $f : [-T,0] \times \mathbb{R}^3 \to [0,1]$ be a classical, radially symmetric solution of* (7.1)*, where Q is the Landau–Coulomb operator* (7.2)*. Assume that $c_0 \leq (-\Delta_v)^{-1} f(0,0) \leq 1$ for some constant $c_0 > 0$. Then, for any values of $R > 0$ and $T > 0$, there are constants $C_1 \geq c_1 > 0$ so that for $(t,v) \in [-T,0] \times B_R$,*

$$c_1 \mathrm{I} \leq \{\bar{a}_{ij}(t,v)\} \leq C_1 \mathrm{I}.$$

The constants c_1 and C_1 depend on c_0, R, and T only.

Proof We first estimate the value of the coefficients at $v = 0$. Recall that

$$\{\bar{a}_{ij}(t,0)\} = \frac{1}{3}(-\Delta)^{-1} f(t,0)\mathrm{I} = c\left(\int_{\mathbb{R}^3} f(t,v)|v|^{-1}\,\mathrm{d}v\right)\mathrm{I}.$$

Let $r_0 > 0$ be so that $r_0^2 < c_0/2$. Let $\varphi(v)$ be a smooth positive function so that $\varphi(v) = |v|^{-1}$ whenever $|v| > 1$ and $0 < \varphi(v) \leq |v|^{-1}$ elsewhere. We define

$$a(t) := \int_{\mathbb{R}^3} f(t,v)\varphi(v)\, dv.$$

Naturally, $a(t) \leq (-\Delta)^{-1} f(t,0)$. Also, using that $0 \leq f(t,v) \leq 1$, we also have $a(t) \geq (-\Delta)^{-1} f(t,0) - r_0^2$. Thus $c_0/2 \leq a(0) \leq 1$.

We compute $a'(t)$ using the integral expression for the Landau operator.

$$a'(t) = \frac{1}{8\pi} \int \varphi(v)\partial_i \int \left(|w|^2\delta_{ij} - w_i w_j\right) |w|^{-3}$$
$$\times \left(f(t,v+w)\partial_j f(t,v) - f(t,v)\partial_j f(t,v+w)\right) dw\, dv.$$

We integrate by parts twice to get

$$= \frac{1}{8\pi} \iint \left(\partial_{ij}\varphi(v) \left(|w|^2\delta_{ij} - w_i w_j\right) |w|^{-3} + 2\partial_i\varphi(v) \frac{w_i}{|w|^3} \right)$$
$$\times f(t,v+w) f(t,v)\, dw\, dv.$$

Taking absolute values, we have

$$|a'(t)| \lesssim \iint \left(\langle v\rangle^{-3}|w|^{-1} + \langle v\rangle^{-2}|w|^{-2} \right) f(t,v+w) f(t,v)\, dw\, dv.$$

In order to estimate $|a'(t)|$, we divide the domain of integration in various subdomains. We start with $|v+w| < 2\langle v\rangle$.

$$I := \iint_{|v+w|<2\langle v\rangle} \left(\langle v\rangle^{-3}|w|^{-1} + \langle v\rangle^{-2}|w|^{-2} \right) f(t,v+w) f(t,v)\, dw\, dv.$$

Using $f(t,v+w) \leq 1$,

$$\lesssim \int f(t,v) \left(\int_{|v+w|<2\langle v\rangle} \langle v\rangle^{-3}|w|^{-1} + \langle v\rangle^{-2}|w|^{-2}\, dw \right) dv,$$
$$\approx \int f(t,v)\langle v\rangle^{-1}\, dv \approx a(t).$$

We continue with one of the terms in the integrand in the subdomain $|v+w| > 2\langle v\rangle$.

$$II := \iint_{|v+w|>2\langle v\rangle} \langle v\rangle^{-2}|w|^{-2} f(t,v+w) f(t,v)\, dw\, dv.$$

Using $f(t, v) \leq 1$ and changing variables $z = v + w$,

$$\lesssim \int f(t,z) \left(\int_{\langle v \rangle < |z|/2} \langle v \rangle^{-2} |v - z|^{-2} \, \mathrm{d}v \right) \mathrm{d}z,$$

$$\lesssim \int f(t,z) \langle z \rangle^{-1} \, \mathrm{d}z \approx a(t).$$

We finish with the second term in the integrand in the subdomain $|v - w| > 2\langle v \rangle$.

$$II := \iint_{|v+w|>2\langle v \rangle} \langle v \rangle^{-3} |w|^{-1} f(t, v+w) f(t,v) \, \mathrm{d}w \, \mathrm{d}v.$$

Changing variables $z = v + w$,

$$\lesssim \int f(t,z) \left(\int_{\langle v \rangle < |z|/2} \langle v \rangle^{-3} |v - z|^{-1} f(t,v) \, \mathrm{d}v \right) \mathrm{d}z.$$

Note that the domain of the inner integral is empty if $|z| < 2$. Thus, $|v - z|^{-1} \lesssim \langle z \rangle^{-1}$.

$$\lesssim \int f(t,z) \langle z \rangle^{-1} \left(\int f(t,v) \langle v \rangle^{-3} \, \mathrm{d}v \right) \mathrm{d}z.$$

For an arbitrary $R > 0$, we divide the inner integral into the two subdomains B_R and $\mathbb{R}^3 \setminus B_R$.

$$\lesssim \int f(t,z) \langle z \rangle^{-1} \left(\int_{B_R} f(t,v) \langle v \rangle^{-3} \, \mathrm{d}v + \int_{\mathbb{R}^3 \setminus B_R} f(t,v) \langle v \rangle^{-3} \, \mathrm{d}v \right) \mathrm{d}z.$$

We use $f(t, v) \leq 1$ in the first term and the value of $a(t)$ to bound the second term.

$$\lesssim \int f(t,z) \langle z \rangle^{-1} \left(\log(1 + R^3) + \langle R \rangle^{-2} a(t) \right) \mathrm{d}z.$$

Choosing $R = \sqrt{a(t)}$, we get

$$\lesssim a(t) \left(1 + \log(1 + a(t))\right).$$

Adding the three terms I, II, and III, we conclude that $|a'(t)| \lesssim a(t)(1 + \log(1+a(t)))$. This ODE, together with the value of $a(0)$, implies that $a(t)$ may grow at most double exponentially as $t \to -\infty$. It has to stay bounded in any interval of time $[-T, 0]$. Moreover, for small values of $a(t)$, the logarithmic correction is irrelevant. The value of $a(t)$ may decay at most exponentially as $t \to -\infty$. This ODE, together with $a(0) \geq c_0/2$, gives us a lower bound for $a(t)$ for $t \in [-T, 0]$.

It is easy to obtain bounds for $(-\Delta)^{-1} f(t, 0)$ from the values of $a(t)$. Indeed, $a(t) \leq (-\Delta)^{-1} f(t, 0) \leq a(t) + r_0^2 \|f\|_{L^\infty}$. Thus, we have already obtained lower and upper bounds for the coefficients $\bar{a}_{ij}(t, 0)$ for $t \in [-T, 0]$. We need to extend these bounds for nonzero values of $v \in B_R$.

For the upper bounds, we observe that $\operatorname{tr} \bar{a}_{ij}(t, v) = (-\Delta)^{-1} f(t, v)$. Therefore

$$
\begin{aligned}
\operatorname{tr} \bar{a}_{ij}(t, v) &= c \int_{\mathbb{R}^3} |v - w|^{-1} f(t, w) \, dw \\
&\lesssim \int_{B_r(v)} |v - w|^{-1} f(w) \, dw + \int_{\mathbb{R}^3 \setminus B_r(v)} |v - w|^{-1} f(w) \, dw \\
&\lesssim r^2 (\sup f) + \frac{|v| + r}{r} (-\Delta)^{-1} f(t, 0).
\end{aligned}
$$

Choosing any arbitrary $r > 0$, we get an upper bound for $\bar{a}_{ij}(t, v)$.

To get a lower bound for $\bar{a}_{ij}(t, v)$, we use that f is radially symmetric in v. Let $\tilde{f} : [-T, 0] \times [0, \infty) \to [0, 1]$ so that $f(t, v) = \tilde{f}(t, |v|)$. The smallest eigenvalue λ of $\bar{a}_{ij}(t, v)$ points in the radial direction. It can be computed from $\tilde{f}$ by the formula

$$
\lambda(t, v) = \frac{1}{3} |v|^{-3} \int_0^{|v|} s^4 \tilde{f}(t, s) \, ds + \frac{1}{3} \int_{|v|}^{\infty} s \tilde{f}(t, s) \, ds.
$$

We choose any $s_0 < |v|$ and get

$$
\begin{aligned}
&\geq \frac{1}{3} |v|^{-3} s_0^3 \int_0^{\infty} s \tilde{f}(t, s) \, ds - \frac{1}{6} |v|^{-3} s_0^5 (\sup \tilde{f}) \\
&\geq \frac{1}{3} |v|^{-3} s_0^3 (-\Delta)^{-1} f(t, 0) - \frac{1}{6} |v|^{-3} s_0^5.
\end{aligned}
$$

This gives us a lower bound if we choose s_0 sufficiently small, and we conclude the proof. □

Going back to the blow-up sequence f_k, because of Lemma 7.4, we have that the coefficients $a^k(t, v)$ will be uniformly elliptic in any bounded set $[-T, 0] \times B_R$. Using standard parabolic estimates (first the theorem by Krylov and Safonov and then Schauder estimates), it is easy to see that the sequence f_k is uniformly smooth on bounded sets. We pass to the limit so that, up to extracting a subsequence, there are a function $f_\infty : (-\infty, 0] \times \mathbb{R}^d \to [0, 1]$ and coefficients $\bar{a}_{ij}^\infty : (-\infty, 0] \times \mathbb{R}^d \to \mathbb{R}$, so that

$$
\begin{aligned}
f_k &\to f_\infty, \\
\partial_t f_k &\to \partial_t f_\infty,
\end{aligned}
$$

$$D_v^2 f_k \to D_v^2 f_\infty,$$
$$\bar{a}_{ij}^k \to \bar{a}_{ij}^\infty,$$

with convergence being uniform over every compact set. Clearly, the function f_∞ solves

$$\partial_t f_\infty = \bar{a}_{ij}^\infty \partial_{ij} f_\infty + f_\infty^2 \qquad \text{in } (-\infty, 0] \times \mathbb{R}^3.$$

From Fatou's lemma, we see that

$$\begin{aligned} \bar{a}_{ij}^\infty &= \lim_{k\to\infty} c \int_{\mathbb{R}^3} (|w|^2 \mathrm{I} - w_i w_j)|w|^{-3} f_k(v-w)\, \mathrm{d}w \\ &\geq c \int_{\mathbb{R}^3} (|w|^2 \mathrm{I} - w_i w_j)|w|^{-3} f_\infty(v-w)\, \mathrm{d}w. \end{aligned}$$

We cannot say that f_∞ is an ancient solution of the Landau–Coulomb equation because the coefficients $\bar{a}_{ij}^\infty$ are not exactly given by the formula (7.2) applied to f_∞. Yet, the inequality above tells us that the diffusion is only *enhanced* in the limit. Intuitively, more diffusion should only improve our regularity estimates.

Admittedly, it is far from clear how to derive a contradiction from this blow-up analysis. It seems reasonable to expect that the only bounded ancient solutions to the Landau–Coulomb equation are stationary Maxwellians. However, even if we start with a solution f that has finite mass, energy and entropy, none of these quantities will be preserved through the sequence and blow-up limit. The final function f_∞ that we obtain is a smooth, bounded function that belongs in addition to $L^\infty_{loc}((-\infty, 0], L^1_{-1}(\mathbb{R}^d))$ (as a consequence of the upper bounds in Lemma 7.4). There is no obvious reason to expect the blow-up limit f_∞ to have finite mass or energy.

We state the corresponding Liouville-type conjecture here.

Conjecture 7.5 *Let* $f : (-\infty, 0] \times \mathbb{R}^3 \to [0, \infty)$ *be a smooth classical solution of* (7.1), *where* Q *is the Landau–Coulomb operator* (7.2). *We make the following assumptions:*

- *For all* $(t, v) \in (-\infty, 0] \times \mathbb{R}^3$, *we have* $0 \leq f(t, v) \leq 1$.
- *For every* $t \in (-\infty, 0]$, *there is some constant* C_t *so that*

$$\int_{\mathbb{R}^3} f(v) \langle v \rangle^{-1}\, \mathrm{d}v \leq C_t.$$

Then f *must be a stationary Maxwellian.*

Note that the second assumption in Conjecture 7.5 is necessary just to make sense of the formula for the coefficients in (7.2). In the radially symmetric case, the constants C_t are related for different values of t through Lemma 7.4.

The blow-up limit f_∞ cannot be a stationary Maxwellian because it would contradict (7.4). That is the basic idea of how the blow-up analysis would rule out the emergence of singularities. However, a positive resolution of Conjecture 7.5 seems to be very difficult with current techniques.

It would be interesting to study solutions to the space homogeneous Landau–Coulomb equation (7.1), with Q given by (7.2), in the class $f \in L^\infty([0, T] \times \mathbb{R}^3) \cap L^\infty([0, T], L^1_{-1}(\mathbb{R}^3))$. These are solutions with possibly infinite mass, energy, and entropy. There is currently no short-time Cauchy theory in this class.

8 Open Problems

In Sect. 7, we analyzed the problem of finding L^∞ bounds in the case of soft potentials and in particular the problem of global existence of smooth solutions for the homogeneous Landau–Coulomb equation. In this section, we quickly review other open problems that are related to the regularity issues reviewed in this chapter.

8.1 Unconditional Regularity Estimates

Observing the conditional regularity results of Theorems 3.2 and 3.1, the obvious question is whether the hydrodynamic bounds (3.1), (3.2), and (3.3) can possibly be proved.

If we somehow established the bounds for the hydrodynamic quantities in (3.1), (3.2), and (3.3), we would conclude that the inhomogeneous Landau and Boltzmann equations have global smooth solutions, for any initial data. This is a remarkable open problem that Cedric Villani describes in the first chapter of his book [95]. If unconditional regularity estimates were true, it seems that they would be very difficult to prove.

At a certain scale, the hydrodynamic quantities associated with solutions of the Landau or Boltzmann equation approximately solve the compressible Euler equation (see [13]). There are some recent results studying the stability of implosion singularities for the compressible Euler equation and producing singularities for the compressible Navier–Stokes equation (see [77] and [78]). Can there be similar implosion singularities for the Boltzmann equation? The idea is to look for a solution that mostly flows toward a single point. The mass and energy densities would be blowing up at that point in a way that resembles the blow-up profile of the compressible Euler equation in [77]. This scenario appears to be more plausible than the unconditional regularity of the previous paragraph. If such an implosion singularity was indeed possible for the Boltzmann and/or Landau equation, then the bounds (3.1), (3.2), and (3.3) would not be true in general.

8.2 The Kinetic Krylov–Safonov Theorem

We explain a kinetic version of De Giorgi–Nash–Moser theorem in Theorem 5.1. It concerns Eq. (5.1), which is a kinetic equation with second-order diffusion in divergence form:

A central result for parabolic equations in non-divergence form is the theorem of Krylov and Safonov [73]. Its statement is analogous to that of De Giorgi, Nash, and Moser, but for diffusion in non-divergence form. It would be natural to expect a result of that kind to hold for kinetic equations as well. It is still an open problem. We state it here as a conjecture.

Conjecture 8.1 *Assume $f : [0,1] \times B_1 \times B_1 \to \mathbb{R}$ is a (classical) solution of an equation of the following form:*

$$f_t + v \cdot \nabla_x f = a_{ij}(t,x,v)\partial_{v_i v_j} f.$$

Here, the coefficients a_{ij} are assumed to be uniformly elliptic in the sense that there are constants $\Lambda \geq \lambda > 0$, such that for all $(t,x,v) \in [-1,0] \times B_1 \times B_1$,

$$\lambda \mathrm{I} \leq \{a_{ij}(t,x,v)\} \leq \Lambda \mathrm{I}.$$

Then, there exist constants $\alpha > 0$ and C, depending only on dimension and the ellipticity parameters λ and Λ, so that

$$\|f\|_{C^\alpha((1/2,1)\times B_{1/2}\times B_{1/2})} \leq C\|f\|_{L^\infty([0,1]\times B_1\times B_1)}.$$

It would also make sense to expect an integro-differential version of Conjecture 8.1 to hold. One can state it for different ellipticity conditions on the kernel. As a starting point, we state the conjecture using a strong notion of fractional ellipticity.

Conjecture 8.2 *Assume $f : [0,1] \times B_1 \times \mathbb{R}^d \to \mathbb{R}$ is a (classical) solution of* (5.2) *for $(t,x,v) \in (0,1] \times B_1 \times B_1$ and some bounded function h. We assume that the kernel K is symmetric $K(t,x,v,v+w) = K(t,x,v,v-w)$ and that there exist constants $\Lambda \geq \lambda > 0$ such that*

$$\lambda|v'-v|^{-d-2s} \leq K(t,x,v,v') \leq \Lambda|v'-v|^{-d-2s}.$$

Then there are constants $\alpha > 0$ and C, depending only on dimension, λ and Λ, so that f satisfies the estimate

$$\|f\|_{C^\alpha((1/2,1)\times B_{1/2}\times B_{1/2})} \leq C\left(\|f\|_{L^\infty([0,1]\times B_1\times\mathbb{R}^d)} + \|h\|_{L^\infty([0,1]\times B_1\times B_1)}\right).$$

The elliptic version of the theorem of Krylov and Safonov is easier to prove for integro-differential equations than it is for classical second-order elliptic equations (see [87]). Based on that, it is possible that Conjecture 8.2 may have a simpler resolution than Conjecture 8.1. As of now, both remain open.

8.3 Bounded Domains

The current versions of the conditional regularity results for Boltzmann and Landau equations in Theorems 3.1 and 3.2 do not allow any form of spatial boundary.

For physical applications, it is natural to have the space variable x confined to some bounded domain. There are different boundary conditions that have been considered in the literature: diffuse reflection, specular reflection, and bounce back reflection.

The boundary effects may have an impact on the regularity of the solutions to kinetic equations. See [46–49, 69, 70] for results concerning the cutoff Boltzmann and Landau equations and solutions near a Maxwellian. It is an interesting research direction to study the possibility of extending some form of Theorem 3.1 to any domain with boundary, for any of the physical boundary conditions.

In the cutoff case, it is known that the boundary conditions produce discontinuities when the domain is not convex (see [69]). Naturally, in the non-cutoff case, the solution is expected to be smooth away from the boundary. However, it is unclear if we should expect uniform smoothness estimates up to the boundary, especially in the nonconvex case.

8.4 Sharper Conditions for Coercivity Estimates

In Theorem 4.3, we present a sufficient condition for a kernel $K : B_2 \times B_2 \to [0, \infty)$ to have coercivity estimates with respect to the $\dot{H}^s$ norm. Our conditions are not sharp. We would like to explore what kernels K have the property that for some $c > 0$,

$$\begin{aligned}\iint_{B_2\times B_2} &|f(v') - f(v)|^2 K(v, v')\,\mathrm{d}v'\,\mathrm{d}v \\ &\geq c \iint_{B_1\times B_1} |f(v') - f(v)|^2 |v' - v|^{-d-2s}\,\mathrm{d}v'\,\mathrm{d}v.\end{aligned}$$

There are simple examples replacing $K(v, v')\,\mathrm{d}v'\,\mathrm{d}v$ for a singular measure that satisfy the coercivity condition. For example, in two dimensions, the operator $(-\partial_1)^{2s} + (-\partial_2)^{2s}$ is clearly coercive. It corresponds to

$$\int_{\mathbb{R}^2}\int_{\mathbb{R}}\Big(|f(v_1+w,v_2)-f(v_1,v_2)|^2$$
$$+|f(v_1,v_2+w)-f(v_1,v_2)|^2\Big)\,|w|^{-1-2s}\,dw\,dv$$
$$\geq c\iint_{\mathbb{R}^d\times\mathbb{R}^d}|f(v')-f(v)|^2|v'-v|^{-d-2s}\,dv'\,dv.$$

What this simple example shows is that a sharp condition for coercivity should allow for singular measures instead of $K(v,v')\,dv'\,dv$, possibly supported in a set of zero Lebesgue measure.

In the context of the Boltzmann equation, Theorem 4.3 allows us to obtain a coercivity estimate for the Boltzmann collision operator in terms of the mass, energy and entropy of f. We would need a sharper version of Theorem 4.3 if we want to replace the upper bound on the entropy with a lower bound on the temperature tensor.

8.5 Conditional Regularity for the Cutoff Boltzmann Equation

It is well known that we should not expect any regularization from the cutoff Boltzmann equation. Yet, one may start with a smooth initial data and study the propagation of its initial regularity.

It seems plausible that if the hydrodynamic bounds (3.1), (3.2), and (3.3) hold, then a solution of the inhomogeneous Boltzmann equation whose initial data is smooth and rapidly decaying would stay smooth and rapidly decaying for all time. Moreover, it is conceivable that the smoothness estimates stay uniform as $t\to\infty$, just like in the non-cutoff case.

Conditional regularity estimates of this type, for the full Boltzmann equation with cutoff, have not yet been studied.

8.6 Renormalized Solutions

For generic initial data f_0, Alexandre and Villani constructed certain kind of global solution f of the non-cutoff Boltzmann equation. This type of generalized solutions are called *renormalized solutions with defect measure*. The uniqueness of solutions is not known within this class.

It is natural to wonder if a localized version of Theorem 3.1 may hold for renormalized solutions with defect measure. That is, if a renormalized solution f satisfies the hydrodynamic inequalities (3.1), (3.2), and (3.3) for almost every (t,x) in some cylinder $(-1,0]\times B_1$, should we expect f to be C^∞ for $(t,x)\in(-1/2,0]\times B_{1/2}$?

8.7 *Conditional Regularity for Solutions Without Rapid Decay in the Moderately Soft Potential Case*

In the case of soft potentials ($\gamma \leq 0$), Theorem 3.1 requires the initial data to have rapid decay. More precisely, we assume that for every $q \geq 0$, there exists a C_q so that $f_0(x, v) \leq C_q \langle v \rangle^{-q}$. The reason for this assumption is that when in the proof of Theorem 3.1 we iteratively compute estimates for all derivatives of f, the decay of our estimates deteriorates every time we take an extra derivative. If we only start with $f_0(x, v) \leq C_q \langle v \rangle^{-q}$ for one particular value of q, after taking a certain number of derivatives, our estimates would not decay at all, and eventually the analysis fails.

It seems difficult to imagine that if $f_0(x, v) \leq C_q \langle v \rangle^{-q}$ for some large value of q, and the hydrodynamic bounds (3.1), (3.2), and (3.3) hold, then f can possibly fail to be C^∞. Yet, this problem is currently open whenever $\gamma \leq 0$.

References

1. R. Alexandre, L. Desvillettes, C. Villani, and B. Wennberg. Entropy dissipation and long-range interactions. *Arch. Ration. Mech. Anal.*, 152(4):327–355, 2000.
2. R. Alexandre, Y. Morimoto, S. Ukai, C.-J. Xu, and T. Yang. The Boltzmann equation without angular cutoff in the whole space: II, Global existence for hard potential. *Anal. Appl. (Singap.)*, 9(2):113–134, 2011.
3. R. Alexandre, Y. Morimoto, S. Ukai, C.-J. Xu, and T. Yang. The Boltzmann equation without angular cutoff in the whole space: I, Global existence for soft potential. *Journal of Functional Analysis*, 262(3):915–1010, 2012.
4. R. Alexandre and C. Villani. On the Boltzmann equation for long-range interactions. *Comm. Pure Appl. Math.*, 55(1):30–70, 2002.
5. Radjesvarane Alexandre and Mouhamad El Safadi. Littlewood-Paley theory and regularity issues in Boltzmann homogeneous equations. I. Non-cutoff case and Maxwellian molecules. *Math. Models Methods Appl. Sci.*, 15(6):907–920, 2005.
6. Radjesvarane Alexandre and Mouhamad Elsafadi. Littlewood-Paley theory and regularity issues in Boltzmann homogeneous equations. II. Non cutoff case and non Maxwellian molecules. *Discrete Contin. Dyn. Syst.*, 24(1):1–11, 2009.
7. Radjesvarane Alexandre, Yoshinori Morimoto, Seiji Ukai, C-J Xu, and Tong Yang. The Boltzmann equation without angular cutoff in the whole space: Qualitative properties of solutions. *Archive for Rational Mechanics and Analysis*, 202(2):599–661, 2011.
8. Radjesvarane Alexandre, Yoshinori Morimoto, Seiji Ukai, Chao-Jiang Xu, and Tong Yang. Smoothing effect of weak solutions for the spatially homogeneous Boltzmann equation without angular cutoff. *Kyoto J. Math.*, 52(3):433–463, 2012.
9. Radjesvarane Alexandre, Yoshinori Morimoto, Seiji Ukai, Chao-Jiang Xu, and Tong Yang. Local existence with mild regularity for the Boltzmann equation. *Kinet. Relat. Models*, 6(4):1011–1041, 2013.
10. R Alonso, Y Morimoto, W Sun, and T Yang. De Giorgi argument for weighted $L^2 \cap L^\infty$ solutions to the non-cutoff Boltzmann equation. *arXiv preprint arXiv:2010.10065*, 2020.
11. Ricardo Alonso. Emergence of exponentially weighted L^p-norms and Sobolev regularity for the Boltzmann equation. *Comm. Partial Differential Equations*, 44(5):416–446, 2019.
12. Ricardo Alonso, Yoshinori Morimoto, Weiran Sun, and Tong Yang. Non-cutoff Boltzmann equation with polynomial decay perturbations. *Rev. Mat. Iberoam.*, 37(1):189–292, 2021.

13. Claude Bardos, François Golse, and David Levermore. Fluid dynamic limits of kinetic equations. I. Formal derivations. *J. Statist. Phys.*, 63(1-2):323–344, 1991.
14. Martin T. Barlow, Richard F. Bass, Zhen-Qing Chen, and Moritz Kassmann. Non-local Dirichlet forms and symmetric jump processes. *Trans. Amer. Math. Soc.*, 361(4):1963–1999, 2009.
15. Richard F Bass and Moritz Kassmann. Harnack inequalities for non-local operators of variable order. *Transactions of the American Mathematical Society*, pages 837–850, 2005.
16. Richard F Bass and Moritz Kassmann. Hölder continuity of harmonic functions with respect to operators of variable order. *Communications in Partial Differential Equations*, 30(8):1249–1259, 2005.
17. Richard F Bass and David A Levin. Harnack inequalities for jump processes. *Potential Analysis*, 17(4):375–388, 2002.
18. Richard F. Bass and David A. Levin. Transition probabilities for symmetric jump processes. *Trans. Amer. Math. Soc.*, 354(7):2933–2953 (electronic), 2002.
19. C. Bjorland, L. Caffarelli, and A. Figalli. Non-local gradient dependent operators. *Adv. Math.*, 230(4–6):1859–1894, 2012.
20. Luis Caffarelli, Chi Hin Chan, and Alexis Vasseur. Regularity theory for parabolic nonlinear integral operators. *J. Amer. Math. Soc.*, 24(3):849–869, 2011.
21. Luis Caffarelli and Luis Silvestre. Regularity theory for fully nonlinear integro-differential equations. *Communications on Pure and Applied Mathematics*, 62(5):597–638, 2009.
22. Luis Caffarelli and Alexis Vasseur. Drift diffusion equations with fractional diffusion and the quasi-geostrophic equation. *Ann. of Math*, 171(3):1903–1930, 2010.
23. Stephen Cameron, Luis Silvestre, and Stanley Snelson. Global a priori estimates for the inhomogeneous Landau equation with moderately soft potentials. *Ann. Inst. H. Poincaré Anal. Non Linéaire*, 35(3):625–642, 2018.
24. Jamil Chaker and Luis Silvestre. Coercivity estimates for integro-differential operators. *Calc. Var. Partial Differential Equations*, 59(4):Paper No. 106, 20, 2020.
25. Héctor Chang Lara and Gonzalo Dávila. Regularity for solutions of nonlocal, nonsymmetric equations. *Ann. Inst. H. Poincaré Anal. Non Linéaire*, 29(6):833–859, 2012.
26. Sanchit Chaturvedi. Stability of vacuum for the Boltzmann equation with moderately soft potentials. *Ann. PDE*, 7(2):Paper No. 15, 104, 2021.
27. Yemin Chen and Lingbing He. Smoothing estimates for Boltzmann equation with full-range interactions: spatially homogeneous case. *Arch. Ration. Mech. Anal.*, 201(2):501–548, 2011.
28. L. Desvillettes and C. Villani. On the trend to global equilibrium for spatially inhomogeneous kinetic systems: the Boltzmann equation. *Invent. Math.*, 159(2):245–316, 2005.
29. Laurent Desvillettes. Regularization properties of the 2-dimensional non-radially symmetric non-cutoff spatially homogeneous Boltzmann equation for Maxwellian molecules. *Transport Theory Statist. Phys.*, 26(3):341–357, 1997.
30. Laurent Desvillettes. Entropy dissipation estimates for the Landau equation in the Coulomb case and applications. *arXiv preprint arXiv:1408.6025*, 2014.
31. Laurent Desvillettes and Clément Mouhot. Stability and uniqueness for the spatially homogeneous Boltzmann equation with long-range interactions. *Arch. Ration. Mech. Anal.*, 193(2):227–253, 2009.
32. Laurent Desvillettes and Cédric Villani. On the spatially homogeneous Landau equation for hard potentials part i: existence, uniqueness and smoothness. *Communications in Partial Differential Equations*, 25(1-2):179–259, 2000.
33. Laurent Desvillettes and Bernt Wennberg. Smoothness of the solution of the spatially homogeneous Boltzmann equation without cutoff. *Comm. Partial Differential Equations*, 29(1-2):133–155, 2004.
34. R. J. DiPerna and P.-L. Lions. On the Cauchy problem for Boltzmann equations: global existence and weak stability. *Ann. of Math. (2)*, 130(2):321–366, 1989.
35. Hongjie Dong, Tianling Jin, and Hong Zhang. Dini and Schauder estimates for nonlocal fully nonlinear parabolic equations with drifts. *Anal. PDE*, 11(6):1487–1534, 2018.

36. Renjun Duan, Shuangqian Liu, Shota Sakamoto, and Robert M. Strain. Global mild solutions of the Landau and non-cutoff Boltzmann equations. *Comm. Pure Appl. Math.*, 74(5):932–1020, 2021.
37. Matthieu Felsinger and Moritz Kassmann. Local Regularity for Parabolic Nonlocal Operators. *Comm. Partial Differential Equations*, 38(9):1539–1573, 2013.
38. I. M. Gamba, V. Panferov, and C. Villani. Upper Maxwellian bounds for the spatially homogeneous Boltzmann equation. *Arch. Ration. Mech. Anal.*, 194(1):253–282, 2009.
39. Irene M. Gamba, Nataša Pavlović, and Maja Tasković. On pointwise exponentially weighted estimates for the Boltzmann equation. *SIAM J. Math. Anal.*, 51(5):3921–3955, 2019.
40. François Golse, Cyril Imbert, Clément Mouhot, and Alexis F. Vasseur. Harnack inequality for kinetic Fokker-Planck equations with rough coefficients and application to the Landau equation. *Ann. Sc. Norm. Super. Pisa Cl. Sci. (5)*, 19(1):253–295, 2019.
41. François Golse, Maria Pia Gualdani, Cyril Imbert, and Alexis Vasseur. Partial regularity in time for the space homogeneous landau equation with coulomb potential. *arXiv preprint arXiv:1906.02841*, 2019.
42. Philip T. Gressman, Joachim Krieger, and Robert M. Strain. A non-local inequality and global existence. *Adv. Math.*, 230(2):642–648, 2012.
43. Philip T. Gressman and Robert M. Strain. Global classical solutions of the Boltzmann equation without angular cut-off. *J. Amer. Math. Soc.*, 24(3):771–847, 2011.
44. Maria Gualdani and Nestor Guillen. On A_p weights and the Landau equation. *Calc. Var. Partial Differential Equations*, 58(1):Paper No. 17, 55, 2019.
45. Maria Pia Gualdani and Nestor Guillen. Estimates for radial solutions of the homogeneous Landau equation with Coulomb potential. *Anal. PDE*, 9(8):1772–1809, 2016.
46. Y. Guo, C. Kim, D. Tonon, and A. Trescases. BV-regularity of the Boltzmann equation in non-convex domains. *Arch. Ration. Mech. Anal.*, 220(3):1045–1093, 2016.
47. Yan Guo. Decay and continuity of the Boltzmann equation in bounded domains. *Arch. Ration. Mech. Anal.*, 197(3):713–809, 2010.
48. Yan Guo, Hyung Ju Hwang, Jin Woo Jang, and Zhimeng Ouyang. The Landau equation with the specular reflection boundary condition. *Arch. Ration. Mech. Anal.*, 236(3):1389–1454, 2020.
49. Yan Guo, Chanwoo Kim, Daniela Tonon, and Ariane Trescases. Regularity of the Boltzmann equation in convex domains. *Invent. Math.*, 207(1):115–290, 2017.
50. Lingbing He. Well-posedness of spatially homogeneous Boltzmann equation with full-range interaction. *Comm. Math. Phys.*, 312(2):447–476, 2012.
51. Christopher Henderson and Stanley Snelson. C^∞ smoothing for weak solutions of the inhomogeneous Landau equation. *Arch. Ration. Mech. Anal.*, 236(1):113–143, 2020.
52. Christopher Henderson, Stanley Snelson, and Andrei Tarfulea. Local existence, lower mass bounds, and a new continuation criterion for the Landau equation. *J. Differential Equations*, 266(2-3):1536–1577, 2019.
53. Christopher Henderson, Stanley Snelson, and Andrei Tarfulea. Local well-posedness of the Boltzmann equation with polynomially decaying initial data. *Kinetic and Related Models*, 13(4):837–867, 2020.
54. Christopher Henderson, Stanley Snelson, and Andrei Tarfulea. Self-generating lower bounds and continuation for the Boltzmann equation. *Calc. Var. Partial Differential Equations*, 59(6):Paper No. 191, 13, 2020.
55. Frédéric Hérau, Daniela Tonon, and Isabelle Tristani. Regularization estimates and Cauchy theory for inhomogeneous Boltzmann equation for hard potentials without cut-off. *Communications in Mathematical Physics*, pages 1–75, 2020.
56. Zhaohui Huo, Yoshinori Morimoto, Seiji Ukai, and Tong Yang. Regularity of solutions for spatially homogeneous Boltzmann equation without angular cutoff. *Kinet. Relat. Models*, 1(3):453–489, 2008.
57. Cyril Imbert, Tianling Jin, and Roman Shvydkoy. Schauder estimates for an integro-differential equation with applications to a nonlocal Burgers equation. *Ann. Fac. Sci. Toulouse Math. (6)*, 27(4):667–677, 2018.

58. Cyril Imbert, Clément Mouhot, and Luis Silvestre. Decay estimates for large velocities in the Boltzmann equation without cutoff. *J. Éc. polytech. Math.*, 7:143–184, 2020.
59. Cyril Imbert, Clément Mouhot, and Luis Silvestre. Gaussian lower bounds for the Boltzmann equation without cutoff. *SIAM J. Math. Anal.*, 52(3):2930–2944, 2020.
60. Cyril Imbert and Luis Silvestre. Regularity for the Boltzmann equation conditional to macroscopic bounds. *EMS Surv. Math. Sci.*, 7(1):117–172, 2020.
61. Cyril Imbert and Luis Silvestre. The weak Harnack inequality for the Boltzmann equation without cut-off. *J. Eur. Math. Soc. (JEMS)*, 22(2):507–592, 2020.
62. Cyril Imbert and Luis Silvestre. The Schauder estimate for kinetic integral equations. *Anal. PDE*, 14(1):171–204, 2021.
63. Cyril Imbert and Luis Silvestre. Global regularity estimates for the Boltzmann equation without cut-off. *Journal of the American Mathematical Society*, to appear.
64. Tianling Jin and Jingang Xiong. Schauder estimates for solutions of linear parabolic integro-differential equations. *Discrete Contin. Dyn. Syst.*, 35(12):5977–5998, 2015.
65. Moritz Kassmann. A priori estimates for integro-differential operators with measurable kernels. *Calculus of Variations and Partial Differential Equations*, 34(1):1–21, 2009.
66. Moritz Kassmann and Ante Mimica. Intrinsic scaling properties for nonlocal operators. *J. Eur. Math. Soc. (JEMS)*, 19(4):983–1011, 2017.
67. Moritz Kassmann, Marcus Rang, and Russell W. Schwab. Integro-differential equations with nonlinear directional dependence. *Indiana Univ. Math. J.*, 63(5):1467–1498, 2014.
68. Moritz Kassmann and Russell W. Schwab. Regularity results for nonlocal parabolic equations. *Riv. Math. Univ. Parma (N.S.)*, 5(1):183–212, 2014.
69. Chanwoo Kim. Formation and propagation of discontinuity for Boltzmann equation in non-convex domains. *Comm. Math. Phys.*, 308(3):641–701, 2011.
70. Chanwoo Kim and Donghyun Lee. The Boltzmann equation with specular boundary condition in convex domains. *Comm. Pure Appl. Math.*, 71(3):411–504, 2018.
71. Takashi Komatsu. Uniform estimates for fundamental solutions associated with non-local Dirichlet forms. *Osaka J. Math.*, 32(4):833–860, 1995.
72. Joachim Krieger and Robert M. Strain. Global solutions to a non-local diffusion equation with quadratic non-linearity. *Comm. Partial Differential Equations*, 37(4):647–689, 2012.
73. N. V. Krylov and M. V. Safonov. A property of the solutions of parabolic equations with measurable coefficients. *Izv. Akad. Nauk SSSR Ser. Mat.*, 44(1):161–175, 239, 1980.
74. Héctor Chang Lara and Gonzalo Dávila. Regularity for solutions of non-local parabolic equations. *Calc. Var. Partial Differential Equations*, 49(1-2):139–172, 2014.
75. Zhang Liqun. The C^α regularity of a class of ultraparabolic equations. In *Third International Congress of Chinese Mathematicians. Part 1, 2*, volume 2 of *AMS/IP Stud. Adv. Math., 42, pt. 1*, pages 619–622. Amer. Math. Soc., Providence, RI, 2008.
76. Jonathan Luk. Stability of vacuum for the Landau equation with moderately soft potentials. *Ann. PDE*, 5(1):Paper No. 11, 101, 2019.
77. Frank Merle, Pierre Raphael, Igor Rodnianski, and Jeremie Szeftel. On smooth self-similar solutions to the compressible Euler equations. *arXiv preprint arXiv:1912.10998*, 2019.
78. Frank Merle, Pierre Raphael, Igor Rodnianski, and Jeremie Szeftel. On the implosion of a three dimensional compressible fluid. *arXiv preprint arXiv:1912.11009*, 2019.
79. R. Mikulevicius and H. Pragarauskas. On the Cauchy problem for integro-differential operators in Hölder classes and the uniqueness of the martingale problem. *Potential Anal.*, 40(4):539–563, 2014.
80. Yoshinori Morimoto and Seiji Ukai. Gevrey smoothing effect of solutions for spatially homogeneous nonlinear Boltzmann equation without angular cutoff. *J. Pseudo-Differ. Oper. Appl.*, 1(1):139–159, 2010.
81. Yoshinori Morimoto, Seiji Ukai, Chao-Jiang Xu, and Tong Yang. Regularity of solutions to the spatially homogeneous Boltzmann equation without angular cutoff. *Discrete Contin. Dyn. Syst.*, 24(1):187–212, 2009.
82. Yoshinori Morimoto and Tong Yang. Local existence of polynomial decay solutions to the Boltzmann equation for soft potentials. *Anal. Appl. (Singap.)*, 13(6):663–683, 2015.

83. Andrea Pascucci and Sergio Polidoro. The Moser's iterative method for a class of ultraparabolic equations. *Commun. Contemp. Math.*, 6(3):395–417, 2004.
84. Russell W. Schwab and Luis Silvestre. Regularity for parabolic integro-differential equations with very irregular kernels. *Anal. PDE*, 9(3):727–772, 2016.
85. Polidoro Sergio. Recent results on Kolmogorov equations and applications. In *Workshop on Second Order Subelliptic Equations and Applications*, volume 3, pages 129–143. GRAFICOM Edizioni, 2004.
86. Joaquim Serra. Regularity for fully nonlinear nonlocal parabolic equations with rough kernels. *Calc. Var. Partial Differential Equations*, 54(1):615–629, 2015.
87. Luis Silvestre. Hölder estimates for solutions of integro-differential equations like the fractional Laplace. *Indiana Univ. Math. J.*, 55(3):1155–1174, 2006.
88. Luis Silvestre. On the differentiability of the solution to the Hamilton–Jacobi equation with critical fractional diffusion. *Advances in mathematics*, 226(2):2020–2039, 2011.
89. Luis Silvestre. A new regularization mechanism for the Boltzmann equation without cut-off. *Comm. Math. Phys.*, 348(1):69–100, 2016.
90. Luis Silvestre. Upper bounds for parabolic equations and the Landau equation. *J. Differential Equations*, 262(3):3034–3055, 2017.
91. Luis Silvestre and Stanley Snelson. Solutions to the non-cutoff Boltzmann equation uniformly near a Maxwellian. *arXiv preprint arXiv:2106.03909*, 2021.
92. Renming Song and Zoran Vondraček. Harnack inequality for some classes of Markov processes. *Math. Z.*, 246(1-2):177–202, 2004.
93. Cédric Villani. On the spatially homogeneous landau equation for Maxwellian molecules. *Mathematical Models and Methods in Applied Sciences*, 8(06):957–983, 1998.
94. Cédric Villani. A review of mathematical topics in collisional kinetic theory. In *Handbook of mathematical fluid dynamics, Vol. I*, pages 71–305. North-Holland, Amsterdam, 2002.
95. Cédric Villani. *Théorème vivant*. Grasset, 2012.
96. WenDong Wang and LiQun Zhang. The C^α regularity of a class of non-homogeneous ultraparabolic equations. *Sci. China Ser. A*, 52(8):1589–1606, 2009.
97. Wendong Wang and Liqun Zhang. The C^α regularity of weak solutions of ultraparabolic equations. *Discrete Contin. Dyn. Syst.*, 29(3):1261–1275, 2011.
98. Haoyu Zhang. Global solutions in $W_k^{\zeta,p} L_x^\infty L_v^2$ for the Boltzmann equation without cutoff. *arXiv preprint arXiv:2008.10269*, 2020.

Part II
Invited Contributions

An Optimization-Based Strategy for Peridynamic-FEM Coupling and for the Prescription of Nonlocal Boundary Conditions

Marta D'Elia, David J. Littlewood, Jeremy Trageser, Mauro Perego, and Pavel B. Bochev

Abstract We describe an optimization-based method for the coupling of a static peridynamic (PD) model and a static classical elasticity model. The approach formulates the coupling as a control problem in which the states are the solutions of the PD and classical equations, the objective is to minimize their mismatch on an overlap of the PD and classical domains, and the controls are virtual volume constraints and boundary conditions applied at the local–nonlocal interface. Our numerical tests performed on three-dimensional geometries illustrate the consistency and accuracy of our method, its numerical convergence, and its applicability to realistic engineering geometries. We demonstrate the coupling strategy as a means to reduce computational expense by confining the nonlocal model to a subdomain of interest, and as a means to transmit local (e.g., traction) boundary conditions applied at a surface to a nonlocal model in the bulk of the domain.

Keywords Peridynamic · Nonlocal models · Classical elasticity · Coupling methods · Optimization · Meshfree method · Finite element method · Boundary conditions

1 Introduction

Nonlocal models have become viable alternatives to classical partial differential equation (PDE) models for certain classes of problems, particularly those in which discontinuities such as cracks are present. The use of these models is increasing in several scientific and engineering applications such as fracture mechanics [34, 59],

M. D'Elia
Computational Science and Analysis, Sandia National Laboratories, , Livermore, CA, USA

D. J. Littlewood (✉) · J. Trageser · M. Perego · P. B. Bochev
Center for Computing Research, Sandia National Laboratories, Albuquerque, NM, USA
e-mail: djlittl@sandia.gov

T. Mengesha, A. J. Salgado (eds.), *A³N²M: Approximation, Applications, and Analysis of Nonlocal, Nonlinear Models*, The IMA Volumes in Mathematics and its Applications 165, https://doi.org/10.1007/978-3-031-34089-5_4

anomalous subsurface transport [5, 20, 48, 49], phase transitions [7, 13, 30], image processing [1, 24, 33], stochastic processes [6, 18, 40, 41], and turbulence [11, 42, 43]. In this chapter, we specifically focus on mechanics applications modeled by peridynamics, a nonlocal extension of continuum mechanics developed to capture discontinuities that result from material failure, as well as other phenomena [56, 60, 61]. However, several applications mentioned above, especially the ones where long-range interactions are limited to finite regions, may benefit from the technique presented in this chapter.

Spatial nonlocal operators are integral operators that embed length scales in their definition; the most general form of a nonlocal operator acting on a vector function $\mathbf{u} : \mathbb{R}^d \to \mathbb{R}^d$ is given by

$$\mathcal{L}[\mathbf{u}](\boldsymbol{x}) = \int_{B_\delta(\boldsymbol{x})} \mathbf{f}(\boldsymbol{x}, \boldsymbol{y}; \mathbf{u})\, d\boldsymbol{y},$$

where $B_\delta(\boldsymbol{x})$ is the ball centered at $\boldsymbol{x}$ of radius δ, usually referred to in the peridynamics literature as the *horizon* or interaction radius. The horizon determines the extent of the nonlocal interactions and embeds the characteristic length scale of the system. Furthermore, since the integrand does not depend on the derivatives of $\mathbf{u}$, the regularity requirements on the solutions of nonlocal problems are minimal (as opposed to the case of PDEs).

While the integral form allows one to capture multiscale behavior and discontinuities, it also poses theoretical and numerical challenges. Theoretical challenges include the lack of a complete nonlocal theory [12, 19, 21, 27], the nontrivial treatment of nonlocal interfaces [2, 8, 55], and nonlocal boundary conditions [26, 28, 31, 68, 70], which must be prescribed in a volumetric region surrounding the domain of interest to guarantee the uniqueness of the solution. Computational challenges are related to the integral form that may require sophisticated quadrature rules, yielding discretized systems whose matrices are dense or even full. Among the works dedicated to improving implementation of nonlocal discretizations and numerical solvers for discretized nonlocal problems, we mention variational methods [4, 9, 17, 22] and meshfree methods [45, 58, 63, 65, 67].

Despite the recent tremendous effort toward improving the efficiency of nonlocal simulations, the usability of nonlocal models is still hindered by their computational cost. This work strives to improve the viability of nonlocal methods for systems in which nonlocal modeling can be restricted to certain subregions within the overall domain. It is often the case that in engineering mechanics problems, the application of a nonlocal model is advantageous only in specific regions of a body, for example, due to the presence of cracks. In these circumstances, it is natural to model the material displacement with a nonlocal description in the proximity of the discontinuity and to use classical PDE models elsewhere, so that the bulk of the computation is concentrated only on select nonlocal regions. Furthermore, by using local models far from these regions, many classical codes that contain an array of features not widely available in peridynamics codes (such as structural elements) can be exploited.

The approach of unifying local and nonlocal models within a single computational framework is known in the literature as *local-to-nonlocal (LtN) coupling* and has been the subject of very active research during the last decade. According to the review paper [23], LtN approaches can be divided into two categories: constant horizon approaches, where the nonlocal region is characterized by a constant value of the horizon and the transition to the local region is abrupt, and variable horizon approaches, characterized by a smooth transition from the nonlocal to the local region by means of a varying horizon. Among the first category, we mention optimization-based methods (OBM) [14–16, 25], partitioned procedures [69], Arlequin approaches [36, 66], morphing methods [35, 39], quasi-nonlocal methods [29, 37], blending methods [50, 51], and splice methods [32]. Among the second category, we mention shrinking horizon methods [62] and partial stress methods [57]. A common feature of all these methods is the fact that the domain of interest is divided into possibly overlapping nonlocal and local regions under the assumption that there exists a local model that accurately describes the behavior of the system when the nonlocal effects vanish. As we describe in more detail below, the most widely used nonlocal models for mechanics converge (pointwise or in norm) to a well-known PDE model [51] as the extent of the nonlocal interaction vanishes, i.e., as $\delta \to 0$.

Building on our previous work on OBM for nonlocal diffusion problems [14, 15, 25] and our preliminary results on OBM for mechanics [16], in this chapter, we propose a physically consistent and non-intrusive OB coupling scheme for the coupling of a state-based peridynamic model and a classical, linear elasticity model. Specifically, we address the LtN coupling of the linear peridynamic solid (LPS) model [56] and the Navier–Cauchy classical model. The main feature of OBMs, which distinguishes these methods from other LtN techniques, is the fact that coupling conditions between local and nonlocal models are prescribed weakly, via optimization, while the local and nonlocal equations act as constraints of the optimization problem. More specifically, OBMs are formulated as optimization problems where the objective functional is the norm of the mismatch between local and nonlocal displacements in the overlap of the local and nonlocal subdomains, the constraints are the local and nonlocal equations in their respective subdomains, and the control variables are the local boundary condition and nonlocal volume constraint on the fictitious interfaces induced by the domain decomposition.

We summarize the main advantages OBMs below:

1. The constraints are fully decoupled, i.e., they can be solved independently, as opposed to other coupling methods such as blending approaches that introduce hybrid models via linear combination of local and nonlocal forces or energies. As a consequence, OBMs may be applied either within a unified computational framework or within a framework in which the local and nonlocal models are evaluated independently, as so-called black boxes.
2. An immediate consequence of 1 is that nonlocal and local discretizations, grids, and software can be different and independent. In fact, the only firm requirement is that the nonlocal and local solutions be compared on the overlap region, such

that a simple evaluation or projection operator suffices. Additional information, for example, evaluation of gradients, are optional depending on the nature of the optimization strategy employed. These facts make OBMs extremely flexible in terms of implementation.
3. OB solutions inherit the mathematical and numerical properties of the constraints. More specifically: when the constraints are well-posed problems, the OB formulation is also well-posed. Moreover, the overall numerical convergence properties of the OB solutions are the same as those of the schemes used for the discretization of the constraints (as illustrated by three-dimensional numerical tests in the current study).
4. OBMs for LtN coupling are amenable to the conversion of local (e.g., surface) boundary conditions to nonlocal volume constraints. It is often the case that volume constraints, necessary for the well-posedness of nonlocal equations, are not readily available. By placing a local model in the vicinity of the boundary, one can use available surface data as boundary conditions for the local model and utilize the coupling scheme to transmit the effect of the boundary conditions to the nonlocal domain. A demonstration of this technique is presented in Sect. 5.

The major contribution of this manuscript is consolidation of OB coupling methods for mechanics problems in three dimensions by demonstrating their consistency and convergence properties and by illustrating their applicability on representative test cases. Additionally, we show how to employ this technique to facilitate the prescription of boundary conditions in the absence of volumetric (nonlocal) data.

Outline of the Paper In Sect. 2, we describe the peridynamic model and its classical counterpart, i.e., the classical linear elasticity equation. In Sect. 3, we introduce the OB coupling scheme and highlight some of its properties. In Sect. 4, we provide details on the discretization techniques used in this chapter, namely, a meshfree approach for the LPS model and the finite element method for the classical model. Finally, in Sect. 5, we illustrate the consistency, accuracy, and applicability of our scheme using numerical examples.

2 The Peridynamics Model and Its Local Counterpart

In this section, we introduce the static, linearized peridynamic equilibrium equation [56] and its corresponding local limit, i.e., the classical Navier–Cauchy equation (NCE) of static elasticity.

Given the bounded body $\Omega \subset \mathbb{R}^3$ with boundary $\partial\Omega = \Gamma$, the peridynamic equilibrium equation at the material point $\boldsymbol{x} \in \Omega$ is given by

$$-\mathcal{L}_n[\mathbf{u}](\boldsymbol{x}) := -\int_\Omega \left\{\mathbf{T}[\boldsymbol{x}]\langle \boldsymbol{x}' - \boldsymbol{x}\rangle - \mathbf{T}[\boldsymbol{x}']\langle \boldsymbol{x} - \boldsymbol{x}'\rangle\right\} dV_{\boldsymbol{x}'} = \mathbf{b}(\boldsymbol{x}). \quad (1)$$

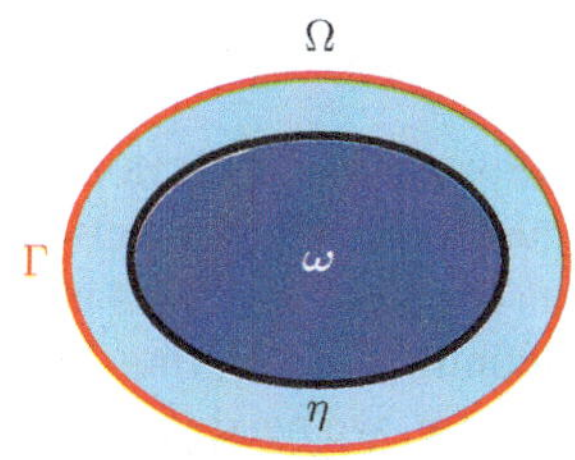

Fig. 1 A domain configuration in two dimensions

Here, $\mathbf{u} : \Omega \times \mathbb{R}^+ \to \mathbb{R}^3$ is the displacement field, $\mathbf{b} : \Omega \times \mathbb{R}^+ \to \mathbb{R}^3$ is a given body force density, and $\mathbf{T} : \Omega \times \mathbb{R}^+ \to \mathbb{R}^{(3,3)}$ is the force state field. The latter represents the force state at $\boldsymbol{x}$ that maps the bond $\langle \boldsymbol{x}' - \boldsymbol{x} \rangle$ to force per unit volume squared. A fundamental assumption in nonlocal modeling is that a material point $\boldsymbol{x}$ interacts with material points in a ball centered in $\boldsymbol{x}$ of radius δ, which we refer to as the *horizon* or nonlocal interaction radius. Formally, this nonlocal neighborhood is given by

$$B_\delta(\boldsymbol{x}) = \{\boldsymbol{x}' \in \Omega : |\boldsymbol{x} - \boldsymbol{x}'| \le \delta\}.$$

As a consequence, we require that the force state field satisfies the following property:

$$\mathbf{T}[\boldsymbol{x}]\langle \boldsymbol{x}' - \boldsymbol{x} \rangle = 0, \quad \forall \boldsymbol{x}' \notin B_\delta(\boldsymbol{x}).$$

Equation (1) is solved in the body $\omega \subset \Omega$, and Dirichlet volume constraints are prescribed in a volumetric layer η surrounding ω. These domains are such that $\Omega = \omega \cup \eta$, see Fig. 1 for a two-dimensional illustration. The thickness of η depends on the definition of $\mathbf{T}$ and guarantees, together with other conditions, that the peridynamics problem is well-posed. According to the linear peridynamic solid (LPS) model considered in this work [56], for all $\boldsymbol{x} \in \Omega$, the force state field $\mathbf{T}$ is defined as

$$\mathbf{T}[\boldsymbol{x}]\langle \boldsymbol{\xi} \rangle = \frac{\kappa(|\boldsymbol{\xi}|)}{m} \left\{ (3K - 5G)\,\theta(\boldsymbol{x})\boldsymbol{\xi} + 15G \frac{\boldsymbol{\xi} \otimes \boldsymbol{\xi}}{|\boldsymbol{\xi}|^2} (\mathbf{u}(\boldsymbol{x} + \boldsymbol{\xi}) - \mathbf{u}(\boldsymbol{x})) \right\}, \tag{2}$$

where $\boldsymbol{\xi} = \boldsymbol{x}' - \boldsymbol{x}$, K is the bulk modulus, G is the shear modulus, and the linearized nonlocal dilatation, $\theta : \Omega \to \mathbb{R}$, is defined as

$$\theta(\boldsymbol{x}) = \frac{3}{m} \int_{B_\delta(\mathbf{0})} \kappa(|\boldsymbol{\zeta}|)\, \boldsymbol{\zeta} \cdot (\mathbf{u}(\boldsymbol{x} + \boldsymbol{\zeta}) - \mathbf{u}(\boldsymbol{x}))\, dV_{\boldsymbol{\zeta}}, \quad \text{with } m = \int_{B_\delta(\mathbf{0})} \kappa(|\boldsymbol{\zeta}|)\, |\boldsymbol{\zeta}|^2\, dV_{\boldsymbol{\zeta}}.$$

The spherical function κ is referred to as the *influence function*; it determines the support of force states and modulates the bond strength [54, 56]. With this choice of $\mathbf{T}$, we define the volumetric layer η as

$$\eta = \{\boldsymbol{x}' \in \Omega : |\boldsymbol{x}' - \boldsymbol{x}| < 2\delta\} \quad \forall \boldsymbol{x} \in \Gamma, \tag{3}$$

whose thickness is double the size of the horizon. As mentioned before, this choice guarantees that the peridynamic operator can be evaluated for any point in ω, up to its boundary $\partial\omega$. This is due to the fact that θ introduces an additional integral in the definition of $\mathcal{L}_n$; thus, we have a double integral over $B_\delta(\mathbf{0}) \times B_\delta(\mathbf{0})$, i.e., for every point $\boldsymbol{x}$, we need values of the displacement in $B_{2\delta}(\boldsymbol{x})$.

Using the linearized LPS force state field in (2), the three-dimensional peridynamic equation is then formulated as follows. Find $\mathbf{u} \in [L^2(\Omega)]^3$ such that

$$\begin{cases} -\mathcal{L}_{\text{LPS}}[\mathbf{u}](\boldsymbol{x}) = \mathbf{b}(\boldsymbol{x}) & \boldsymbol{x} \in \omega \\ \mathbf{u}(\boldsymbol{x}) = \mathbf{g}(\boldsymbol{x}) & \boldsymbol{x} \in \eta. \end{cases} \tag{4}$$

Here, $\mathbf{g} \in [L^2(\eta)]^3$ is the volumetric Dirichlet data and $\mathcal{L}_{\text{LPS}}$ is the nonlocal operator $\mathcal{L}_n$ corresponding to the choice of $\mathbf{T}$ as in (2), i.e.,

$$\begin{aligned} \mathcal{L}_{\text{LPS}}[\mathbf{u}](\boldsymbol{x}) := \int_{B_\delta(\mathbf{0})} \frac{\kappa(|\boldsymbol{\xi}|)}{m} \Big\{ (3K - 5G)(\theta(\boldsymbol{x}) + \theta(\boldsymbol{x} + \boldsymbol{\xi}))\boldsymbol{\xi} \\ + 30G \, \frac{\boldsymbol{\xi} \otimes \boldsymbol{\xi}}{|\boldsymbol{\xi}|^2} (\mathbf{u}(\boldsymbol{x} + \boldsymbol{\xi}) - \mathbf{u}(\boldsymbol{x})) \Big\} \, dV_{\boldsymbol{\xi}}. \end{aligned} \tag{5}$$

The linearized LPS model (5) has two important properties that one can exploit to choose a viable candidate for the local model and perform consistency and converge tests for the coupling method. First, the limit for $\delta \to 0$ (i.e., when the nonlocal interactions vanish) is the classical NCE of static elasticity [53]:

$$-\mathcal{L}_{\text{NC}}[\mathbf{u}](\boldsymbol{x}) := -\left[\left(K + \frac{1}{3} G \right) \nabla(\nabla \cdot \mathbf{u})(\boldsymbol{x}) + G \, \nabla^2 \mathbf{u}(\boldsymbol{x}) \right] = \mathbf{b}(\boldsymbol{x}), \tag{6}$$

where K, G, and $\mathbf{b}$ are defined as in (2), or equivalently,

$$\begin{aligned} &-\nabla \cdot \boldsymbol{\sigma}[\mathbf{u}](\boldsymbol{x}) = \mathbf{b}(\boldsymbol{x}), \quad \text{where} \\ &\boldsymbol{\sigma}[\mathbf{u}] = \lambda(\nabla \cdot \mathbf{u})\mathbf{I} + \mu(\nabla \mathbf{u} + \nabla \mathbf{u}^T) \\ &(\lambda, \mu) = \left(K - \tfrac{2G}{3}, \, G \right), \end{aligned} \tag{7}$$

where $\mathbf{I}$ is the identity tensor. Note that the latter equation is the classical linear elasticity equation in terms of the Lamé constants (λ, μ). This implies that the NC model is a fair approximation of the peridynamics model for sufficiently regular solutions, and, as a consequence, it can be used as the local model in the coupling strategy.

Furthermore, the linearized LPS model equates to the classical NCE model when the displacement field is quadratic [53, Proposition 1]. This property provides analytical solutions for the LPS model for the quadratic patch test and numerical convergence tests presented in Sect. 5.1.

3 Optimization-Based LtN Formulation

To construct the continuous formulation of the coupling procedure, we first introduce a partitioning of the domain Ω into a nonlocal subdomain Ω_n and a local subdomain Ω_l, with boundary Γ_l, such that $\Omega_n = \omega_n \cup \eta_n$ and $\Omega_n \cap \Omega_l = \Omega_o \neq \emptyset$. A two-dimensional illustration of this strategy is given in Fig. 2. As is common in LtN coupling methods, we assume that the linearized LPS model (5) provides an accurate description of the material behavior in Ω_n, whereas the local NC model gives a fair representation in the remainder of the domain. In OBM approaches to coupling, the main idea is to formulate the coupling as an optimization problem where the difference between the nonlocal and the local displacements is minimized on the overlap Ω_o by tuning their values on the virtual interfaces induced by the partition, i.e., the virtual interaction volume η_c and the virtual boundary Γ_c. Let $\eta_D = \eta \cap \eta_n$ and $\Gamma_D = \Gamma \cap \Gamma_l$ be the physical interaction volume and boundary where we prescribe the given Dirichlet data. We define the virtual control volume and boundary as $\eta_c = \eta_n \setminus \eta_D$ and $\Gamma_c = \Gamma_l \setminus \Gamma_D$. We then pose the peridynamics problem on the nonlocal domain ω_n and the classical NC problem on the local domain Ω_l. This results in the following systems of equations:

$$\begin{cases} -\mathcal{L}_{\text{LPS}}[\mathbf{u}_n](\boldsymbol{x}) = \mathbf{b}(\boldsymbol{x}) & \boldsymbol{x} \in \omega_n \\ \mathbf{u}_n(\boldsymbol{x}) = \mathbf{g}(\boldsymbol{x}) & \boldsymbol{x} \in \eta_D \\ \mathbf{u}_n(\boldsymbol{x}) = \boldsymbol{v}_n(\boldsymbol{x}) & \boldsymbol{x} \in \eta_c \end{cases} \qquad \begin{cases} -\mathcal{L}_{\text{NC}}[\mathbf{u}_l](\boldsymbol{x}) = \mathbf{b}(\boldsymbol{x}) & \boldsymbol{x} \in \Omega_l \\ \mathbf{u}_l(\boldsymbol{x}) = \mathbf{g}(\boldsymbol{x}) & \boldsymbol{x} \in \Gamma_D \\ \mathbf{u}_l(\boldsymbol{x}) = \boldsymbol{v}_l(\boldsymbol{x}) & \boldsymbol{x} \in \Gamma_c. \end{cases} \tag{8}$$

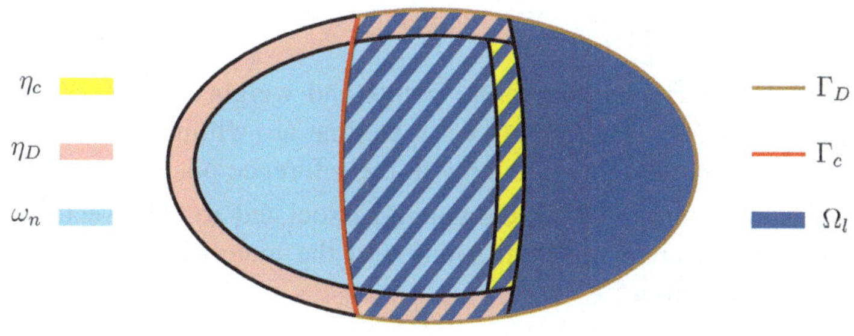

Fig. 2 An example LtN domain configuration in two dimensions

Here, $\boldsymbol{v}_n \in [L^2(\eta_c)]^3$ and $\boldsymbol{v}_l \in [H^{1/2}(\Gamma_c)]^3$ are the unknown volume constraint and boundary condition. The systems (8) act as constraints to the optimization problem, while $(\boldsymbol{v}_n, \boldsymbol{v}_l)$ act as control variables. As a result, we aim at solving the following optimization problem:

$$\min_{\mathbf{u}_n, \mathbf{u}_l, \boldsymbol{v}_n, \boldsymbol{v}_l} \mathcal{J}(\mathbf{u}_n, \mathbf{u}_l) = \frac{1}{2} \int_{\Omega_o} |\mathbf{u}_n - \mathbf{u}_l|^2 \, d\boldsymbol{x} \quad \text{subject to (8).} \tag{9}$$

Upon solution of (9), we denote the optimal controls by $\boldsymbol{v}_n^*$ and $\boldsymbol{v}_l^*$, and the corresponding optimal displacements by $\mathbf{u}_n^* = \mathbf{u}_n(\boldsymbol{v}_n^*)$ and $\mathbf{u}_l^* = \mathbf{u}_l(\boldsymbol{v}_l^*)$. Then, the optimal coupled solution is defined as

$$\mathbf{u}^* = \begin{cases} \mathbf{u}_n^* \; \boldsymbol{x} \in \Omega_n \\ \mathbf{u}_l^* \; \boldsymbol{x} \in \Omega_l \setminus \Omega_o. \end{cases} \tag{10}$$

Remark 1 The well-posedness of the minimization (9) has been rigorously studied in [15] and [25] for nonlocal Poisson's problems with Dirichlet and Neumann conditions in a multi-dimensional setting. Further considerations on the well-posedness in the context of peridynamics models can be found in [16].

4 Numerical Solution of the Optimization-Based LtN Formulation

For the linearized LPS model described in Sect. 2, we utilize the meshfree approach of Silling and Askari [61]. We discretize the nonlocal domain using a set of material points $\{\boldsymbol{x}_i\}_{i=1}^{N_n} \subset \Omega_n$ and associate with each material point $\boldsymbol{x}_i$ a volume V_i such that $\sum_{i=1}^{N_n} V_i = \Omega_n$. For every point $\boldsymbol{x}_i$, we approximate the operator $\mathcal{L}_{\text{LPS}}$ as follows:

$$\mathcal{L}_{\text{LPS}}^h[\boldsymbol{x}_i] := \sum_{j \in \mathcal{F}_i} \left\{ \mathbf{T}[\boldsymbol{x}_i]\langle \boldsymbol{x}_j - \boldsymbol{x}_i \rangle - \mathbf{T}[\boldsymbol{x}_j]\langle \boldsymbol{x}_i - \boldsymbol{x}_j \rangle \right\} V_j^{(i)}, \tag{11}$$

where $\boldsymbol{x}_i$ and $V_j^{(i)}$ serve as quadrature points and weights and $\mathcal{F}_i$ represents the *family* of $\boldsymbol{x}_i$, i.e., the set of all points $\boldsymbol{x}_j$ in Ω_n that are within a distance of δ from $\boldsymbol{x}_i$. Here, the $\boldsymbol{x}_j$ is chosen to coincide with the reference position of the jth node, and the quadrature weight[1] $V_j^{(i)}$ is the volume associated $\boldsymbol{x}_j$. The vector of degrees of freedom of the discrete nonlocal solution at the material points is denoted by $\vec{\boldsymbol{u}}_n = [\vec{u}_n^1, \vec{u}_n^2, \vec{u}_n^3]$, with $\vec{u}_n^k \in \mathbb{R}^{N_n}$.

[1] Details regarding the computation of $V_j^{(i)}$ can be found in [53].

We discretize the NC model in (6) by the finite element (FE) method using continuous piecewise linear basis functions. As it is standard, the FE implementation is not discussed in detail. We denote the vector of degrees of freedom of the local discrete solution by $\vec{\boldsymbol{u}}_l = [\vec{u}_l^1, \vec{u}_l^2, \vec{u}_l^3]$, with $\vec{u}_l^k \in \mathbb{R}^{N_l}$, where N_l is the number of degrees of freedom of each spatial component over the FE computational mesh.

Although the peridynamics equation and the NC equation do not directly interact, the difference of the corresponding solutions must be computed on the overlap region Ω_o. To this end, we introduce the nonlocal and local selection matrices that allow us to compute an approximation of the cost functional. Let $S_n \in \mathbb{R}^{N_o,N_n}$ be the matrix that selects the components of $\vec{u}_n^k$ in Ω_o and $S_l \in \mathbb{R}^{N_o,N_l}$ be the operator that evaluates $\vec{u}_l^k$ at the material points $\{\boldsymbol{x}_i\} \in \Omega_o$. Formally, given the FE basis functions $\{\phi_j\}_{i=1}^{N_l}$, the selection matrices are defined as

$$(S_n)_{ij} := \delta_{ij} \quad \text{and} \quad (S_l)_{ij} := \phi_j(\boldsymbol{x}_i), \quad \forall \boldsymbol{x}_i \in \Omega_o.$$

Then, we define the discrete functional as

$$J_d(\vec{\boldsymbol{u}}_n, \vec{\boldsymbol{u}}_l) = \frac{1}{2} \sum_{i=1}^{N_o} \sum_{k=1}^{3} |(S_n \vec{u}_n^k)_i - (S_l \vec{u}_l^k)_i|^2 \, \widetilde{V}_i, \tag{12}$$

where $\widetilde{V}_i$ is the volume associated with the ith material point, properly scaled.

5 Numerical Tests

In this section, we report the results of several numerical tests performed on three-dimensional geometries. The purpose of these experiments is, first, to illustrate the consistency and accuracy of our method via patch tests and numerical convergence studies, and, second, to demonstrate the applicability of our approach for realistic engineering geometries. We also show that our coupling approach can be used to circumvent the nontrivial prescription of "nonlocal boundary conditions" or volume constraints, when only surface data are available. We start by providing some details on the software used for the simulations. These are followed by consistency and convergence tests with manufactured solutions. We then consider examples of coupling in the presence of cracks and provide an example of prescribing boundary conditions via OB coupling.

The simulations presented in this work were performed using *Peridigm* [44, 46] and *Albany* [3, 47], which are open-source codes developed primarily at Sandia National Laboratories. *Peridigm* is a peridynamics code for solid mechanics based on the meshfree approach of Silling and Askari [38, 61], and *Albany* is a FE code for the simulation of several physical systems governed by PDEs. Both *Peridigm* and *Albany* are C++ codes designed for use on large-scale parallel computing platforms.

For the current study, *Peridigm* and *Albany* were coupled directly, resulting in a single executable in which the peridynamic and FE contributions are treated in a unified fashion and the implicit solver acts on a single, monolithic system. Routines in the *Peridigm* code base are used to evaluate peridynamic forces and the corresponding entries in the tangent stiffness matrix. The *Albany* code handles most remaining aspects of the computation, including FE assembly for the NC equation, computation of the discrete functional and its derivative, and solution of the state and adjoint systems necessary for the numerical solution of the optimization problem, conducted by means of the LBFGS algorithm. The *Peridigm* and *Albany* codes, and the software infrastructure developed in this study to couple them, rely heavily on several *Trilinos* [64] packages, including *ROL* for solution of the optimization problem, *Epetra* for the management of parallel data structures, *Intrepid* for FE assembly, and *Ifpack* and *AztecOO* for the preconditioning and solution of the linear systems.

5.1 Patch Tests and Convergence Study

We consider analytic solutions of the coupling problem in order to perform patch tests and a convergence study. We utilize linear displacement fields for the patch test and quadratic displacement fields for the convergence study. As discussed above, it can be shown (see [53]) that Eqs. (4) and (6) are equivalent for quadratic displacements, and thus the known analytical solution to the local problem also holds for the nonlocal problem under this condition. In our tests, we let $\Omega^+ = [0, 1]^3$ and $\Omega = [2\delta, 1 - 2\delta]^3$, so that the thickness of the external layer where the volume constraints are prescribed is twice the size of the horizon. Specifically, for $\boldsymbol{x} = (x, y, z)$, we consider the following data sets:

I: Linear
$\mathbf{u}_{lin}(\boldsymbol{x}) = (x, 0, 0)$, $\mathbf{b}_{lin}(\boldsymbol{x}) = \mathbf{0}$, $\mathbf{g}_{lin}(\boldsymbol{x}) = \mathbf{u}_{lin}(\boldsymbol{x})$.
II: Quadratic
$\mathbf{u}_q(\boldsymbol{x}) = (x^2, 0, 0)$,
$\mathbf{b}_q(\boldsymbol{x}) = (-304.49, 0, 0)$, obtained by substitution of $\mathbf{u}$ in (6),
$\mathbf{g}_q(\boldsymbol{x}) = \mathbf{u}_q(\boldsymbol{x})$,
$(\lambda, \mu) = (109.62, 73.08)$ (Lamé parameters representative of stainless steel).

To assess the accuracy of the coupled solution, we consider the ℓ^2 norm of the difference between the vectors of the computed displacements and the analytic displacements $\vec{\boldsymbol{u}}^*$ at the nonlocal and local degrees of freedom, i.e.,

$$\begin{aligned} \text{error}_n &= \|\vec{\boldsymbol{u}}_n - \vec{\boldsymbol{u}}_*\|_{\ell^2} \\ \text{error}_l &= \|\vec{\boldsymbol{u}}_l - \vec{\boldsymbol{u}}_*\|_{\ell^2}, \end{aligned} \tag{13}$$

where $\vec{\boldsymbol{u}}^*$ is either $\mathbf{u}_{lin}$ or $\mathbf{u}_q$ evaluated at the appropriate degrees of freedom.

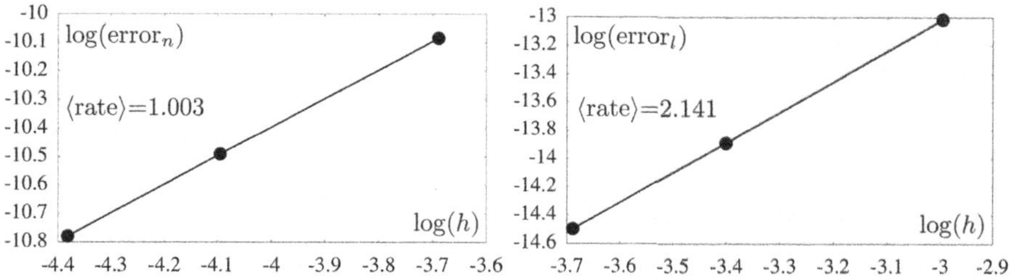

Fig. 3 Convergence plots for the nonlocal (left) and local (right) portions of the coupled model

We perform a linear patch test to verify the consistency of the method; to this end, we prescribe **b** and **g** as in case I above[2]. Numerical results indicate a perfect match of the nonlocal and local solutions on the overlap and show that the coupled solution is ϵ-machine accurate with respect to $\mathbf{u}_{lin}$. This happens because linear solutions can be approximated exactly by both the peridynamic and FE discretizations.

Results of the convergence tests are reported in Fig. 3. Here, we prescribe **b** and **g** as in case II above. For decreasing values of h, we report the errors corresponding to the nonlocal and local solutions, according to (13), using a log–log scale. We observe a linear rate of convergence for the nonlocal solution and a quadratic rate of convergence for the local solution, see the average rate, ⟨rate⟩, reported in the figure. This is in line with theoretical results for the meshfree nonlocal discretization utilized in the nonlocal region [52, 53], and for the piecewise linear finite element approximation utilized in the local region [10, 15]. We point out that, by using a higher-order discretization method that approximates quadratic displacements exactly, one would obtain solutions that are ϵ-machine accurate by construction.

5.2 *Geometries with Cracks*

Peridynamic models are advantageous for modeling cracks because the governing equations remain valid in the presence of material discontinuities. We investigate the performance of the OB coupling scheme for this class of problems using a three-dimensional bar with a crack. We assign material parameters that are representative of steel alloys, as shown in Table 1.

[2] For more numerical tests regarding linear and quadratic patch tests in three-dimensional settings, we refer the interested reader to [16].

Table 1 Material parameters for the bar and compact tension experiments

Parameter	Prenotched bar	Compact tension
Horizon	1.00 mm	0.30 mm
Poisson's ratio	0.3	0.27
Bulk modulus	140.0 GPa	150.0 GPa

5.2.1 Prenotched Rectangular Bar

This experiment considers a prenotched bar described by the geometry $\Omega := [-18, 18] \times [-4, 4] \times [-2, 2]$, where the unit of length is mm. See Fig. 4 for an illustration of a cross-section along the XY plane. The local solution domain, boundary regions, and control regions are given by

$$\begin{aligned}\Omega_l &:= ((-18, -2) \times [-4, 4] \times [-2, 2]) \cup ((2, 18) \times [-4, 4] \times [-2, 2])\\ \Gamma_D &:= (\{-18\} \times [-4, 4] \times [-2, 2]) \cup (\{18\} \times [-4, 4] \times [-2, 2])\\ \Gamma_c &:= (\{-2\} \times [-4, 4] \times [-2, 2]) \cup (\{2\} \times [-4, 4] \times [-2, 2])\,.\end{aligned} \tag{14}$$

The nonlocal solution domain and control regions are given by

$$\begin{aligned}\omega_n &:= (-3.5, 3.5) \times [-4, 4] \times [-2, 2]\\ \eta_c &:= ([-5, -3.5] \times [-4, 4] \times [-2, 2]) \cup ([3.5, 5] \times [-4, 4] \times [-2, 2])\,.\end{aligned} \tag{15}$$

In addition, a prenotch region P is described by

$$P := \{0\} \times [1, 4] \times [-2, 2]. \tag{16}$$

The prenotch is modeled in the numerical experiment by omitting all bonds crossing P. Our choice of domain decomposition is dictated by the fact that the nonlocal model is advantageous in the vicinity of the prenotch, while the local model is sufficient far from it. The nonlocal and local computational domains are further illustrated in Fig. 5, where material points in the nonlocal model are rendered as spheres and the hexahedral FE mesh corresponds to the local model.

A prescribed displacement boundary condition is applied in the x direction on the local boundary region Γ_D, specifically -0.05 mm at $x = -18.0$ mm and 0.05 mm at $x = 18.0$ mm, resulting in tensile loading. Stress-free conditions are employed on the remaining surfaces, i.e., a homogeneous Neumann volume constraint and boundary condition for the nonlocal and local problems, respectively. To eliminate rigid body modes, additional zero displacement boundary conditions are applied in the y direction along the edges defined by $x = -18.0$ mm , $y = -4.0$ mm and $x = 18.0$ mm, $y = -4.0$ mm, and in the z direction along the edges defined by $x = -18.0$ mm, $z = -2.0$ mm and $x = 18.0$ mm, $z = -2.0$ mm. The results of the experiment are shown in Fig. 5. As expected, the influence of the crack on the displacement is restricted predominantly to the nonlocal region, and the OB coupling provides a smooth transition between the local and nonlocal models.

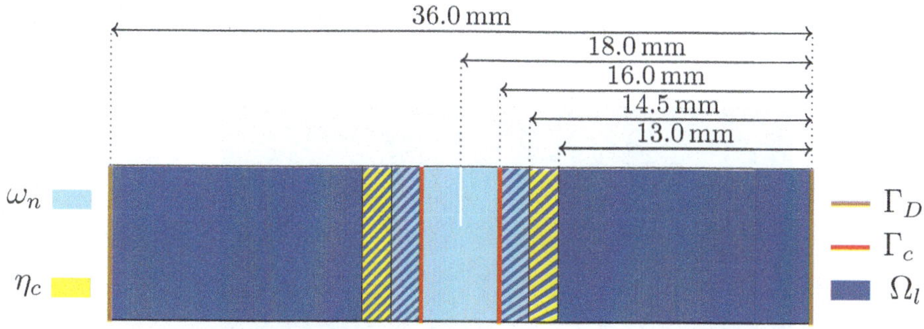

Fig. 4 XY cross-section of the geometry for the prenotched bar experiment in which Dirichlet boundary conditions are imposed. The nonlocal model is applied in the vicinity of the prenotch, and the local model is applied elsewhere

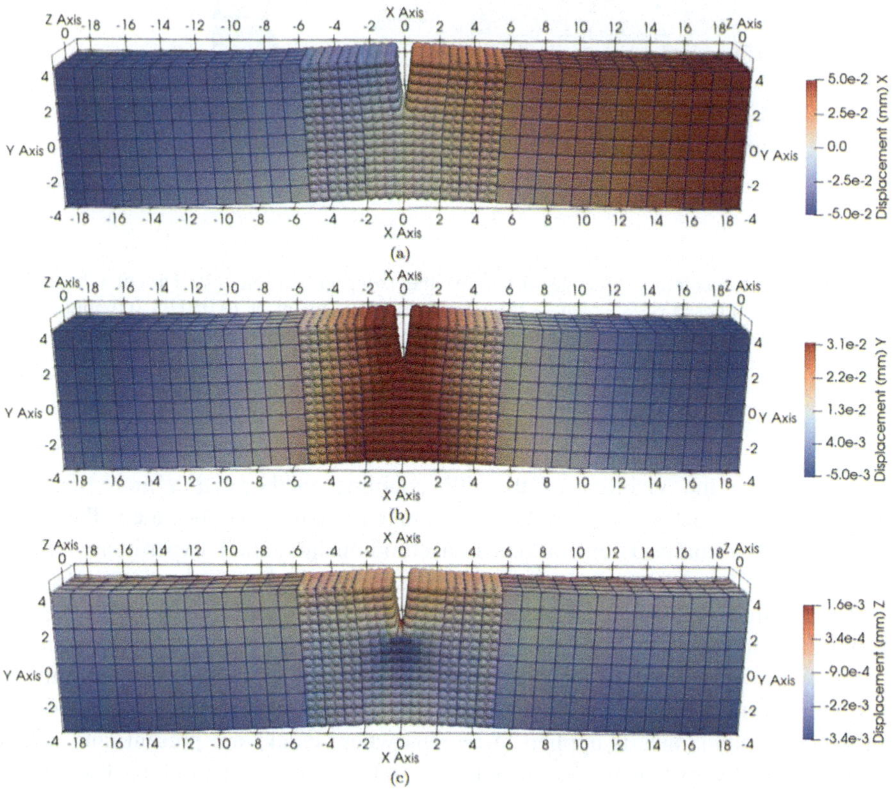

Fig. 5 Results of the bar experiment with prescribed displacements of -0.05 mm at $x = -18.0$ mm and 0.05 mm at $x = 18.0$ mm (see Fig. 4). Displacements are magnified by a factor of 15 for the purpose of illustration. (**a**) displacement in x. (**b**) displacement in y. (**c**) displacement in z

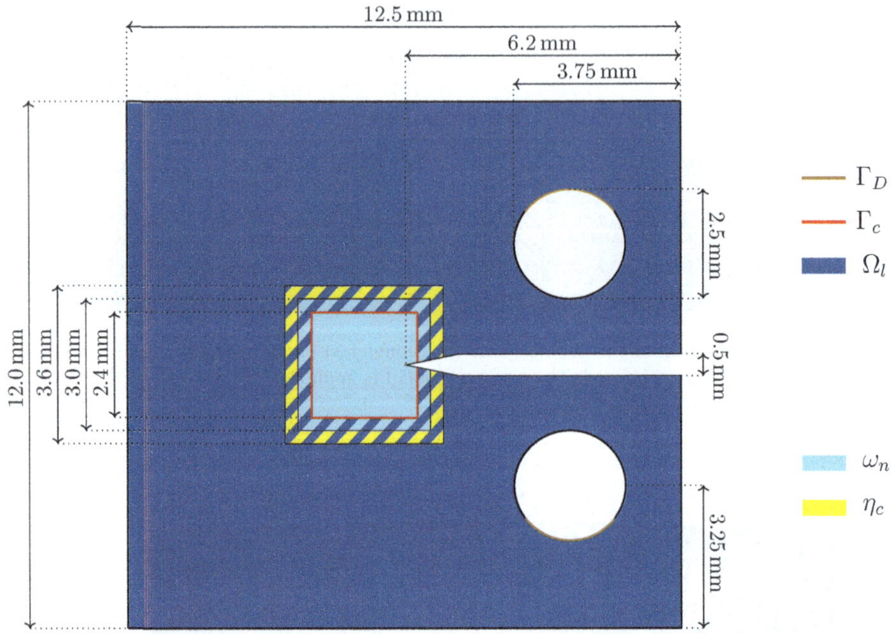

Fig. 6 XY cross-section of the geometry for the compact tension test. Dirichlet boundary conditions are imposed on Γ_D, where we impose a displacement of $\langle 0.0, 0.03, 0.0\rangle$ mm in the top hole and a displacement of $\langle 0.0, -0.03, 0.0\rangle$ mm in the bottom hole

5.2.2 Compact Tension Experiment

To investigate the viability of the OB coupling method on a more realistic engineering geometry, we consider the compact tension test specimen illustrated in Fig. 6. The compact tension test is a common laboratory experiment for the evaluation of material properties, such as fracture toughness, that are related to material failure. We employ a compact tension test geometry based on the ASTM E399-20 standard, although we reduce the thickness to decrease computational expense. As in the case of the prenotched bar in Sect. 5.2.1, the compact tension test simulation demonstrates the ability of the coupling strategy to permit models in which peridynamics is applied to regions where cracks are present, and a less computationally expensive classical FE model is applied elsewhere. The local–nonlocal interface is placed sufficiently far from the crack tip such that the material response in the overlap region is smooth and well-behaved. Prescribed displacement boundary conditions are applied at the holes, which is representative of the physical laboratory experiment in which pins connected to a material testing machine are used to load the specimen. It is noted that in this configuration, loading is transmitted

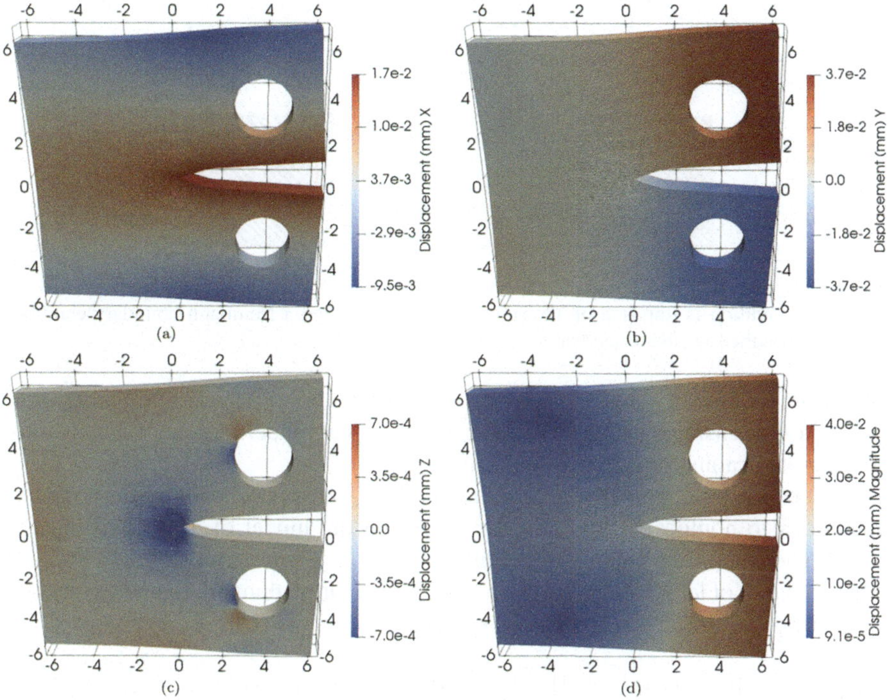

Fig. 7 Compact tension experiment with Dirichlet boundary conditions. (**a**) Displacement in x. (**b**) Displacement in y. (**c**) Displacement in z. (**d**) Magnitude of displacement.

to the peridynamic region only through the coupling interface; nonzero user-prescribed boundary conditions are confined to the local model. The results of the simulation are presented in Fig. 7. The deformations are consistent with the applied loading, and no unphysical artifacts are apparent in the overlap region.

The simulations carried out in the present study are static, i.e., the coupled *Peridigm–Albany* code solves for the equilibrium configuration corresponding to a single set of prescribed boundary conditions. Extending the code capabilities to enable quasi-static simulations that include stable crack propagation emanating from the prenotch is a subject of future work.

5.2.3 Conversion of Local to Nonlocal Boundary Conditions in a Prenotched Bar

A notable use case for LtN coupling is the conversion of local boundary conditions into nonlocal volume constraints. This is important because surface traction boundary conditions, commonly applied in traditional FE models to represent pressure loading, or similar, cannot be applied directly to a nonlocal model. Instead, to

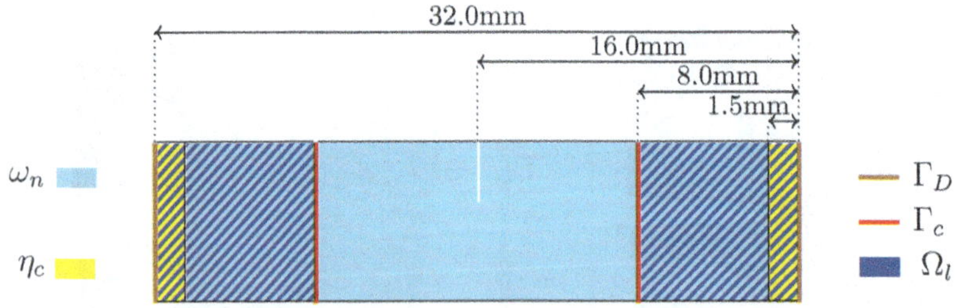

Fig. 8 XY cross-section geometry for the prenotched bar experiment in which a Dirichlet boundary condition is imposed on the right end of the bar and a Neumann boundary condition is imposed on the left side of the bar

ensure uniqueness of solution for the nonlocal problem, tractions must be converted from two-dimensional surface loads for use in a local model into three-dimensional volume constraints for use in a nonlocal model.

We demonstrate the OB coupling method for conversion of traction boundary conditions to nonlocal volume constraints using the model of a bar illustrated in Fig. 8. The computational domain is defined as $\Omega := [-16, 16] \times [-4, 4] \times [-2, 2]$, where the unit of length is millimeters. The local solution domain, boundary regions, and control regions are, respectively, given by

$$\begin{aligned}
\Omega_l &:= ((-16, -8) \times [-4, 4] \times [-2, 2]) \cup ((8, 16) \times [-4, 4] \times [-2, 2]) \\
\Gamma_D &:= (\{-16\} \times [-4, 4] \times [-2, 2]) \cup (\{16\} \times [-4, 4] \times [-2, 2]) \\
\Gamma_c &:= (\{-8\} \times [-4, 4] \times [-2, 2]) \cup (\{8\} \times [-4, 4] \times [-2, 2]) \,.
\end{aligned} \tag{17}$$

The nonlocal solution domain and control regions are, respectively, given by

$$\begin{aligned}
\omega_n &:= (-14.5, 14.5) \times [-4, 4] \times [-2, 2] \\
\eta_c &:= ([-16, -14.5] \times [-4, 4] \times [-2, 2]) \cup ([14.5, 16] \times [-4, 4] \times [-2, 2]) \,.
\end{aligned} \tag{18}$$

In addition, a prenotch region P is described by

$$P := \{0\} \times [1, 4] \times [-2, 2]. \tag{19}$$

In the numerical experiment, all bonds crossing P are omitted from the simulation. On the local boundary region Γ_D, a prescribed traction of $\tau = -1700.0$ MPa in the x direction at $x = -16.0$ mm and a prescribed displacement of 0.0 mm in the x direction at $x = 16.0$ mm are imposed. To eliminate rigid body modes, additional zero displacement boundary conditions are applied in the y direction along the edge defined by $x = 16.0$ mm, $y = -4.0$ mm and in the z direction along the edge defined by $x = 16.0$ mm, $z = -2.0$ mm.

In contrast to the prenotched bar experiment in Sect. 5.2.1, the nonlocal region is not restricted to the vicinity of the prenotch and instead encompasses the entire bar. This approach allows us to observe the specific nonlocal volume constraints that correspond to the local boundary conditions at the ends of the bar. Of particular interest are the nonlocal volume constraints that correspond to the traction boundary condition applied to the face at $x = -16.0$ mm. The nonlocal volume constraints, which are virtual constraints determined by solution of the OB coupling problem, give insight into how nonlocal volume constraints could be applied in a purely nonlocal simulation to reproduce the effect of traction loading in a comparable local simulation. This conversion technique is similar to the approaches introduced in [26, 28].

The results for this experiment are shown in Figs. 9, 10, and 11. Displacement results for both the local and nonlocal domains are shown in Fig. 9 and are consistent

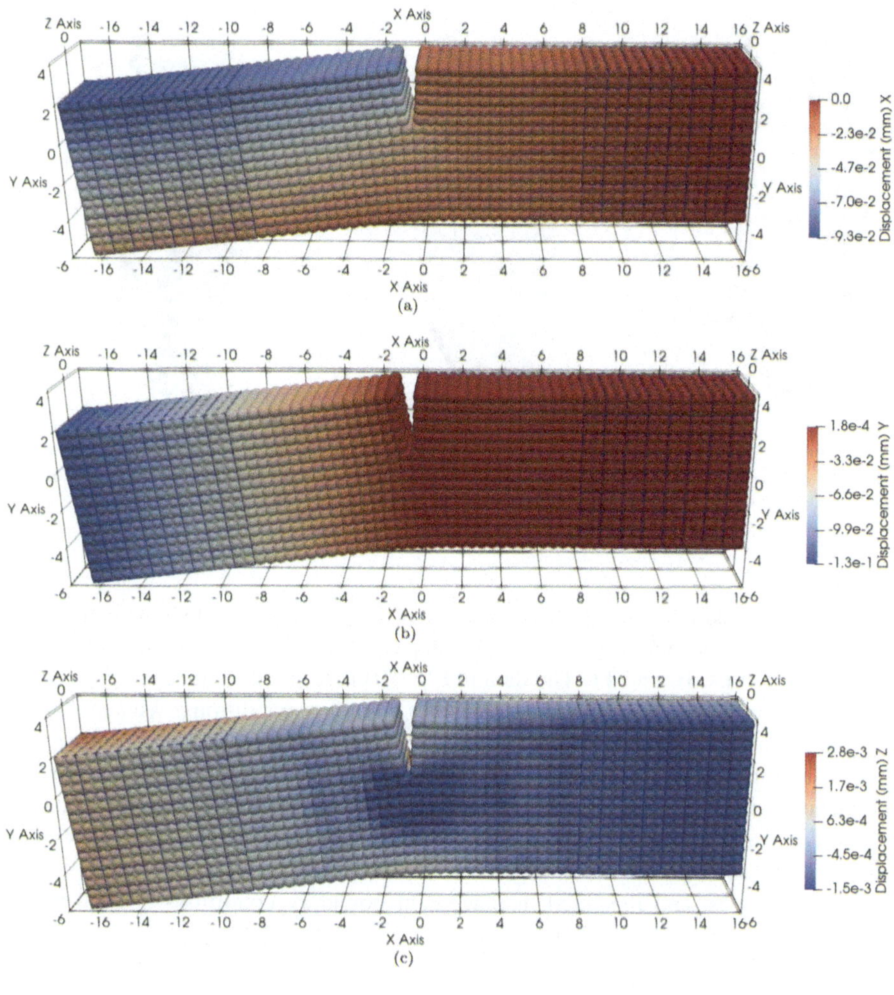

Fig. 9 Bar with Neumann–Dirichlet Boundary Conditions. Displacements are magnified by a factor of 15 to emphasize displacements. (**a**) Displacement in x. (**b**) Displacement in y. (**c**) Displacement in z

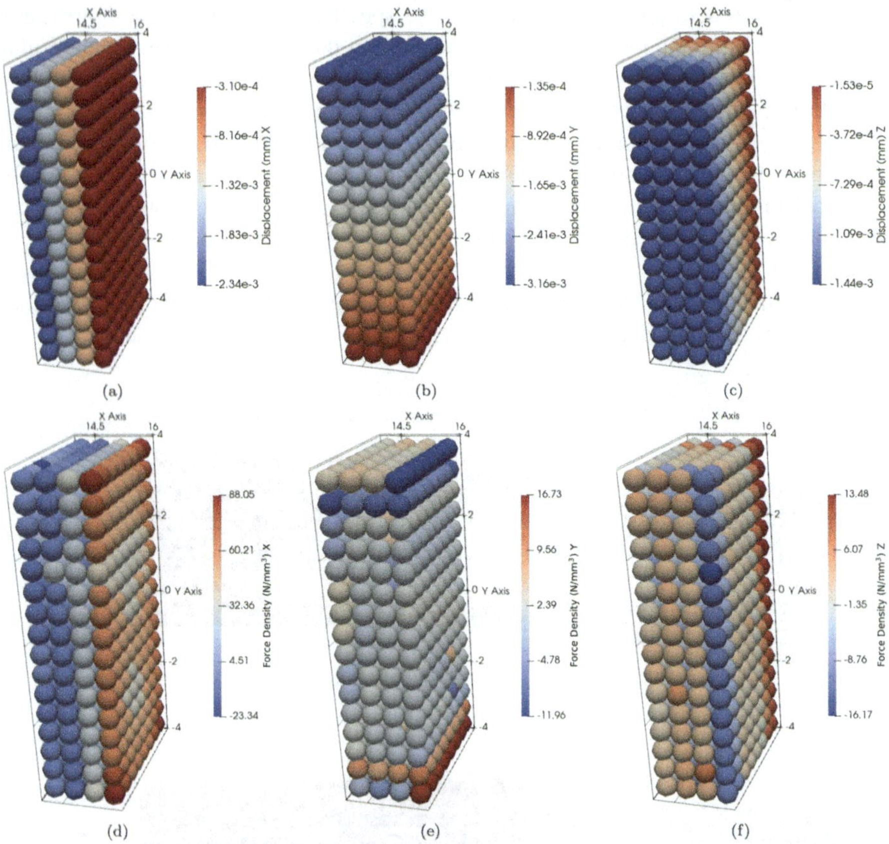

Fig. 10 Nonlocal boundary conditions corresponding to the local Dirichlet boundary condition. (**a**) Displacement in x. (**b**) Displacement in y. (**c**) Displacement in z. (**d**) Force density in x. (**e**) Force density in y. (**f**) Force density in z

with the expected behavior. Displacement and force density results for the nonlocal model in the direct vicinity of the boundary conditions at the ends of the bar are presented in Figs. 10 and 11. The data in Fig. 10 correspond to the end of the bar at $x = 16.0$ mm, adjacent to the local Dirichlet boundary conditions. We see that the displacements in the nonlocal model vary approximately linearly over this region. This is an intuitive result, and it suggests that a fair approximation of the prescribed displacement volume constraints for a comparable purely nonlocal model could be determined simply by assuming a linear displacement field in this region. It is noted that the force density results in Fig. 10 reflect the influence of the applied loading, the free surfaces, and the fixed displacement boundary conditions applied to the local model to prevent rigid body modes.

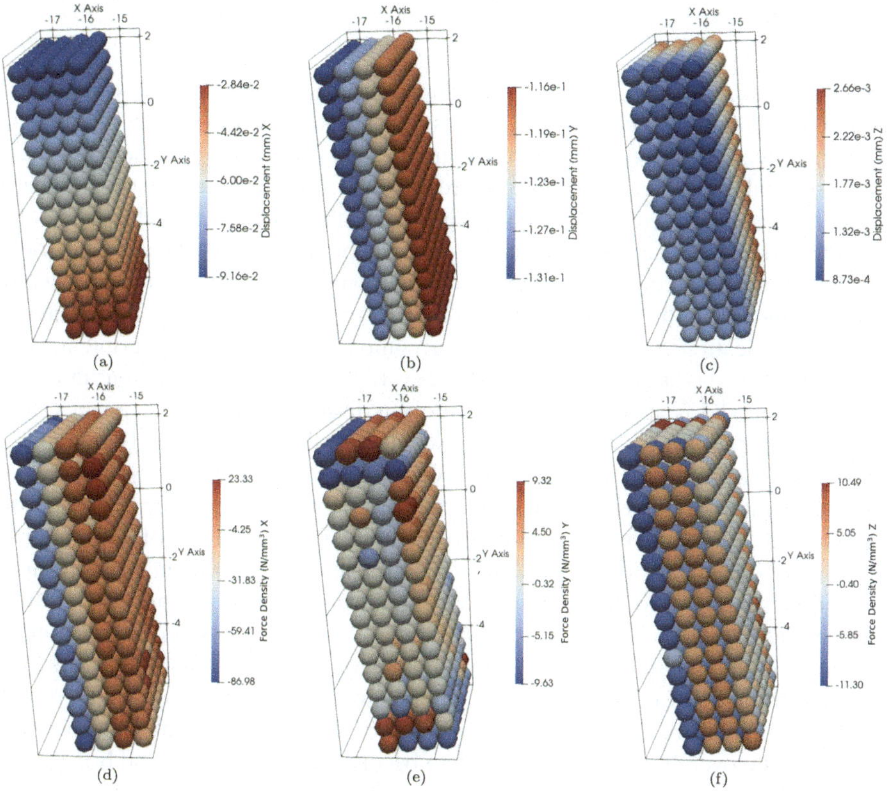

Fig. 11 Nonlocal boundary conditions corresponding to the local Neumann boundary condition. (**a**) Displacement in x. (**b**) Displacement in y. (**c**) Displacement in z. (**d**) Force density in x. (**e**) Force density in y. (**f**) Force density in z

Results in Fig. 11 show the displacements and force densities for the nonlocal model directly adjacent to the traction boundary condition that is applied to the local model. The displacement field is again approximately linear, which is consistent with the expected behavior. The force density field is much more complex, however, which is a reflection of the nonlocal interactions between material points in the peridynamic model. These interactions are dictated by the nonlocal material model, the geometry (e.g., free surfaces), the discretization, and the value of the horizon. A goal of the current study is to illustrate the complexity of determining prescribed force density volume constraints for a purely nonlocal model that approximate the effect of traction boundary conditions in a comparable local model and to provide a systematic approach for doing so via LtN coupling.

6 Summary

In this chapter, we presented an optimization-based approach for local-to-nonlocal coupling and demonstrated its effectiveness on a series of computational examples. The objective function for the optimization problem is the difference between local and nonlocal solutions in an overlap region, the controls are virtual boundary conditions on the local and nonlocal models at the coupling interface, and the models themselves act as constraints on the optimization problem. Numerical examples on statics problems, carried out by coupling the *Peridigm* and *Albany* codes, demonstrate the accuracy of the coupling method, its convergence behavior, and its application to the geometry of a compact tension test experiment. The coupling method is shown to be a viable means for carrying out simulations in which a nonlocal model is applied in the vicinity of a discontinuity, such as a crack, and a less computationally expensive local model is applied elsewhere. The coupling approach is also shown to enable the conversion of local boundary conditions, including tractions, to prescribed displacements or force densities for the definition of nonlocal volume constraints. Future work includes application of the coupling scheme for quasi-static problems that include stable crack growth and an investigation of its use for approximating nonlocal volume constraints in transient dynamics simulations. Another natural extension of this work includes its application to nonlinear problems; in this setting, the optimization-based formulation is the same, and the proposed framework can be applied straightforwardly as long as robust nonlinear solvers are provided for the solution of the local and nonlocal problems. However, the overall optimization problem becomes more complex and brings new computational challenges that might require the adoption of more efficient and robust optimization algorithms.

7 Declarations

This article has been authored by an employee of National Technology & Engineering Solutions of Sandia, LLC under Contract No. DE-NA0003525 with the U.S. Department of Energy (DOE). The employee owns all right, title, and interest in and to the article and is solely responsible for its contents. The United States Government retains, and the publisher, by accepting the article for publication, acknowledges that the United States Government retains a non-exclusive, paid-up, irrevocable, world-wide license to publish or reproduce the published form of this article or allow others to do so, for United States Government purposes. The DOE will provide public access to these results of federally sponsored research in accordance with the DOE Public Access Plan https://www.energy.gov/downloads/doe-public-access-plan.

Funding M. D'Elia, P. Bochev, and M. Perego were partially supported by the U.S. Department of Energy, Office of Advanced Scientific Computing Research under the Collaboratory on Mathematics and Physics-Informed Learning Machines for Multiscale and Multiphysics Problems (PhILMs) project. M. D'Elia and J. Trageser were also supported by the Sandia National Laboratories Laboratory-directed Research and Development (LDRD) program, project 218318.

Conflicts of Interest The authors have no conflicts of interest to disclose.

Availability of Data and Material All the experiments can be reproduced by using the geometries, parameters, and input data indicated in the text.

Code Availability The code used in the work is open source. However, the specific optimization-based code has not been disclosed. Interested users may contact the corresponding author.

Authors' Contributions Marta D'Elia provided the main idea and prototyping and wrote several parts of the paper. David Littlewood provided the building blocks of the coupling code and performed preliminary tests; he also contributed to writing the numerical section. Jeremy Trageser conducted all the experiments in the paper and contributed to writing parts of the numerical section. Mauro Perego provided support for the Albany code, for its coupling with Peridigm, and for the implementation of the optimization algorithm. Pavel Bochev provided the main idea and its mathematical foundations.

Ethics Approval No concerns.

Consent to Participate Not applicable.

Consent for Publication Not applicable.

References

1. A. A. Buades, B. Coll, and J.M. Morel. Image denoising methods. A new nonlocal principle. *SIAM Review*, 52:113–147, 2010.
2. B. Alali and M. Gunzburger. Peridynamics and material interfaces. *Journal of Elasticity*, 120(2):225–248, 2015.
3. The Albany Project Team. *The Albany Project Website*, https://github.com/gahansen/Albany.
4. E. Aulisa, G. Capodaglio, A. Chierici, and M. D'Elia. Efficient quadrature rules for finite element discretizations of nonlocal equations. *Num. Meth. PDE*, 2021, Accepted.
5. D.A. Benson, S.W. Wheatcraft, and M.M. Meerschaert. Application of a fractional advection-dispersion equation. *Water Resources Research*, 36(6):1403–1412, 2000.
6. N. Burch, M. D'Elia, and R. Lehoucq. The exit-time problem for a Markov jump process. *The European Physical Journal Special Topics*, 223:3257–3271, 2014.
7. O. Burkovska and M. Gunzburger. On a nonlocal Cahn-Hilliard model permitting sharp interfaces. *Mathematical Models and Methods in Applied Sciences*, 2021. in print.
8. G. Capodaglio, M. D'Elia, P. Bochev, and M. Gunzburger. An energy-based coupling approach to nonlocal interface problems. *Computers and Fluids*, 207:104593, 2019.

9. G. Capodaglio, M. D'Elia, M. Gunzburger, P. Bochev, M. Klar, and C. Vollmann. A general framework for substructuring-based domain decomposition methods for models having nonlocal interactions. *Num. Meth. PDE*, 2021, Accepted.
10. Xi Chen and M. Gunzburger. Continuous and discontinuous finite element methods for a peridynamics model of mechanics. *Computer Methods in Applied Mechanics and Engineering*, 200(9-12):1237–1250, 2011.
11. Patricio Clark Di Leoni, Tamer A. Zaki, George Karniadakis, and Charles Meneveau. Two-point stress–strain-rate correlation structure and non-local eddy viscosity in turbulent flows. *Journal of Fluid Mechanics*, 914:A6, 2021.
12. O. Defterli, M D'Elia, Q. Du, M. Gunzburger, R. Lehoucq, and M. M. Meerschaert. Fractional diffusion on bounded domains. *Fractional Calculus and Applied Analysis*, 18(2):342–360, 2015.
13. A.H. Delgoshaie, D.W. Meyer, P. Jenny, and H. Tchelepi. Non-local formulation for multiscale flow in porous media. *Journal of Hydrology*, 531(1):649–654, 2015.
14. M. D'Elia and P. Bochev. Optimization-based coupling of nonlocal and local diffusion models. In R. Lipton, editor, *Proceedings of the Fall 2014 Materials Research Society Meeting*, MRS Symposium Proceedings, Boston, MA, 2014. Cambridge University Press.
15. M. D'Elia and P. Bochev. Formulation, analysis and computation of an optimization-based local-to-nonlocal coupling method. *Results in Applied Mathematics*, 9:100129, 2021.
16. M. D'Elia, P. Bochev, D. Littlewood, and M. Perego. Optimization-based coupling of local and nonlocal models: Applications to peridynamics. *Chapter in Handbook of nonlocal continuum mechanics for materials and structures*, 2017.
17. M. D'Elia, Q. Du, C. Glusa, M. Gunzburger, Xiaochuan Tian, and Zhi Zhou. Numerical methods for nonlocal and fractional models. *Acta Numerica*, 29:1–124, 2020.
18. M. D'Elia, Q. Du, M. Gunzburger, and R. Lehoucq. Nonlocal convection-diffusion problems on bounded domains and finite-range jump processes. *Computational Methods in Applied Mathematics*, 29:71–103, 2017.
19. M. D'Elia, C. Flores, X. Li, P. Radu, and Y. Yu. Helmholtz-Hodge decompositions in the nonlocal framework: well-posedness analysis and applications. *Journal of Peridynamics and Nonlocal Modeling*, 2(4):401–418, 2020.
20. M. D'Elia and M. Gulian. Analysis of anisotropic nonlocal diffusion models: Well-posedness of fractional problems for anomalous transport. *arXiv preprint arXiv:2101.04289*, 2021.
21. M. D'Elia, M. Gulian, H. Olson, and G. E. Karniadakis. Towards a unified theory of fractional and nonlocal vector calculus. *Fractional Calculus and Applied Analysis*, 24(5):1301–1355, 2021.
22. M. D'Elia, M. Gunzburger, and C. Vollman. A cookbook for finite element methods for nonlocal problems, including quadrature rule choices and the use of approximate neighborhoods. *M3AS*, 2020. in print.
23. M. D'Elia, Xingjie Li, Pablo Seleson, Xiaochuan Tian, and Yue Yu. A review of local-to-nonlocal coupling methods in nonlocal diffusion and nonlocal mechanics. *arXiv preprint arXiv:1912.06668*, 2019. to appear in the Journal of Peridynamics and Nonlocal Modeling.
24. M. D'Elia, J.C. De los Reyes, and A. Trujillo. Bilevel parameter optimization for nonlocal image denoising model. *Journal of Mathematical Imaging and Vision*, 2021. in print.
25. M. D'Elia, M. Perego, P. Bochev, and D.J. Littlewood. A coupling strategy for nonlocal and local diffusion models with mixed volume constraints and boundary conditions. *Computers & Mathematics with Applications*, 71(11):2218–2230, 2016.
26. M. D'Elia and Y. Yu. On the prescription of boundary conditions for nonlocal Poisson's and peridynamics models. *arXiv preprint arXiv:2107.04450*, 2021.
27. Marta D'Elia, Mamikon Gulian, Tadele Mengesha, and James M Scott. Connections between nonlocal operators: from vector calculus identities to a fractional Helmholtz decomposition. *arXiv preprint arXiv:2112.05317*, 2021.
28. Marta D'Elia, Xiaochuan Tian, and Yue Yu. A physically consistent, flexible, and efficient strategy to convert local boundary conditions into nonlocal volume constraints. *SIAM Journal on Scientific Computing*, 42(4):A1935–A1949, 2020.

29. Q. Du, X.H. Li, J. Lu, and X. Tian. A quasi-nonlocal coupling method for nonlocal and local diffusion models. *SIAM Journal on Numerical Analysis*, 56(3):1386–1404, 2018.
30. P. Fife. *Some nonclassical trends in parabolic and parabolic-like evolutions*, chapter Vehicular Ad Hoc Networks, pages 153–191. Springer-Verlag, New York, 2003.
31. Mikil Foss, Petronela Radu, and Yue Yu. Convergence analysis and numerical studies for linearly elastic peridynamics with Dirichlet-type boundary conditions. *preprint*, 2021.
32. U. Galvanetto, T. Mudric, A. Shojaei, and M. Zaccariotto. An effective way to couple FEM meshes and peridynamics grids for the solution of static equilibrium problems. *Mechanics Research Communications*, 76:41–47, 2016.
33. G. Gilboa and S. Osher. Nonlocal linear image regularization and supervised segmentation. *Multiscale Model. Simul.*, 6:595–630, 2007.
34. Youn Doh Ha and Florin Bobaru. Characteristics of dynamic brittle fracture captured with peridynamics. *Engineering Fracture Mechanics*, 78(6):1156–1168, 2011.
35. F. Han, G. Lubineau, Y. Azdoud, and A. Askari. A morphing approach to couple state-based peridynamics with classical continuum mechanics. *Computer methods in applied mechanics and engineering*, 301:336–358, 2016.
36. Fei Han and Gilles Lubineau. Coupling of nonlocal and local continuum models by the Arlequin approach. *International Journal for Numerical Methods in Engineering*, 89(6):671–685, 2012.
37. X.H. Li and J. Lu. Quasinonlocal coupling of nonlocal diffusions. *SIAM Journal on Numerical Analysis*, 55(5):2394–2415, 2017.
38. D.J. Littlewood. Roadmap for peridynamic software implementation. Technical Report SAND2015-9013, Sandia National Laboratories, Albuquerque, NM and Livermore, CA, 2015.
39. G. Lubineau, Y. Azdoud, F. Han, C. Rey, and A. Askari. A morphing strategy to couple non-local to local continuum mechanics. *Journal of the Mechanics and Physics of Solids*, 60(6):1088–1102, 2012.
40. M.M. Meerschaert and A. Sikorskii. *Stochastic models for fractional calculus*. Studies in Mathematics, Gruyter, 2012.
41. R. Metzler and J. Klafter. The random walk's guide to anomalous diffusion: a fractional dynamics approach. *Physics Reports*, 339(1):1–77, 2000.
42. G. Pang, M. D'Elia, M. Parks, and G. E. Karniadakis. nPINNs: nonlocal Physics-Informed Neural Networks for a parametrized nonlocal universal Laplacian operator. Algorithms and Applications. *Journal of Computational Physics*, 422:109760, 2020.
43. G. Pang, L. Lu, and G. E. Karniadakis. fPINNs: Fractional physics-informed neural networks. *SIAM Journal on Scientific Computing*, 41:A2603–A2626, 2019.
44. M.L. Parks, D.J. Littlewood, J.A. Mitchell, and S.A. Silling. Peridigm Users' Guide v1.0.0. Technical Report SAND2012-7800, Sandia National Laboratories, Albuquerque, NM and Livermore, CA, 2012.
45. M. Pasetto. *Enhanced Meshfree Methods for Numerical Solution of Local and Nonlocal Theories of Solid Mechanics*. PhD thesis, UC San Diego, 2019.
46. The Peridigm Project Team. *The Peridigm Project Website, github.com/peridigm*.
47. A.G. Salinger, R.A. Bartlett, Q. Chen, X. Gao, G.A. Hansen, I. Kalashnikova, A. Mota, R.P. Muller, E. Nielsen, J.T. Ostien, R.P. Pawlowski, E.T. Phipps, and W. Sun. Albany: A component–based partial differential equation code built on Trilinos. Technical Report SAND2013-8430J, Sandia National Laboratories, Albuquerque, NM and Livermore, CA, 2013.
48. R. Schumer, D.A. Benson, M.M. Meerschaert, and B. Baeumer. Multiscaling fractional advection-dispersion equations and their solutions. *Water Resources Research*, 39(1):1022–1032, 2003.
49. R. Schumer, D.A. Benson, M.M. Meerschaert, and S.W. Wheatcraft. Eulerian derivation of the fractional advection-dispersion equation. *Journal of Contaminant Hydrology*, 48:69–88, 2001.
50. P. Seleson, S. Beneddine, and S. Prudhomme. A force-based coupling scheme for peridynamics and classical elasticity. *Computational Materials Science*, 66:34–49, 2013.

51. P. Seleson, Y.-D. Ha, and S. Beneddine. Concurrent coupling of bond-based peridynamics and the Navier equation of classical elasticity by blending. *International Journal for Multiscale Computational Engineering*, 13(2):91–113, 2015.
52. P. Seleson and D. J. Littlewood. Numerical tools for effective meshfree discretizations of peridynamic models. *In George Z. Voyiadjis, editor, Handbook of Nonlocal Continuum Mechanics for Materials and Structures*, 2019.
53. P. Seleson and D.J. Littlewood. Convergence studies in meshfree peridynamic simulations. *Computers & Mathematics with Applications*, 71(11):2432–2448, 2016.
54. P. Seleson and M.L. Parks. On the role of the influence function in the peridynamic theory. *International Journal for Multiscale Computational Engineering*, 9(6):689–706, 2011.
55. Pablo Seleson, M. Gunzburger, and Michael L Parks. Interface problems in nonlocal diffusion and sharp transitions between local and nonlocal domains. *Computer Methods in Applied Mechanics and Engineering*, 266:185–204, 2013.
56. S. Silling, M. Epton, O. Weckner, J. Xu, and E. Askari. Peridynamic states and constitutive modeling. *Journal of Elasticity*, 88:151–184, 2007.
57. S. Silling, D. Littlewood, and P. Seleson. Variable horizon in a peridynamic medium. *Journal of Mechanics of Materials and Structures*, 10(5):591–612, 2015.
58. S. A Silling and E. Askari. A meshfree method based on the peridynamic model of solid mechanics. *Computers & structures*, 83(17-18):1526–1535, 2005.
59. S.A. Silling. Reformulation of elasticity theory for discontinuities and long-range forces. *Journal of the Mechanics and Physics of Solids*, 48:175–209, 2000.
60. S.A. Silling. Reformulation of elasticity theory for discontinuities and long-range forces. *Journal of the Mechanics and Physics of Solids*, 48(1):175–209, 2000.
61. S.A. Silling and E. Askari. A meshfree method based on the peridynamic model of solid mechanics. *Computers and Structures*, 83(17-18):1526–1535, 2005.
62. Y. Tao, X. Tian, and Q. Du. Nonlocal models with heterogeneous localization and their application to seamless local-nonlocal coupling. *Multiscale Modeling & Simulation*, 17(3):1052–1075, 2019.
63. Nathaniel Trask, Huaiqian You, Yue Yu, and Michael L Parks. An asymptotically compatible meshfree quadrature rule for nonlocal problems with applications to peridynamics. *Computer Methods in Applied Mechanics and Engineering*, 343:151–165, 2019.
64. The Trilinos Project Team. *The Trilinos Project Website, trilinos.github.io*.
65. H. Wang, K. Wang, and T. Sircar. A direct $\mathcal{O}(N \log^2 N)$ finite difference method for fractional diffusion equations. *Journal of Computational Physics*, 229(21):8095–8104, 2010.
66. X. Wang, S.S. Kulkarni, and A. Tabarraei. Concurrent coupling of peridynamics and classical elasticity for elastodynamic problems. *Computer Methods in Applied Mechanics and Engineering*, 344:251–275, 2019.
67. Xiao Xu, Christian Glusa, Marta D'Elia, and John T. Foster. A FETI approach to domain decomposition for meshfree discretizations of nonlocal problems. *Computer Methods in Applied Mechanics and Engineering*, 387:114148, 2021.
68. Huaiqian You, XinYang Lu, Nathaniel Task, and Yue Yu. An asymptotically compatible approach for Neumann-type boundary condition on nonlocal problems. *ESAIM: Mathematical Modelling and Numerical Analysis*, 54(4):1373–1413, 2020.
69. Y. Yu, F.F. Bargos, H. You, M.L. Parks, M.L. Bittencourt, and G.E. Karniadakis. A partitioned coupling framework for peridynamics and classical theory: Analysis and simulations. *Computer Methods in Applied Mechanics and Engineering*, 340:905–931, 2018.
70. Yue Yu, Huaiqian You, and Nathaniel Trask. An asymptotically compatible treatment of traction loading in linearly elastic peridynamic fracture. *Computer Methods in Applied Mechanics and Engineering*, 377:113691, 2021.

Nonlocal Diffusion Models with Consistent Local and Fractional Limits

Qiang Du, Xiaochuan Tian, and Zhi Zhou

Abstract For some spatially nonlocal diffusion models with a finite range of nonlocal interactions measured by a positive parameter δ, we review their formulation defined on a bounded domain subject to various conditions that correspond to some inhomogeneous data. We consider their consistency to similar inhomogeneous boundary value problems of classical partial differential equation (PDE) models as the nonlocal interaction kernel gets localized in the local $\delta \to 0$ limit, and at the same time, for rescaled fractional-type kernels, to corresponding inhomogeneous nonlocal boundary value problems of fractional equations in the global $\delta \to \infty$ limit. Such discussions help to delineate issues related to nonlocal problems defined on a bounded domain with inhomogeneous data.

Keywords Nonlocal models · Nonlocal diffusion · Peridynamics · Fractional PDEs · Well-posedness · Inhomogeneous data

AMS Subject Classifications 47G10, 46E35, 35R11, 35J25

The research of Q. Du was supported in part by NSF grant DMS-2012562 and DMS-1937254. The research of X. Tian was supported in part by NSF grant DMS-2111608. The research of Z. Zhou is partially supported by Hong Kong Research Grants Council (No. 15303122) and an internal grant of Hong Kong Polytechnic University (Project ID: P0031041, Programme: ZZKS)

Q. Du (✉)
Department of Applied Physics and Applied Mathematics, and Data Science Institute, Columbia University, New York, NY, USA
e-mail: qd2125@columbia.edu

X. Tian
Department of Mathematics, University of California, San Diego, CA, USA
e-mail: xctian@ucsd.edu

Z. Zhou
Department of Applied Mathematics, Hong Kong Polytechnic University, Hong Kong, China
e-mail: zhizhou@polyu.edu.hk

T. Mengesha, A. J. Salgado (eds.), *A^3N^2M: Approximation, Applications, and Analysis of Nonlocal, Nonlinear Models*, The IMA Volumes in Mathematics and its Applications 165, https://doi.org/10.1007/978-3-031-34089-5_5

1 Introduction

There has been much interest in nonlocal models [4, 10, 20, 28, 46]. A characterization of nonlocal models can be found in [29], together with a study on the close connections between nonlocal modeling and other mathematical subjects such as homogenization, model reduction, and coarse graining.

Motivated by studies from mechanics to image analysis and from traffic flows of autonomous vehicles to anomalous diffusion, a growing area of research is the study on models with nonlocal interactions that span a finite range [28, 30], measured by the horizon parameter $\delta \in \mathbb{R}$, as we will describe in detail later. It should be noted that in such models, the interaction range is generically nonzero (thus nonlocal) but, for problems on a bounded domain, may not be on the same scale as the size of the whole domain. In mathematical modeling, nonlocal models with a finite δ, simply referred to as nonlocal models here, can play a number of roles [28]:

- They can either complement or serve as an alternative to traditional PDE-based local models, and they can account for nonlocal interactions explicitly and remain valid for not only smooth but also singular solutions. Examples include peridynamic (PD) models for fracture mechanics [36, 65] and nonlocal models of anomalous diffusion and transport [4, 30, 56, 57].
- They can serve as tools to better understand and simulate existing local PDE models. Popular numerical methods such as smoothed particle hydrodynamics (SPH) and vortex blob methods are good examples of discretizing PDEs through nonlocal smoothing/averaging [18, 35, 37, 58].
- They can serve as bridges to build connections between continuum PDEs, discrete algebraic, graph and network models, metric geometry, and fractional differential equations [11, 15, 16, 27, 28, 74, 75].

The main goal here is related to the bridging roles mentioned above. That is, the discussion is centered on connecting local PDEs with fractional PDEs through nonlocal equations with a finite range of interactions via a unified description. In particular, we reexamine some concepts presented in [30] on nonlocal diffusion models with a finite δ and make further investigations so that the theory can be consistent with both the local and fractional limits. While the related issues have been subject to earlier investigations, see, e.g., [20, 30, 75], most known results are stated for problems with homogeneous boundary data (or nonlocal constraints). By presenting problems in more general settings that involve inhomogeneous data, we also provide some insight on how to connect some of the relevant concepts concerning nonlocal boundary conditions (volumetric constraints [30]).

2 Local, Nonlocal, and Fractional Models

For illustration, we consider problems defined on a bounded, open domain $\Omega \subset \mathbb{R}^d$ with a Lipschitz boundary $\partial\Omega$. We adopt a set of notations that is largely similar to those used in [20] but with necessary modifications in some cases.

Let the closure of Ω be denoted by $\overline{\Omega} = \Omega \cup \partial\Omega$. The complements are denoted by $\Omega_c = \mathbb{R}^d \setminus \overline{\Omega}$ and $\overline{\Omega}_c = \mathbb{R}^d \setminus \Omega$, respectively. Note that $\overline{\Omega}_c = \overline{\mathbb{R}^d \setminus \Omega} = \overline{\Omega_c}$.

Given the domain Ω and a positive parameter $\delta > 0$, we define an *interaction layer of width* δ by

$$\Omega_{\mathcal{I}_\delta} := \{ \boldsymbol{y} \in \Omega_c \mid \operatorname{dist}(\boldsymbol{y}, \Omega) < \delta \}. \tag{2.1}$$

We let $\overline{\Omega_{\mathcal{I}_\delta}}$ denote the closure of $\Omega_{\mathcal{I}_\delta}$. $\widehat{\Omega}_\delta = \Omega \cup \Omega_{\mathcal{I}_\delta}$ and $\overline{\Omega}_\delta = \overline{\Omega \cup \Omega_{\mathcal{I}_\delta}} = \overline{\Omega} \cup \overline{\Omega_{\mathcal{I}_\delta}}$. In this chapter, we are interested in some special cases and limiting regimes

$$\delta = 0, \quad 0 \leftarrow \delta \ll \operatorname{diam}\Omega, \quad \delta \approx \operatorname{diam}\Omega, \quad \operatorname{diam}\Omega \ll \delta \to \infty, \quad \delta = \infty. \tag{2.2}$$

Note that $\Omega_{\mathcal{I}_0} = \emptyset$ and we let $\overline{\Omega_{\mathcal{I}_0}} = \partial\Omega$, while $\Omega_{\mathcal{I}_\infty} = \Omega_c$ and $\overline{\Omega_{\mathcal{I}_\infty}} = \overline{\Omega}_c$.

We also define some domains for any pair of points $(\boldsymbol{x}, \boldsymbol{y})$. These include $\Omega^2 = \Omega \times \Omega$, which denotes the tensor product of any domain Ω with itself and

$$\widetilde{\Omega}_\delta = \widehat{\Omega}_\delta^2 \setminus \Omega_{\mathcal{I}_\delta}^2 \quad \text{and} \quad \widetilde{\Omega}_\infty = (\mathbb{R}^d)^2 \setminus \Omega_c^2.$$

2.1 Local Diffusion Models

We first consider a classical Poisson problem for an unknown solution $u \colon \Omega \subset \mathbb{R}^d \to \mathbb{R}$,

$$\begin{cases} -\mathcal{L}_0\, u(\boldsymbol{x}) := -(\Delta u)(\boldsymbol{x}) = F(\boldsymbol{x}, u(\boldsymbol{x})), & \forall \boldsymbol{x} \in \Omega, \\ \mathcal{B}_0\, u(\boldsymbol{x}) = g_0(\boldsymbol{x}), & \forall \boldsymbol{x} \in \overline{\Omega_{\mathcal{I}_0}} = \partial\Omega, \end{cases} \tag{2.3}$$

where $F(\boldsymbol{x}, u)$ and $g_0(\boldsymbol{x})$ denote given functions defined on $\Omega \times \mathbb{R}$ and $\partial\Omega$, respectively. Here, we elect to work with the Laplacian Δ rather than more general diffusion operators. A special case is that $F(\boldsymbol{x}, u)$ is independent of u, thus leading to a standard linear PDE on Ω. For the boundary conditions operator $\mathcal{B}_0$, we have the choices

$$\mathcal{B}_0\, u(\boldsymbol{x}) = \begin{cases} u(\boldsymbol{x}), & \boldsymbol{x} \in \partial\Omega^D \text{ (Dirichlet BC)}, \\ \dfrac{\partial u}{\partial \mathbf{n}}(\boldsymbol{x}) & \boldsymbol{x} \in \partial\Omega^N \text{ (Neumann BC)}, \end{cases} \tag{2.4}$$

where $\partial\Omega^D$ and $\partial\Omega^N$ are subset of $\partial\Omega$ such that $\partial\Omega^D \cup \partial\Omega^N = \partial\Omega$ with $\partial\Omega^D \cap \partial\Omega^N$ having zero $d-1$-dimensional measure. Note that $\mathcal{B}_0$ can also take on other forms (e.g.,

$$\mathcal{B}_0\, u(\boldsymbol{x}) = \frac{\partial u}{\partial \mathbf{n}}(\boldsymbol{x}) - \alpha u(\boldsymbol{x}), \quad \forall \boldsymbol{x} \in \partial\Omega^N \tag{2.5}$$

for a positive constant $\alpha > 0$) to yield other (e.g., Robin type) boundary conditions. In the case $\partial\Omega^D = \partial\Omega$, we have a pure Dirichlet problem, while $\partial\Omega^N = \partial\Omega$ leads to a pure Neumann problem. In the latter case, compatibility conditions may be needed on the data, and an additional constraint (e.g., having a zero mean over Ω) on the solutions may be imposed to ensure uniqueness.

Remark 2.1 Problem (2.3), which is named the local diffusion problem here, is a well-studied local PDE model used for steady-state normal diffusion.

2.2 *Fractional Diffusion Models*

Fractional PDEs are PDEs with fractional-order partial derivatives of unknown solutions. They are nonlocal integral models with a global or infinite range of interactions. Here, we consider a nonlocal fractional diffusion model associated with the *integral fractional Laplacian* $\mathcal{L}^s_\infty = -(-\Delta)^s$, which, for $u\colon \mathbb{R}^d \to \mathbb{R}$ and $s \in (0, 1)$, is given by

$$-\mathcal{L}^s_\infty\, u(\boldsymbol{x}) = (-\Delta)^s\, u(\boldsymbol{x}) = \int_{\mathbb{R}^d} (u(\boldsymbol{x}) - u(\boldsymbol{y}))\gamma_s(|\boldsymbol{y}-\boldsymbol{x}|)d\boldsymbol{y}, \qquad \forall \boldsymbol{x} \in \mathbb{R}^d, \tag{2.6}$$

where the kernel function is defined by

$$\gamma_s(|\boldsymbol{z}|) = \frac{C_{d,s}}{|\boldsymbol{z}|^{d+2s}}, \quad \text{with} \quad C_{d,s} = \frac{2^{2s} s\Gamma(s+d/2)}{\pi^{d/2}\Gamma(1-s)}, \tag{2.7}$$

where Γ denotes the gamma function. Note that the singular integral in (2.6) should be interpreted in the principal value sense.

The *fractional diffusion problem* considered here is then given by

$$\begin{cases} -\mathcal{L}^s_\infty\, u(\boldsymbol{x}) = F(\boldsymbol{x}, u(\boldsymbol{x})), & \forall \boldsymbol{x} \in \Omega, \\ \mathcal{B}^s_\infty\, u(\boldsymbol{x}) = g_\infty(\boldsymbol{x}), & \forall \boldsymbol{x} \in \Omega_{\mathcal{I}_\infty} = \Omega_c, \end{cases} \tag{2.8}$$

where $F(\boldsymbol{x}, u)$ is as before and $g_\infty(\boldsymbol{x})$ is a given data defined for $\boldsymbol{x} \in \Omega_{\mathcal{I}_\infty}$. The operator $\mathcal{B}^s_\infty$ is defined by

$$\mathcal{B}_\infty^s u(\boldsymbol{x}) = \begin{cases} u(\boldsymbol{x}), & \boldsymbol{x} \in \Omega_{\mathcal{I}_\infty}^D \text{ (Dirichlet VC)}, \\ \mathcal{N}_\infty^s u(\boldsymbol{x}), & \boldsymbol{x} \in \Omega_{\mathcal{I}_\infty}^N \text{ (Neumann VC)}, \end{cases} \tag{2.9}$$

where $\Omega_{\mathcal{I}_\infty}^D$ and $\Omega_{\mathcal{I}_\infty}^N$ are two disjoint subdomains of $\Omega_{\mathcal{I}_\infty}$ such that $\overline{\Omega_{\mathcal{I}_\infty}^D \cup \Omega_{\mathcal{I}_\infty}^N} = \overline{\Omega}_c$. $\mathcal{N}_\infty^s$ is a suitably defined nonlocal operator, for example, given by Dipierro et al. [26],

$$\mathcal{N}_\infty^s u(\boldsymbol{x}) = -\int_\Omega (u(\boldsymbol{y}) - u(\boldsymbol{x}))\gamma_s(|\boldsymbol{y} - \boldsymbol{x}|)d\boldsymbol{y}, \qquad \forall \boldsymbol{x} \in \Omega_{\mathcal{I}_\infty} = \Omega_c. \tag{2.10}$$

In the above definition, $\mathcal{N}_\infty^s$ is defined as an integral over the domain Ω only, instead of the whole space $\mathbb{R}^d$.

Remark 2.2 Note that the constraint $\mathcal{B}_\infty^s u = g_\infty$ is applied on a domain having a nonzero volume in $\mathbb{R}^d$, thus representing a nonlocal boundary condition that is also called a *volume constraint* (VC) [30].

The integral fractional Laplacian is sometimes called the whole space fractional Laplacian as the integral ranges the whole space $\mathbb{R}^d$. It is a special case of the nonlocal Laplacian defined later (see (2.19) and related discussions in Sect. 2.3). It is also equivalent to the Fourier representation [76]

$$(-\Delta)^s u(\boldsymbol{x}) = \frac{1}{(2\pi)^d} \int_{\mathbb{R}^d} |\boldsymbol{\xi}|^{2s} (u, \mathrm{e}^{-\mathrm{i}\boldsymbol{\xi}\cdot\boldsymbol{x}})\, \mathrm{e}^{\mathrm{i}\boldsymbol{\xi}\cdot\boldsymbol{x}} d\boldsymbol{\xi} = \mathcal{F}^{-1}(|\boldsymbol{\xi}|^{2s} \mathcal{F}\{u\}(\boldsymbol{\xi}))(\boldsymbol{x}), \tag{2.11}$$

where $\mathcal{F}$ denotes the Fourier transform. Formally, if we let $s \to 1^-$ in the above, it recovers the Fourier spectral representation of the classical local Laplacian $-\Delta$. Indeed, it is well known that $(-\Delta)^s u(x) \to -\Delta u(x)$ as $s \to 1^-$ for any function u with sufficient regularity; see, e.g., [68, Theorems 3 and 4] and [25, Proposition 4.4].

Remark 2.3 The fractional Laplacian (2.6) is often used to model superdiffusion. In probabilistic terms, in contrast to the local Laplacian that corresponds to the normal diffusion described by Brownian motion, the fractional Laplacian can be connected to superdiffusion associated with Lévy flights in which the length of particle jumps follows a heavy-tailed power law distribution, reflecting the long-range interactions between particles, see e.g., [26, 56, 66].

The problem (2.8) is called a fractional diffusion model here to distinguish it from other diffusion models. It is given in the distribution sense so that the equations are only specified over an open domain Ω. More rigorous descriptions of (2.8) will be discussed later.

Other variants of the operators $\mathcal{L}_\infty^s$ and $\mathcal{N}_\infty^s$ are discussed in Sect. 2.4. For notation simplicity, we drop the superscript s in $\mathcal{L}_\infty^s$, $\mathcal{B}_\infty^s$, and $\mathcal{N}_\infty^s$ whenever there is no ambiguity, that is, $\mathcal{L}_\infty^s = \mathcal{L}_\infty$, $\mathcal{B}_\infty^s = \mathcal{B}_\infty$, and $\mathcal{N}_\infty^s = \mathcal{N}_\infty$. Meanwhile, similar to the local case, one can also consider other volume constraints such as $\mathcal{B}_\infty u(\boldsymbol{x}) = \mathcal{N}_\infty u(\boldsymbol{x}) - \alpha u(\boldsymbol{x})$ for $\boldsymbol{x} \in \Omega_c$ to mimic a Robin-type constraint.

2.3 *Nonlocal Diffusion Models*

To distinguish from the local and fractional counterparts, we consider a class of special nonlocal diffusion models here associated with a finite range of interactions, measured by the so-called horizon parameter (interaction radius) $\delta > 0$.

For $\delta > 0$, we consider the following nonlocal problem for a scalar-valued unknown function $u(\boldsymbol{x})$ defined on $\widehat{\Omega}_\delta$,

$$\begin{cases} -\mathcal{L}_\delta u = F(\boldsymbol{x}, u(\boldsymbol{x})), & \forall \boldsymbol{x} \in \Omega, \\ \mathcal{B}_\delta u = g_\delta(\boldsymbol{x}), & \forall \boldsymbol{x} \in \Omega_{\mathcal{I}_\delta}, \end{cases} \tag{2.12}$$

with F and g_δ given and a *nonlocal diffusion operator* defined by

$$\mathcal{L}_\delta u(\boldsymbol{x}) := \int_{\mathbb{R}^d} (u(\boldsymbol{y}) - u(\boldsymbol{x}))\gamma_\delta(\boldsymbol{x}, \boldsymbol{y}) d\boldsymbol{y}, \qquad \forall \boldsymbol{x} \in \mathbb{R}^d, \tag{2.13}$$

where the kernel function $\gamma_\delta(\boldsymbol{x}, \boldsymbol{y})$ is symmetric, i.e.,

$$\gamma_\delta(\boldsymbol{x}, \boldsymbol{y}) = \gamma_\delta(\boldsymbol{y}, \boldsymbol{x}), \quad \forall \boldsymbol{x}, \boldsymbol{y} \in \mathbb{R}^d,$$

non-negative, and with a compact support in $\boldsymbol{y} - \boldsymbol{x}$ in $B_\delta(\mathbf{0})$, the δ-ball of the origin, that is,

$$\gamma_\delta(\boldsymbol{x}, \boldsymbol{y}) = 0 \text{ whenever } |\boldsymbol{y} - \boldsymbol{x}| > \delta.$$

Again, the integral in (2.13) (and the one in (2.19), which will be introduced shortly) should be understood in the sense of principal value throughout the paper. Moreover, due to the compact support of γ_δ, for $\boldsymbol{x} \in \Omega$, we can rewrite (2.13) as

$$\mathcal{L}_\delta u(\boldsymbol{x}) := \int_{\widehat{\Omega}_\delta} (u(\boldsymbol{y}) - u(\boldsymbol{x}))\gamma_\delta(\boldsymbol{x}, \boldsymbol{y}) d\boldsymbol{y}, \qquad \forall \boldsymbol{x} \in \Omega. \tag{2.14}$$

Analogous to (2.4), we have the choice of volume constraints

$$\mathcal{B}_\delta u(\boldsymbol{x}) = \begin{cases} u(\boldsymbol{x}) & \boldsymbol{x} \in \Omega^D_{\mathcal{I}_\delta} \text{ (Dirichlet VC)}, \\ \mathcal{N}_\delta u(\boldsymbol{x}), & \boldsymbol{x} \in \Omega^N_{\mathcal{I}_\delta} \text{ (Neumann VC)}, \end{cases} \tag{2.15}$$

where $\Omega^D_{\mathcal{I}_\delta}$ and $\Omega^N_{\mathcal{I}_\delta}$ are two disjoint subdomains of $\Omega_{\mathcal{I}_\delta}$ such that $\overline{\Omega^D_{\mathcal{I}_\delta} \cup \Omega^N_{\mathcal{I}_\delta}} = \overline{\Omega_{\mathcal{I}_\delta}}$. The operator $\mathcal{N}_\delta$ can be defined, for example, analogously to (2.10) by

$$\mathcal{N}_\delta\, u(\boldsymbol{x}) := -\int_\Omega (u(\boldsymbol{y}) - u(\boldsymbol{x}))\gamma_\delta(\boldsymbol{x}, \boldsymbol{y})d\boldsymbol{y}, \quad \forall \boldsymbol{x} \in \Omega_{\mathcal{I}_\delta}. \tag{2.16}$$

Note that generally $\mathcal{N}_\delta\, u$ only needs to be defined and applied on $\Omega^N_{\mathcal{I}_\delta}$. Its definition on $\Omega_{\mathcal{I}_\delta}$ is given for the convenience of notation only.

Rather than considering kernels in the general form, we focus on those translation-invariant (i.e., functions of $\boldsymbol{y} - \boldsymbol{x}$) and radial symmetric (i.e., depending only on $|\boldsymbol{y} - \boldsymbol{x}|$). In particular, we consider a more specific rescaled form given by

$$\gamma_\delta(\boldsymbol{x}, \boldsymbol{y}) = C_d(\delta)\gamma\left(\frac{|\boldsymbol{y} - \boldsymbol{x}|}{\delta}\right) \tag{2.17}$$

for a kernel $\gamma(|\cdot|)$ that satisfies

$$\begin{cases} \gamma(s) \geq 0, \text{ and nonincreasing}, \forall s \in [0, 1), \quad \gamma(s) = 0, \forall s \geq 1, \\ \widehat{\gamma}(|z|) = |z|^2\gamma(|z|) \in L^1_{\text{loc}}(\mathbb{R}^d). \end{cases} \tag{2.18}$$

For such kernels, we see that $\mathcal{L}_\delta$ can also be written as

$$\mathcal{L}_\delta\, u(\boldsymbol{x}) := C_d(\delta)\int_{\mathbb{R}^d} (u(\boldsymbol{y}) - u(\boldsymbol{x}))\gamma\left(\frac{|\boldsymbol{y} - \boldsymbol{x}|}{\delta}\right)d\boldsymbol{y}, \qquad \forall \boldsymbol{x} \in \mathbb{R}^d. \tag{2.19}$$

The scaling factor $C_d(\delta)$ is of special interest in this chapter and is defined differently depending on the regimes that we study. We give two examples below:

- For $0 < \delta \ll \text{diam}\,(\Omega)$, $C_d(\delta)$ is chosen to satisfy

$$\int_{\mathbb{R}^d} |\boldsymbol{y} - \boldsymbol{x}|^2\gamma_\delta(\boldsymbol{x}, \boldsymbol{y})d\boldsymbol{y} = \delta^{d+2}C_d(\delta)\int_{B_1(\mathbf{0})} |\boldsymbol{\xi}|^2\gamma(|\boldsymbol{\xi}|)d\boldsymbol{\xi} = 2d, \tag{2.20}$$

  so that as $\delta \to 0$, the local limit of $\mathcal{L}_\delta$ becomes the standard Laplacian $\mathcal{L}_0 = \Delta$ [28, 30, 54, 55].
- For $\delta \gg \text{diam}\,(\Omega)$, the factor and kernels are chosen to satisfy

$$C_d(\delta) = \frac{C_{d,s}}{\delta^{d+2s}} \quad \text{for} \quad \gamma(t) = t^{-d-2s}\mathbb{1}_{(0,1)}, \tag{2.21}$$

  so that $C_d(\delta)\gamma(t/\delta) = C_{d,s}t^{-d-2s}\mathbb{1}_{(0,\delta)}$, that is, $\gamma_\delta(\boldsymbol{x}, \boldsymbol{y})$ is a truncation of the fractional kernel γ_s defined in (2.7)

$$\gamma_\delta(\boldsymbol{x},\boldsymbol{y}) = \begin{cases} \gamma_s(|\boldsymbol{y}-\boldsymbol{x}|) = \dfrac{C_{d,s}}{|\boldsymbol{y}-\boldsymbol{x}|^{d+2s}}, & \forall\, \boldsymbol{y}-\boldsymbol{x} \in B_\delta(\mathbf{0}), \\ 0, & \text{otherwise}. \end{cases} \tag{2.22}$$

Consequently, we have $\mathcal{L}_\delta \to \mathcal{L}^s_\infty = -(-\Delta)^s$ as $\delta \to \infty$ [12, 22, 75].

Remark 2.4 Although $\mathcal{L}_\delta\, u(\boldsymbol{x})$ is defined for any $\boldsymbol{x} \in \mathbb{R}^d$, independent of any spatial domain, for problem (2.12), we only need $\mathcal{L}_\delta\, u(\boldsymbol{x})$ for $\boldsymbol{x} \in \Omega$. Then, with a compactly supported γ_δ, the integral in (2.19) can be equivalently defined over $\widehat{\Omega}_\delta$ instead, i.e.,

$$\mathcal{L}_\delta\, u(\boldsymbol{x}) := \int_{\widehat{\Omega}_\delta} (u(\boldsymbol{y}) - u(\boldsymbol{x}))\gamma_\delta\,(\boldsymbol{x},\boldsymbol{y})d\boldsymbol{y}, \qquad \forall \boldsymbol{x} \in \Omega. \tag{2.23}$$

Remark 2.5 The operator $\mathcal{L}_\delta$ defined by (2.19), with a kernel scaled by (2.20) or (2.22), is a nonlocal analog of the partial differential operator $\mathcal{L}_0 = \Delta$ or the fractional operator $\mathcal{L}^s_\infty = -(-\Delta)^s$. Likewise, the nonlocal model (2.12) is a nonlocal analog of the local diffusion model (2.3) and the fractional diffusion model (2.8). Nonlocal diffusion problems have also been extensively studied in recent years, and we refer to the literature review given in [30] and [20]. In addition, one can find additional references in a few more recent works [23, 45, 60, 78].

Remark 2.6 We note that, different from [20, 31], the operators $\mathcal{L}_\delta$ and $\mathcal{N}_\delta$, involving the respective domains of integration in their definitions, are adopted as in [28, 33] to be consistent with the fractional version, see, e.g., [26, 43], given by (2.10). We also consider their variants, as well as the variants of the fractional operators later in the paper.

Remark 2.7 Concerning the operator $\mathcal{B}_\delta$ in (2.15), one may consider more general forms. For example, with $\Omega_{\mathcal{I}_\delta} = \Omega^D_{\mathcal{I}_\delta}$, we may have

$$\mathcal{B}_\delta\, u(\boldsymbol{x}) = \mathcal{N}_\delta\, u(\boldsymbol{x}) - \mathcal{D}_\delta u(\boldsymbol{x}),\ \forall \boldsymbol{x} \in \Omega_{\mathcal{I}_\delta}, \tag{2.24}$$

where the linear operator $\mathcal{D}_\delta$ is given by

$$\mathcal{D}_\delta\, u(\boldsymbol{x}) = \int_{\Omega_{\mathcal{I}_\delta}} u(\boldsymbol{y})\sigma_\delta(\boldsymbol{x},\boldsymbol{y})d\boldsymbol{y},\ \forall \boldsymbol{x} \in \Omega_{\mathcal{I}_\delta} = \Omega^D_{\mathcal{I}_\delta}, \tag{2.25}$$

where the $\sigma_\delta(\boldsymbol{x},\boldsymbol{y})$ can be a bounded and symmetric kernel such that as $\delta \to 0$,

$$\int_{\Omega_{\mathcal{I}_\delta}} \mathcal{D}_\delta\, u(\boldsymbol{x})v(\boldsymbol{x})d\boldsymbol{x} \to \int_{\partial\Omega} u(\boldsymbol{x})v(\boldsymbol{x})d\boldsymbol{x}, \tag{2.26}$$

for any pair of smoothly defined functions u and v on Ω. In this case, we get a problem analogous to the local Robin boundary value problem.

Remark 2.8 While we focus on nonlocal diffusion operators defined by (2.19) with a symmetric kernel for illustration, we note that more general nonlocal diffusion-convection operators with a nonsymmetric kernel have also been studied, see, for instance, [21, 32, 48, 70, 71].

Note that for the pure Neumann case, i.e., $\Omega^N_{\mathcal{I}_\delta} = \Omega_{\mathcal{I}_\delta}$, we impose a constraint on the solution u to have mean 0 over $\widehat{\Omega}_\delta$ and the equations in (2.12) are modified to

$$\begin{cases} -\mathcal{L}_\delta\, u = F(\boldsymbol{x}, u(\boldsymbol{x})) - \lambda_N, & \forall \boldsymbol{x} \in \Omega, \\ \mathcal{N}_\delta\, u = g^\delta_N(\boldsymbol{x}) - \lambda_N, & \forall \boldsymbol{x} \in \Omega_{\mathcal{I}_\delta}, \end{cases} \tag{2.27}$$

where λ_N is a constant Lagrange multiplier satisfying

$$\int_\Omega F(\boldsymbol{x}, u(\boldsymbol{x}))\, d\boldsymbol{x} + \int_{\Omega_{\mathcal{I}_\delta}} g^\delta_N(\boldsymbol{x}) = \lambda_N |\widehat{\Omega}_\delta|, \tag{2.28}$$

where $|\widehat{\Omega}_\delta|$ denotes the nonzero measure of $\widehat{\Omega}_\delta$.

2.4 *Regional Fractional and Nonlocal Laplacians*

An operator related to the fractional Laplacian, but defined with respect to the domain Ω, is the *regional fractional Laplacian* given by Bogdan et al. [7] and Chen and Kim [14]

$$(-\Delta)^s_\Omega\, u(\boldsymbol{x}) = \int_\Omega (u(\boldsymbol{x}) - u(\boldsymbol{y}))\gamma_s(\boldsymbol{x}, \boldsymbol{y}) d\boldsymbol{y}, \quad \forall \boldsymbol{x} \in \Omega. \tag{2.29}$$

It can be used to define a problem given by

$$(-\Delta)^s_\Omega\, u(\boldsymbol{x}) = F(\boldsymbol{x}, u(\boldsymbol{x})), \qquad \forall \boldsymbol{x} \in \Omega, \tag{2.30}$$

which is another approach to generalize the local Poisson Neumann problem to the nonlocal fractional case [26, Section 7]. Note that in comparison with the fractional Laplacian $\mathcal{L}^s_\infty$ defined in (2.6), the only difference is that the regional operator in (2.29) involves an integral over Ω instead of $\mathbb{R}^d$.

Similarly, a nonlocal problem on Ω related to the regional fractional Laplacian is

$$-\mathcal{L}_{\delta,\Omega}\, u(\boldsymbol{x}) = F(\boldsymbol{x}, u(\boldsymbol{x})), \qquad \forall \boldsymbol{x} \in \Omega, \tag{2.31}$$

where the *regional nonlocal diffusion operator* $\mathcal{L}_{\delta,\Omega}$ is given by

$$\mathcal{L}_{\delta,\Omega}\, u(\boldsymbol{x}) = \int_{\Omega} (u(\boldsymbol{y}) - u(\boldsymbol{x}))\gamma_\delta\, (\boldsymbol{x}, \boldsymbol{y}) d\boldsymbol{y}, \qquad \forall \boldsymbol{x} \in \Omega. \tag{2.32}$$

Then (2.30) is a special case of (2.31) by choosing special fractional kernels as in (2.21) and setting $\delta = \infty$. Hence, for kernels of the form (2.21), we may see $\mathcal{L}_{\delta,\Omega}$ as a "regional" nonlocal Laplacian with respect to the domain Ω.

Indeed, let $\Omega = \Omega_1 \cup \Omega_2$ with $\Omega_1 \cap \Omega_2 = \emptyset$, the problem (2.31), by a simple change of notations, can be equivalently written as the nonlocal problem:

$$\begin{cases} -\mathcal{L}_{\delta,\Omega}\, u(\boldsymbol{x}) = F(\boldsymbol{x}, u(\boldsymbol{x})) & \forall \boldsymbol{x} \in \Omega_1, \\ \mathcal{N}_{\delta,\Omega}\, u(\boldsymbol{x}) = G(\boldsymbol{x}, u(\boldsymbol{x})) & \forall \boldsymbol{x} \in \Omega_2. \end{cases} \tag{2.33}$$

Here, a nonlocal operator $\mathcal{N}_{\delta,\Omega}$ is introduced, which is defined by

$$\mathcal{N}_{\delta,\Omega}\, u(\boldsymbol{x}) := -\mathcal{L}_{\delta,\Omega}\, u(\boldsymbol{x}) = \int_{\Omega} (u(\boldsymbol{x}) - u(\boldsymbol{y}))\gamma_\delta(\boldsymbol{x}, \boldsymbol{y}) d\boldsymbol{y}, \quad \forall \boldsymbol{x} \in \Omega_2, \tag{2.34}$$

and $G(\boldsymbol{x}, u(\boldsymbol{x})) = F(\boldsymbol{x}, u(\boldsymbol{x}))$ on Ω_2. $\mathcal{N}_{\delta,\Omega}$ and $G(\boldsymbol{x}, u(\boldsymbol{x}))$ only carry symbolic meanings in this case, and the equivalence of (2.31) and (2.33) holds independently of how Ω is decomposed into Ω_1 and Ω_2. For example, a special choice of the domain Ω_2, for δ sufficiently small, is the interior δ-layer of Ω given by

$$\Omega_{\mathcal{I}_\delta^0} = \{\boldsymbol{x} \in \Omega \mid \operatorname{dist}(\boldsymbol{x}, \partial\Omega) < \delta\}. \tag{2.35}$$

Remark 2.9 Note that the definition (2.34) of $\mathcal{N}_{\delta,\Omega}$ has been adopted in discussions in [20, 31], except for the use of different notations for the geometric domains. It can be seen that $\mathcal{N}_{\delta,\Omega}$ adopts the same definition as $-\mathcal{L}_{\delta,\Omega}$ except that while $-\mathcal{L}_{\delta,\Omega}\, u(\boldsymbol{x})$ is applied on Ω_1, $\mathcal{N}_{\delta,\Omega}\, u(\boldsymbol{x})$ inherits the definition on Ω_2.

Remark 2.10 We note that the reformulation (2.33) of (2.31) resembles more closely, in appearance, the conventional way of presenting the problem as an equation coupled with a boundary condition (or say volume constraint). Yet, the equivalence of the formulations also highlights the fact that for nonlocal problems, the notion of boundary conditions can be superfluous. A more general way to understand the formulation is that a nonlocal model is largely described by the law of nonlocal interactions that could be altered due to the presence of a boundary.

Remark 2.11 We also note that the nonlocal diffusion operator $\mathcal{L}_\delta$ reformulated in (2.23), with the integral being over $\widehat{\Omega}_\delta$ instead of $\mathbb{R}^d$ for compactly supported kernels of an interaction range δ, may have the appearance of being associated with a "regional" nonlocal diffusion operator defined on the combined domain $\widehat{\Omega}_\delta$. However, this association can only be meaningful in the case where Eq. (2.23), rather than being confined to Ω, is extended to all $\boldsymbol{x} \in \widehat{\Omega}_\delta$.

Let us examine some limiting cases formally. First, with the kernel $\gamma_\delta(\boldsymbol{x}, \boldsymbol{y})$ properly scaled with respect to the horizon parameter δ as in (2.20), we see that

(2.31), thus (2.33), is expected to converge as $\delta \to 0$ to a Neumann problem of the local diffusion operator with the corresponding boundary condition determined by the limiting behavior of F on the interior δ-layer $\Omega_{\mathcal{I}_\delta^0}$ defined in (2.35).

Likewise, with $\gamma_\delta(\boldsymbol{x}, \boldsymbol{y})$ chosen as in (2.22), then (2.33) is expected to recover exactly the regional fractional Laplacian Neumann problem (2.30) if $\delta > \text{diam}(\Omega)$. We thus see the regional nonlocal Laplacian again provides a bridge between the local Laplacian and the regional fractional Laplacian.

We note that spectral decomposition is another approach to define fractional differential operators and nonlocal integral operators. Generically, such spectrally represented operators might not yield operators with a finite range of interactions spatially. In many cases, the resulting nonlocal kernels for their spatial integral formulations might not have simple analytical representations in real space. We thus do not consider further in this direction, and interested readers can find additional discussions in [1, 5, 8, 13, 19, 43, 59, 62, 67].

2.5 Nonlocal Flux and Nonlocal Green's Identities

In [20, 31], the operator $\mathcal{N}_{\delta,\Omega}$ is also used to provide a nonlocal Green's formula that shares more symbolic resemblance to the classical Green's formula for the local Laplacian and for defining a nonlocal flux. We provide a more rigorous statement and derivation here for clarity.

Proposition 2.12 *Given the nonlocal operators defined by a rescaled kernel* (2.17) *with the conditions* (2.18) *and* (2.20) *satisfied, for functions u and v in* $C^2(\overline{\Omega \cup \Omega'})$, *we have the nonlocal Green's first identity given by*

$$\begin{aligned}\frac{1}{2}\int_{(\Omega\cup\Omega')^2} &(v(\boldsymbol{y}) - v(\boldsymbol{x}))(u(\boldsymbol{y}) - u(\boldsymbol{x}))\gamma_\delta(\boldsymbol{x}, \boldsymbol{y})d\boldsymbol{y}d\boldsymbol{x}\\ &= -\int_{\Omega\cup\Omega'} v(\boldsymbol{x})\mathcal{L}_{\delta,\Omega\cup\Omega'}\, u(\boldsymbol{x})d\boldsymbol{x}\\ &= -\int_{\Omega} v(\boldsymbol{x})\mathcal{L}_{\delta,\Omega\cup\Omega'}\, u(\boldsymbol{x})d\boldsymbol{x} + \int_{\Omega'} v(\boldsymbol{x})\mathcal{N}_{\delta,\Omega\cup\Omega'}\, u(\boldsymbol{x})d\boldsymbol{x}.\end{aligned} \tag{2.36}$$

This also readily implies the nonlocal Green's second identity

$$\begin{aligned}\int_{\Omega} &v(\boldsymbol{x})\mathcal{L}_{\delta,\Omega\cup\Omega'}\, u(\boldsymbol{x})d\boldsymbol{x} - \int_{\Omega} u(\boldsymbol{x})\mathcal{L}_{\delta,\Omega\cup\Omega'} v(\boldsymbol{x})d\boldsymbol{x}\\ &= -\int_{\Omega'} u(\boldsymbol{x})\mathcal{N}_{\delta,\Omega\cup\Omega'} v(\boldsymbol{x})d\boldsymbol{x} + \int_{\Omega'} v(\boldsymbol{x})\mathcal{N}_{\delta,\Omega\cup\Omega'}\, u(\boldsymbol{x})d\boldsymbol{x}.\end{aligned} \tag{2.37}$$

Proof To prove Proposition 2.12, the key is to show that the integrals are well-defined so that the classical Fubini theorem can be applied to get (2.36). For $u \in$

$C^2(\overline{\Omega \cup \Omega'})$, we now check that $\mathcal{L}_{\delta,\Omega\cup\Omega'}\, u \in L^1(\Omega \cup \Omega')$, which is a sufficient condition to achieve this. To simplify the notation, we use Ω to replace $\Omega \cup \Omega'$ so that

$$\mathcal{L}_{\delta,\Omega}\, u(\boldsymbol{x}) = \int_{\Omega} (u(\boldsymbol{y}) - u(\boldsymbol{x}))\gamma_\delta(\boldsymbol{x}, \boldsymbol{y})d\boldsymbol{y}.$$

Then it suffices to show that $\mathcal{L}_{\delta,\Omega}\, u \in L^1(\Omega)$ for all $u \in C^2(\overline{\Omega})$.

By the regularity of u, we have for $\boldsymbol{x} \in \Omega$,

$$\begin{aligned}
\mathcal{L}_{\delta,\Omega}\, u(\boldsymbol{x}) &= \int_{\Omega\setminus B_{\mathrm{dist}(\boldsymbol{x},\partial\Omega)}(\boldsymbol{x})} (u(\boldsymbol{y}) - u(\boldsymbol{x}))\gamma_\delta(\boldsymbol{x}, \boldsymbol{y})d\boldsymbol{y} \\
&\quad + \int_{B_{\mathrm{dist}(\boldsymbol{x},\partial\Omega)}(\boldsymbol{x})} (u(\boldsymbol{y}) - u(\boldsymbol{x}))\gamma_\delta(|\boldsymbol{y} - \boldsymbol{x}|)d\boldsymbol{y} \\
&\le C \int_{\Omega\setminus B_{\mathrm{dist}(\boldsymbol{x},\partial\Omega)}(\boldsymbol{x})} |\boldsymbol{y} - \boldsymbol{x}|\gamma_\delta(|\boldsymbol{y} - \boldsymbol{x}|)d\boldsymbol{y} \\
&\quad + C \int_{B_{\mathrm{dist}(\boldsymbol{x},\partial\Omega)}(\boldsymbol{x})} |\boldsymbol{y} - \boldsymbol{x}|^2\gamma_\delta(|\boldsymbol{y} - \boldsymbol{x}|)d\boldsymbol{y} \\
&\le C \left(\int_{\mathrm{dist}(\boldsymbol{x},\partial\Omega)<|\boldsymbol{z}|<\delta} |\boldsymbol{z}|\gamma_\delta(|\boldsymbol{z}|)d\boldsymbol{z} + 1 \right),
\end{aligned}$$

where the constant C depends on $\|u\|_{C^2(\overline{\Omega})}$. Thus, the task remains to show that

$$w(\boldsymbol{x}) := \int_{\mathrm{dist}(\boldsymbol{x},\partial\Omega)<|\boldsymbol{z}|<\delta} |\boldsymbol{z}|\gamma_\delta(|\boldsymbol{z}|)d\boldsymbol{z} \in L^1(\Omega)\,.$$

Let Ω^ϵ denote an inner layer of width $\epsilon > 0$ surrounding $\partial\Omega$, where $\epsilon > 0$ is a small fixed number. Since it is obvious that $w \in L^1(\Omega\setminus\Omega^\epsilon)$, we only need to show $w \in L^1(\Omega^\epsilon)$. Define for $t \in (0, \epsilon)$,

$$\partial\Omega_t := \{\boldsymbol{x} \in \Omega \mid \mathrm{dist}(\boldsymbol{x}, \partial\Omega) = t\}.$$

Then, there exists a general constant $C > 0$,

$$\begin{aligned}
\int_{\Omega^\epsilon} \int_{\mathrm{dist}(\boldsymbol{x},\partial\Omega)<|\boldsymbol{z}|<\delta} |\boldsymbol{z}|\gamma_\delta(|\boldsymbol{z}|)d\boldsymbol{z} &\le C \int_0^\epsilon \int_{\partial\Omega_t} \int_{t<r<\delta} r^d\gamma_\delta(r)drd\boldsymbol{x}dt \\
&\le C \int_0^\epsilon \int_{t<r<\delta} r^d\gamma_\delta(r)drdt
\end{aligned}$$

$$= C \int_0^\delta \int_0^{\min(r,\epsilon)} r^d \gamma_\delta(r) dt dr$$

$$\leq C \int_0^\delta r^{d+1} \gamma_\delta(r) dr < \infty.$$

Thus, we get the desired result that $\mathcal{L}_{\delta,\Omega}\, u \in L^1(\Omega)$. □

Remark 2.13 Note that by density argument, one can show that the above identities hold in the closure of $C^2(\overline{\Omega \cup \Omega'})$ with respect to suitable norms.

Remark 2.14 In [31] and [28], the identity (2.37) is stated as for $\Omega = \Omega_1 \cup \Omega_2$,

$$\int_\Omega v(\boldsymbol{x}) \mathcal{L}_{\delta,\Omega}\, u(\boldsymbol{x}) d\boldsymbol{x} = \int_\Omega u(\boldsymbol{x}) \mathcal{L}_{\delta,\Omega} v(\boldsymbol{x}) d\boldsymbol{x}. \tag{2.38}$$

This relation, as a consequence of Fubini's theorem, can be seen as a nonlocal version of integration by parts.

It is worth pointing out that a similar role as $\mathcal{N}_{\delta,\Omega}$ can be played by the operator $\mathcal{N}_\delta$ defined earlier in (2.16). In fact, corresponding to the function u defined on a union $\Omega \cup \Omega'$ of domains Ω and Ω', let $\mathcal{N}_\delta\, u(\boldsymbol{x})$ be defined on Ω' in the same way as (2.16), the so-called nonlocal flux of u between the domain Ω and another domain Ω', which is induced from $\mathcal{N}_\delta$, can be specified as

$$\mathcal{F}(\Omega, \Omega') = \int_{\Omega'} \int_\Omega (u(\boldsymbol{x}) - u(\boldsymbol{y})) \gamma_\delta(\boldsymbol{x}, \boldsymbol{y})\, d\boldsymbol{y}\, d\boldsymbol{x} = \int_{\Omega'} \mathcal{N}_\delta\, u(\boldsymbol{x})\, d\boldsymbol{x}.$$

Then for a symmetric kernel $\gamma_\delta(\boldsymbol{x}, \boldsymbol{y})$, i.e., $\gamma_\delta(\boldsymbol{x}, \boldsymbol{y}) = \gamma_\delta(\boldsymbol{y}, \boldsymbol{x})$, we have the following properties:

- Anti-symmetry:

$$\mathcal{F}(\Omega, \Omega') = -\mathcal{F}(\Omega', \Omega) = -\int_\Omega \int_{\Omega'} (u(\boldsymbol{y}) - u(\boldsymbol{x})) \gamma_\delta(\boldsymbol{y}, \boldsymbol{x})\, d\boldsymbol{x}\, d\boldsymbol{y}$$

  holds formally (assuming that Fubini's theorem is applicable).
- Additivity:

$$\mathcal{F}(\Omega_1, \Omega') + \mathcal{F}(\Omega_2, \Omega') = \mathcal{F}(\Omega, \Omega') \quad \text{for } \Omega = \Omega_1 \cup \Omega_2,\ \Omega_1 \cap \Omega_2 = \emptyset.$$

Remark 2.15 With $\mathcal{F}(\Omega_1, \Omega_1) = 0$, we see that

$$\mathcal{F}(\Omega_1, \Omega_2) = \mathcal{F}(\Omega_1, \Omega_2) + \mathcal{F}(\Omega_2, \Omega_2) = \mathcal{F}(\Omega_1 \cup \Omega_2, \Omega_2)$$

$$= \int_{\Omega_2} \int_\Omega (u(\boldsymbol{x}) - u(\boldsymbol{y})) \gamma_\delta(\boldsymbol{x}, \boldsymbol{y})\, d\boldsymbol{y}\, d\boldsymbol{x} = \int_{\Omega_2} \mathcal{N}_{\delta,\Omega}\, u(\boldsymbol{x})\, d\boldsymbol{x}.$$

This also implies that $\mathcal{N}_\delta$ induces the same flux functional as $\mathcal{N}_{\delta,\Omega}$ for any symmetric kernel.

Next, let us state the nonlocal Green's first identity given by

$$\begin{aligned}\frac{1}{2}\int_{(\Omega\cup\Omega')^2\setminus\Omega'^2}&(v(\boldsymbol{y})-v(\boldsymbol{x}))(u(\boldsymbol{y})-u(\boldsymbol{x}))\gamma_\delta(\boldsymbol{x},\boldsymbol{y})d\boldsymbol{y}d\boldsymbol{x}\\ &=-\int_\Omega v(\boldsymbol{x})\mathcal{L}_{\delta,\Omega\cup\Omega'}\,u(\boldsymbol{x})d\boldsymbol{x}+\int_{\Omega'}v(\boldsymbol{x})\mathcal{N}_\delta\,u(\boldsymbol{x})d\boldsymbol{x}. \qquad (2.39)\end{aligned}$$

In particular, if $\Omega'=\Omega_{\mathcal{I}_\delta}$, then with $\widetilde{\Omega}_\delta=\widehat{\Omega}_\delta^2\setminus\Omega_{\mathcal{I}_\delta}^2$, we have

$$\begin{aligned}\frac{1}{2}\int_{\widetilde{\Omega}_\delta}&(v(\boldsymbol{y})-v(\boldsymbol{x}))(u(\boldsymbol{y})-u(\boldsymbol{x}))\gamma_\delta(\boldsymbol{x},\boldsymbol{y})d\boldsymbol{y}d\boldsymbol{x}\\ &=-\int_\Omega v(\boldsymbol{x})\mathcal{L}_\delta\,u(\boldsymbol{x})d\boldsymbol{x}+\int_{\Omega_{\mathcal{I}_\delta}}v(\boldsymbol{x})\mathcal{N}_\delta\,u(\boldsymbol{x})d\boldsymbol{x}. \qquad (2.40)\end{aligned}$$

Note that the identity (2.40) is different from (2.36) as given in [20, 31], but it is a more natural nonlocal analog of the (generalized) local Green's first identity

$$\int_\Omega\nabla v(\boldsymbol{x})\cdot\nabla u(\boldsymbol{x})d\boldsymbol{x}=-\int_\Omega v(\boldsymbol{x})\Delta u(\boldsymbol{x})d\boldsymbol{x}+\int_{\partial\Omega}v(\boldsymbol{x})\frac{\partial u}{\partial\mathbf{n}}(\boldsymbol{x})d\boldsymbol{x}, \qquad (2.41)$$

as well as the fractional version given by

$$\begin{aligned}\frac{1}{2}\int_{\widetilde{\Omega}_\infty}&(v(\boldsymbol{y})-v(\boldsymbol{x}))(u(\boldsymbol{y})-u(\boldsymbol{x}))\gamma_s(|\boldsymbol{y}-\boldsymbol{x}|)d\boldsymbol{y}d\boldsymbol{x}\\ &=-\int_\Omega v(\boldsymbol{x})\mathcal{L}_\infty^s\,u(\boldsymbol{x})d\boldsymbol{x}+\int_{\Omega_c}v(\boldsymbol{x})\mathcal{N}_\infty^s\,u(\boldsymbol{x})d\boldsymbol{x}, \qquad (2.42)\end{aligned}$$

where $\widetilde{\Omega}_\infty=(\mathbb{R}^d)^2\setminus\Omega_c^2$.

For smooth function u, by simple Taylor expansion, we have $\mathcal{L}_\delta\,u(\boldsymbol{x})\to\Delta u(\boldsymbol{x})$ in Ω if we choose the kernels according to (2.20). In this case, from the first Green's identity (2.40), we may formally let $\delta\to 0$ to get that

$$\int_{\Omega_{\mathcal{I}_\delta}}v(\boldsymbol{x})\mathcal{N}_\delta\,u(\boldsymbol{x})d\boldsymbol{x}\to\int_{\partial\Omega}v(\boldsymbol{x})\frac{\partial u}{\partial\mathbf{n}}d\boldsymbol{x}. \qquad (2.43)$$

This offers some justification to the statement that the local version (2.41) of (2.42) can be seen as the $\delta\to 0$ limit of (2.40). In this case, from the above formal argument, we expect that the Neumann operator $\mathcal{N}_\delta$ converges to the local one in the distribution sense. Meanwhile, the fractional version can be seen as the $\delta \to \infty$ limit if the kernel is given according to (2.22). Furthermore, as $\delta \to 0$, formally we may also note that

$$\int_{\Omega_{\mathcal{I}_\delta}} v(\boldsymbol{x})\mathcal{N}_{\delta,\widehat{\Omega}_\delta}\, u(\boldsymbol{x})d\boldsymbol{x} \to \int_{\partial\Omega} v(\boldsymbol{x})\frac{\partial u}{\partial \mathbf{n}}d\boldsymbol{x}. \tag{2.44}$$

Indeed, the difference between the left-hand side integrals of (2.43) and (2.44) satisfies

$$\begin{aligned}&\int_{\Omega_{\mathcal{I}_\delta}} v(\boldsymbol{x})\int_{\Omega_{\mathcal{I}_\delta}} (u(\boldsymbol{x})-u(\boldsymbol{y}))\gamma_\delta(|\boldsymbol{x}-\boldsymbol{y}|)d\boldsymbol{y}d\boldsymbol{x}\\ &\quad=\frac{1}{2}\int_{\Omega_{\mathcal{I}_\delta}}\int_{\Omega_{\mathcal{I}_\delta}} (u(\boldsymbol{x})-u(\boldsymbol{y}))(v(\boldsymbol{x})-v(\boldsymbol{y}))\gamma_\delta(|\boldsymbol{x}-\boldsymbol{y}|)d\boldsymbol{y}d\boldsymbol{x}=O(\delta)\end{aligned}$$

for smooth functions u and v defined on the whole space. This observation shows the formal consistency using the different Neumann operators in the local limit. However, we note that consistency to the fractional limit, unlike the case of $\mathcal{N}_\delta$, is not shared by the nonlocal analog presented in [20, 31] based on the operator $\mathcal{N}_{\delta,\widehat{\Omega}_\delta}$.

Likewise, the nonlocal analog of

$$\int_\Omega [v(\boldsymbol{x})\Delta u(\boldsymbol{x}) - u(\boldsymbol{x})\Delta v(\boldsymbol{x})]\, d\boldsymbol{x} = \int_{\partial\Omega}\left[v(\boldsymbol{x})\frac{\partial u}{\partial \mathbf{n}}(\boldsymbol{x}) - u(\boldsymbol{x})\frac{\partial v}{\partial \mathbf{n}}(\boldsymbol{x})\right] d\boldsymbol{x}$$

is the nonlocal Green's second identity given by

$$\int_\Omega [v(\boldsymbol{x})\mathcal{L}_\delta\, u(\boldsymbol{x}) - u(\boldsymbol{x})\mathcal{L}_\delta\, v(\boldsymbol{x})]d\boldsymbol{x} = \int_{\Omega_{\mathcal{I}_\delta}} [v(\boldsymbol{x})\mathcal{N}_\delta\, u(\boldsymbol{x}) - u(\boldsymbol{x})\mathcal{N}_\delta\, v(\boldsymbol{x})]d\boldsymbol{x}. \tag{2.45}$$

The factional analog can be easily obtained in the limit $\delta \to \infty$:

$$\int_\Omega [v(\boldsymbol{x})\mathcal{L}^s_\infty\, u(\boldsymbol{x}) - u(\boldsymbol{x})\mathcal{L}^s_\infty\, v(\boldsymbol{x})]d\boldsymbol{x} = \int_{\Omega_c} [v(\boldsymbol{x})\mathcal{N}^s_\infty\, u(\boldsymbol{x}) - u(\boldsymbol{x})\mathcal{N}^s_\infty\, v(\boldsymbol{x})]d\boldsymbol{x}. \tag{2.46}$$

Remark 2.16 Rigorous proofs of the nonlocal Green's identities and their connections to the local and fractional analog in their full generality depend on the characterization of the trace spaces, and we will leave the discussions to future works, though we note some related works in the fractional case [43] and recent works on the nonlocal trace theorems in [38].

3 Well-posed Inhomogeneous Nonlocal Variational Problems

We use linear problems where $F(\boldsymbol{x}, u) = f(\boldsymbol{x})$ as illustrations for well-posed nonlocal models with a horizon parameter δ and inhomogeneous nonlocal constraints. We consider weak formulations corresponding to (2.12) and make connections to a minimization principle. We pay particular attention to the results that are independent of δ so that the local and fractional cases can then become special cases, which enforces the generality of the nonlocal models.

3.1 Function Spaces and Norms

Given a generic domain Ω, a parameter $\delta > 0$, and the extended domain $\widehat{\Omega}_\delta$, let us first define an *"energy" space*

$$V_\delta(\widehat{\Omega}_\delta) = \{v(\boldsymbol{x}) \in L^2(\widehat{\Omega}_\delta) \colon \|v\|_{V_\delta(\widehat{\Omega}_\delta)} < \infty\}, \tag{3.1}$$

where $\|\cdot\|_{V_\delta(\widehat{\Omega}_\delta)}$ denotes an *"energy" norm* induced by the inner product

$$\langle\cdot,\cdot\rangle_{V_\delta(\widehat{\Omega}_\delta)} = (\cdot,\cdot)_{V_\delta(\widehat{\Omega}_\delta)} + (\cdot,\cdot)_{\widehat{\Omega}_\delta}$$

with $(\cdot,\cdot)_{V_\delta(\widehat{\Omega}_\delta)}$ being defined as

$$(u, v)_{V_\delta(\widehat{\Omega}_\delta)} = \frac{1}{2}\int_{\widetilde{\Omega}_\delta} (v(\boldsymbol{y}) - v(\boldsymbol{x}))(u(\boldsymbol{y}) - u(\boldsymbol{x}))\gamma_\delta(\boldsymbol{x}, \boldsymbol{y})\, d\boldsymbol{y} d\boldsymbol{x}\,, \tag{3.2}$$

for any $u, v \in V_\delta(\widehat{\Omega}_\delta)$. Here $\widetilde{\Omega}_\delta = \widehat{\Omega}_\delta^2 \setminus \Omega^2_{\mathcal{I}_\delta}$.

Remark 3.1 A different choice of the nonlocal space and its norm, defined over a domain $\widehat{\Omega}$, corresponds to the inner product

$$\langle u, v\rangle_{V(\widehat{\Omega})} = (u, v)_{V(\widehat{\Omega})} + (u, v)_{\widehat{\Omega}}, \tag{3.3}$$

where

$$(u, v)_{V(\widehat{\Omega})} = \frac{1}{2}\int_{\widehat{\Omega}}\int_{\widehat{\Omega}} (u(\boldsymbol{y}) - u(\boldsymbol{x}))(v(\boldsymbol{y}) - v(\boldsymbol{x}))\gamma_\delta(\boldsymbol{x}, \boldsymbol{y}) d\boldsymbol{y} d\boldsymbol{x}. \tag{3.4}$$

Such space and norm can be associated with the regional version of the nonlocal and fractional versions of diffusion discussed in Sect. 2.4. When $\widehat{\Omega}$ is replaced by $\widehat{\Omega}_\delta$, the main difference with the norm of $V_\delta(\widehat{\Omega}_\delta)$ induced from (3.2) is the inclusion of additional nonlocal interactions in $\Omega^2_{\mathcal{I}_\delta}$ for the norm in $V(\widehat{\Omega}_\delta)$. Note that this difference disappears for functions constrained to have zero values on $\Omega_{\mathcal{I}_\delta}$.

The space $V_\delta(\widehat{\Omega}_\delta)$ is by definition a subspace of $L^2(\widehat{\Omega}_\delta)$ and can be shown, for suitably chosen kernels such as those specified by (2.20) or (2.22), to be a Hilbert space.

The *constrained energy space* is a closed subspace of $V_\delta(\widehat{\Omega}_\delta)$ defined by

$$V_{c,\delta}(\widehat{\Omega}_\delta) = \left\{ v(\boldsymbol{x}) \in V_\delta(\widehat{\Omega}_\delta) \colon v(\boldsymbol{x}) = 0 \text{ for } \boldsymbol{x} \in \Omega^D_{\mathcal{I}_\delta} \right\}. \tag{3.5}$$

Note that if $\Omega^D_{\mathcal{I}_\delta} = \emptyset$, we may pick a nontrivial subset K of $\widehat{\Omega}_\delta$ to define the space by space $V_{c,\delta}(\widehat{\Omega}_\delta)$ by

$$V_{c,\delta}(\widehat{\Omega}_\delta) = \left\{ v(\boldsymbol{x}) \in V_\delta(\widehat{\Omega}_\delta) : \int_K v(\boldsymbol{x}) d\boldsymbol{x} = 0 \right\}. \tag{3.6}$$

Some natural choices include $K = \Omega$ and $K = \widehat{\Omega}_\delta$. The dual space of $V_{c,\delta}(\widehat{\Omega}_\delta)$ is denoted by $V^\star_\delta(\widehat{\Omega}_\delta)$ with $(\cdot,\cdot)_{\widehat{\Omega}_\delta}$ being the duality pairing that, for functions vanishing on $\Omega^D_{\mathcal{I}_\delta}$, is also used to denote the equivalent $L^2(\Omega \cup \Omega^N_{\mathcal{I}_\delta})$ inner product.

For convenience, we also define the energy seminorm

$$|v|_{V_{c,\delta}(\widehat{\Omega}_\delta)} = \left(\frac{1}{2} \int_{\widetilde{\Omega}_\delta} (v(\boldsymbol{y}) - v(\boldsymbol{x}))^2 \gamma_\delta(\boldsymbol{x},\boldsymbol{y}) d\boldsymbol{y} d\boldsymbol{x} \right)^{1/2}, \tag{3.7}$$

which can be a full norm on $V_{c,\delta}(\widehat{\Omega}_\delta)$ as shown later.

Remark 3.2 A natural question is to ask what assumptions on $\Omega^D_{\mathcal{I}_\delta}$ and $\Omega^N_{\mathcal{I}_\delta}$ and g^δ_D and g^δ_N are needed to recover the appropriate local limit. For example, under what assumptions on $\Omega^D_{\mathcal{I}_\delta}$ and $\Omega^N_{\mathcal{I}_\delta}$, one can have bounded extension of functions $V_{c,\delta}(\widehat{\Omega}_\delta)$ with respect to the energy, in the same form but over some larger domain, independent of δ for δ small. Sometimes, it is possible that the above extension cannot be made: for example, if $\Omega^D_{\mathcal{I}_\delta}$ is an inner layer surrounded by $\Omega^N_{\mathcal{I}_\delta}$. For an attempt to put assumptions on $\Omega^D_{\mathcal{I}_\delta}$, $\Omega^N_{\mathcal{I}_\delta}$, g^δ_D, and g^δ_N, we refer to the discussion in Sect. 4.2.

Compactness and Equivalent Norms The work of Bourgain, Brezis and Mironecu [9] characterizes the limiting space of the energy space $V_\delta(\widehat{\Omega}_\delta)$ as $\delta \to 0$ when the kernel γ_δ satisfies (2.18) and (2.20). Compactness is gained through taking the limit $\delta \to 0$. The original result in [9] requires additionally that $\gamma(t)t^2$ is a nonincreasing function for $t \in (0, 1)$. It was found later that the assumption could be replaced with $\gamma(t)$ being nonincreasing instead, see, e.g., discussions in [54, 73] and [34]. Ponce [61] showed a different proof of the Bourgain–Brezis–Mironescu compactness result that removed the nonincreasing assumption on γ for $d \geq 2$. We quote the compactness result below.

Lemma 3.1 *Assume that* $\delta_n \to 0$ *and the kernels* $\{\gamma_{\delta_n}\}_{n\in\mathbb{N}_+}$ *satisfy* (2.18) *and* (2.20). *Let* $\Omega_1 \subset \mathbb{R}^d$ *be a bounded Lipschitz domain. If* $\{v_n\} \subset L^2(\Omega_1)$ *is a bounded sequence such that*

$$\int_{\Omega_1}\int_{\Omega_1} \gamma_{\delta_n}(\boldsymbol{x},\boldsymbol{y})(v_n(\boldsymbol{y})-v_n(\boldsymbol{x}))^2 d\boldsymbol{x}d\boldsymbol{y} \le C_0,$$

then $\{v_n\}$ *is relatively compact in* $L^2(\Omega_1)$ *and any limit function* v *belongs to* $H^1(\Omega_1)$ *with*

$$\|v\|_{H^1(\Omega_1)} \le C_0.$$

Remark 3.3 It is possible to have more general compactness results for a sequence of kernels $\{\gamma_{\delta_n}\}_{n\in\mathbb{N}_+}$ converging to a limit γ_δ, as $\delta_n \to \delta$, under suitable conditions on these kernels and the corresponding energy spaces $V_{c,\delta_n}(\widehat{\Omega}_{\delta_n})$ and $V_{c,\delta}(\widehat{\Omega}_\delta)$. We refer to [34, 73].

Remark 3.4 Note that by classical results of fractional Sobolev spaces, for a kernel γ_δ that satisfies (2.22), we have the pre-compactness of $V_{c,\delta}(\widehat{\Omega}_\delta)$ in $L^2(\widehat{\Omega})$ for $\delta > 0$.

Using the compactness result, we can prove a general Poincaré inequality [28, 30, 51, 52] for kernels satisfying either (2.18)–(2.20) or (2.22) as $\delta \to 0$ and $\delta \to \infty$, respectively.

Lemma 3.2 *There exists a constant* $C > 0$, *independent of* δ, *such that*

$$\|v\|_{L^2(\widehat{\Omega}_\delta)} \le C|v|_{V_{c,\delta}(\widehat{\Omega}_\delta)}, \quad \forall\, v \in V_{c,\delta}(\widehat{\Omega}_\delta). \tag{3.8}$$

Lemma 3.2 unifies the classical version of the Poincaré inequality for the usual energy function spaces associated with the local Laplacian, the fractional, and the nonlocal versions.

Remark 3.5 We see that $|\cdot|_{V_{c,\delta}(\widehat{\Omega}_\delta)}$ is a norm on $V_{c,\delta}(\widehat{\Omega}_\delta)$ equivalent to the norm $\|\cdot\|_{V_{c,\delta}(\widehat{\Omega}_\delta)}$ uniformly in δ for kernels specified in (2.20) and (2.22) as $\delta \to 0$ and $\delta \to \infty$, respectively. Correspondingly, we may view $V_{c,\delta}(\widehat{\Omega}_\delta)$ as a Hilbert space equipped with the inner product

$$(u,v)_{V_{c,\delta}(\widehat{\Omega}_\delta)} = \frac{1}{2}\int_{\widetilde{\Omega}_\delta} (v(\boldsymbol{y})-v(\boldsymbol{x}))(u(\boldsymbol{y})-u(\boldsymbol{x}))\gamma_\delta(\boldsymbol{x},\boldsymbol{y})d\boldsymbol{y}d\boldsymbol{x} \tag{3.9}$$

for all $u, v \in V_{c,\delta}(\widehat{\Omega}_\delta)$.

For more detailed discussions of the conditions on the kernels and rigorous proofs of the properties on these spaces, we refer to [28, 54, 55].

For any subdomain $\Omega' \subset \Omega_{\mathcal{I}_\delta}$ with a nonzero volume (i.e., d-dimensional measure), we define the space $V_t(\Omega')$,

$$V_t(\Omega') = \{v = w|_{\Omega'} \text{ for some } w \in V_\delta(\widehat{\Omega}_\delta)\}, \tag{3.10}$$

which involves restrictions to the domain Ω' having nonzero volume in $\mathbb{R}^d$. A norm on $V_t(\Omega')$ can be defined by

$$\|v\|_{V_t(\Omega')} = \inf\{\|w\|_{V_\delta(\widehat{\Omega}_\delta)} \mid w \in V_\delta(\widehat{\Omega}_\delta),\, w|_{\Omega'} = v.\} \tag{3.11}$$

Remark 3.6 One can view $V_t(\Omega')$, induced by $V_\delta(\widehat{\Omega}_\delta)$, as a nonlocal analogue of the trace space $H^{1/2}(S)$ induced by $H^1(\Omega)$ for $S \subset \partial\Omega$. Some rigorous characterizations of the trace spaces for special classes of kernels can be found in [38].

We are particularly interested in the spaces $V_t(\Omega^D_{\mathcal{I}_\delta})$ and $V_t(\Omega^N_{\mathcal{I}_\delta})$ and their dual spaces $V_t^\star(\Omega^D_{\mathcal{I}_\delta})$ and $V_t^\star(\Omega^N_{\mathcal{I}_\delta})$ corresponding to cases $\Omega' = \Omega^D_{\mathcal{I}_\delta}$ and $\Omega' = \Omega^N_{\mathcal{I}_\delta}$, respectively. We use $\langle g^\delta_N, v\rangle_{\Omega^N_{\mathcal{I}_\delta}}$ to denote the inner product of g^δ_N and v in $L^2(\Omega^N_{\mathcal{I}_\delta})$.

3.2 Weak Formulations and Well-Posedness

For illustration, we consider the case $F(\boldsymbol{x}, u) = -q(u)+f(\boldsymbol{x})$, where $q(u) = Q'(u)$ with $Q = Q(u) = \alpha|u|^m/m$ with some $m \in (2, \min\{4, 2d/(d-2)\})$ and $\alpha \geq 0$. We note that the problem reduces to a linear one if $\alpha = 0$. For any $u \in L^m(\Omega)$, we define a linear functional

$$\mathcal{Q}(u, v) = \int_\Omega q(u(\boldsymbol{x}))v(\boldsymbol{x})d\boldsymbol{x}, \quad \forall v \in L^m(\Omega).$$

Following [28], we consider a weak formulation of problem (2.12) with $\mathcal{B}_\delta\, u = g_\delta$. Let g^δ_D and g^δ_N denote the restriction of g_δ on $\Omega^D_{\mathcal{I}_\delta}$ and $\Omega^N_{\mathcal{I}_\delta}$, respectively, for domains that have nonzero volume (d-dimensional measure). We define

$$\widehat{f}(\boldsymbol{x}) = \begin{cases} f(\boldsymbol{x}), & \boldsymbol{x} \in \Omega, \\ 0, & \boldsymbol{x} \in \Omega^D_{\mathcal{I}_\delta}, \\ g^\delta_N(\boldsymbol{x}), & \boldsymbol{x} \in \Omega^N_{\mathcal{I}_\delta}. \end{cases} \tag{3.12}$$

We assume that $\widehat{f} \in V^\star_\delta(\widehat{\Omega}_\delta)$, which also implies $g^\delta_N \in V^\star_t(\Omega^N_{\mathcal{I}_\delta})$. Moreover, we assume that $g^\delta_D \in V_t(\Omega^D_{\mathcal{I}_\delta})$. By the definition of $V_t(\Omega^D_{\mathcal{I}_\delta})$, we can define an extension of g^δ_D on $\widehat{\Omega}_\delta$, denoted by w^δ_D, such that

$$w^\delta_D \in V_\delta(\widehat{\Omega}_\delta), \quad w^\delta_D\,|_{\Omega^D_{\mathcal{I}_\delta}} = g^\delta_D, \quad \|w^\delta_D\|_{V_\delta(\widehat{\Omega}_\delta)} \leq C\|g^\delta_D\|_{V_t(\Omega^D_{\mathcal{I}_\delta})}, \tag{3.13}$$

for some constant $C > 0$, independent of δ. Without loss of generality, we can set $C = 2$. To deal with the special nonlinear term, we make an additional assumption that w_D^δ also satisfies

$$\alpha w_D^\delta \in \alpha L^m(\Omega), \qquad \mathcal{Q}\left(w_D^\delta, w_D^\delta\right) \le \tilde{q}\left(\|g_D^\delta\|_{V_t(\Omega_{\mathcal{I}_\delta}^D)}\right), \tag{3.14}$$

for a given smooth function $\tilde{q} = \tilde{q}(r)$. Here we use the abbreviation $\alpha L^m(\Omega)$ as the space of all functions v such that $\alpha v \in L^m(\Omega)$, i.e., $\alpha L^m(\Omega)$ can include all functions if $\alpha = 0$ but $\alpha L^m(\Omega) = L^m(\Omega)$ if $\alpha \neq 0$.

The weak formulation can then be stated as follows: given $g_D^\delta \in V_t(\Omega_{\mathcal{I}_\delta}^D)$, $\widehat{f} \in V_\delta^\star(\widehat{\Omega}_\delta)$, and w_D^δ satisfying (3.13) and (3.14), find u such that $u - w_D^\delta \in V_{c,\delta}(\widehat{\Omega}_\delta) \cap \alpha L^m(\Omega)$ and

$$\mathcal{A}_\delta(u, v) + \mathcal{Q}(u, v) = (\widehat{f}, v)_{\widehat{\Omega}_\delta}, \quad \forall v \in V_{c,\delta}(\widehat{\Omega}_\delta), \tag{3.15}$$

where the bilinear form $\mathcal{A}_\delta(\cdot, \cdot)$ is given by

$$\mathcal{A}_\delta(u, v) = (u, v)_{V_{c,\delta}(\widehat{\Omega}_\delta)}, \quad \forall\, u, v \in V_\delta(\widehat{\Omega}_\delta), \tag{3.16}$$

and the linear functional

$$(\widehat{f}, v)_{\widehat{\Omega}_\delta} = \int_\Omega f(\boldsymbol{x}) v(\boldsymbol{x}) d\boldsymbol{x} + \int_{\Omega_{\mathcal{I}_\delta}^N} g_N^\delta(\boldsymbol{x}) v(\boldsymbol{x}) d\boldsymbol{x}\,. \tag{3.17}$$

Note that if $\Omega_{\mathcal{I}_\delta}^D = \Omega_{\mathcal{I}_\delta}$, then for any $v \in V_{c,\delta}(\widehat{\Omega}_\delta)$, we have

$$\begin{aligned}\mathcal{A}_\delta(u, v) = (u, v)_{V_{c,\delta}(\widehat{\Omega}_\delta)} &= \frac{1}{2}\int_{\widetilde{\Omega}_\delta} (v(\boldsymbol{y}) - v(\boldsymbol{x}))(u(\boldsymbol{y}) - u(\boldsymbol{x}))\gamma_\delta(\boldsymbol{x}, \boldsymbol{y}) d\boldsymbol{y} d\boldsymbol{x} \\ &= \frac{1}{2}\int_{\widehat{\Omega}_\delta}\int_{\widehat{\Omega}_\delta} (v(\boldsymbol{y}) - v(\boldsymbol{x}))(u(\boldsymbol{y}) - u(\boldsymbol{x}))\gamma_\delta(\boldsymbol{x}, \boldsymbol{y}) d\boldsymbol{y} d\boldsymbol{x}.\end{aligned} \tag{3.18}$$

The weak formulation is equivalent to a minimization problem: Given $g_D^\delta \in V_t(\Omega_{\mathcal{I}_\delta}^D)$, $\widehat{f} \in V_\delta^\star(\widehat{\Omega}_\delta)$, and w_D^δ satisfying (3.13) and (3.14), find u that minimizes

$$\min_{v \in w_D^\delta + V_{c,\delta}(\widehat{\Omega}_\delta) \cap \alpha L^m(\Omega)} E(v) := \frac{1}{2}\mathcal{A}_\delta(v, v) + \frac{1}{m}\mathcal{Q}(v, v) - (\widehat{f}, v)_{\widehat{\Omega}_\delta}. \tag{3.19}$$

Theorem 3.3 *The problem* (3.15) *has a unique solution* $u \in w_D^\delta + V_{c,\delta}(\widehat{\Omega}_\delta) \cap \alpha L^m(\Omega)$, *for any given data* $\widehat{f}$, g_D^δ, *and* w_D^δ *satisfying the conditions specified earlier. Moreover, there is a constant* $C > 0$, *independent of* δ, *such that the following a priori estimates for the solution* u *are satisfied:*

$$\|u\|^2_{V_\delta(\widehat{\Omega}_\delta)} + \mathcal{Q}(u,u) \le C\left[\|\widehat{f}\|^2_{V^*_\delta(\widehat{\Omega}_\delta)} + \|g^\delta_D\|^2_{V_t(\Omega^D_{\mathcal{I}_\delta})} + \tilde{q}\left(\|g^\delta_D\|_{V_t(\Omega^D_{\mathcal{I}_\delta})}\right)\right]. \tag{3.20}$$

Proof Using the Poincaré inequality in Lemma 3.2, we see that the functional is coercive and continuous in the set $w^\delta_D + V_{c,\delta}(\widehat{\Omega}_\delta) \cap \alpha L^m(\Omega)$ with respect to the norm $\|\cdot\|_{L^2_0(\Omega)} + \alpha\|\cdot\|_{L^m(\Omega)}$. Thus, there exists a minimizing sequence that remains uniformly bounded in $w^\delta_D + V_{c,\delta}(\widehat{\Omega}_\delta)\cap\alpha L^m(\Omega)$. One can take a uniformly bounded and weakly convergent sequence $v_n = w_n + w^\delta_D$, $w_n \in V_{c,\delta}(\widehat{\Omega}_\delta) \cap \alpha L^m(\Omega)$ such that

$$\lim_{n\to\infty} E(v_n) = \inf_{v\in w^\delta_D + V_{c,\delta}(\widehat{\Omega}_\delta)\cap\alpha L^m(\Omega)} E(v) > -\infty.$$

Denote u the limit of v_n, then $u \in w^\delta_D + V_{c,\delta}(\widehat{\Omega}_\delta) \cap \alpha L^m(\Omega)$, and we may use the weak lower semicontinuity of the functional $E = E(v)$ to get

$$E(u) \le \underline{\lim}_{n\to\infty} E(v_n) = \inf_{v\in w^\delta_D + V_{c,\delta}(\widehat{\Omega}_\delta)\cap\alpha L^m(\Omega)} E(v),$$

which gives u as the minimizer of $E(v)$ in $w^\delta_D + V_{c,\delta}(\widehat{\Omega}_\delta) \cap \alpha L^m(\Omega)$. One can easily see that the weak formulation is simply the Euler–Lagrange equation. We thus have the existence of the solution. Since $E(v)$ is strictly convex, we also get the uniqueness of the solution. The a priori estimates can be obtained from the weak formulation. Indeed,

$$\mathcal{A}_\delta\left(u, u - w^\delta_D\right) + \mathcal{Q}\left(u, u - w^\delta_D\right) = \left(\widehat{f}, u - w^\delta_D\right)_{\widehat{\Omega}_\delta}.$$

Then Young's inequality implies

$$\begin{aligned}
\frac{1}{\alpha}\mathcal{Q}\left(u, u - w^\delta_D\right) &= \int_\Omega u^m(x)dx - \int_\Omega u^{m-1}(x) w^\delta_D(x)dx \\
&\ge \int_\Omega u^m(x)dx - \int_\Omega \frac{m-1}{m} u^m(x) + \frac{1}{m}\left(w^\delta_D\right)^m(x)dx \\
&= \frac{1}{m}\int_\Omega u^m(x)dx - \frac{1}{m}\int_\Omega \left(w^\delta_D\right)^m(x)dx.
\end{aligned}$$

Note that for the linear case with $\alpha = 0$, we can also use the Lax–Milgram theorem to show the well-posedness of the problem (3.15). □

Remark 3.7 (A Pure Nonlocal Neumann Problem) For the case $\Omega^N_{\mathcal{I}_\delta} = \Omega_{\mathcal{I}_\delta}$ and $\alpha = 0$, that is, a pure linear Neumann problem, due to the compatibility condition imposed on $V_{c,\delta}(\widehat{\Omega}_\delta)$, we can recover the original strong form (2.12) from the linear functional $(\widehat{f}, v)_{\widehat{\Omega}}$ with the following compatibility condition:

$$\int_{\widehat{\Omega}_\delta} \widehat{f}(\boldsymbol{x})d\boldsymbol{x} = \int_{\Omega} f(\boldsymbol{x})d\boldsymbol{x} + \int_{\Omega_{\mathcal{I}_\delta}} g_N^\delta(\boldsymbol{x})d\boldsymbol{x} = 0. \tag{3.21}$$

Meanwhile, for the nonlinear problem (2.27), the condition is equivalently (2.28). Similarly, a compatibility condition is also required for the "regional" nonlocal diffusion problem given in (2.31) or the equivalent form in (2.33):

$$\int_{\Omega} F(\boldsymbol{x}, u(\boldsymbol{x}))d\boldsymbol{x} = \int_{\Omega_1} F(\boldsymbol{x}, u(\boldsymbol{x}))d\boldsymbol{x} + \int_{\Omega_2} G(\boldsymbol{x}, u(\boldsymbol{x}))d\boldsymbol{x} = 0, \tag{3.22}$$

where Ω_1 and Ω_2 represent a decomposition of Ω as in (2.33). Eq. (3.22) follows easily from the symmetry assumption on the nonlocal interaction kernel.

Remark 3.8 From the discussion here, we see that in the weak formulation, different parts of the data $\widehat{f}$ play different roles, in particular when considering the limiting regimes of $\delta \to 0$ or $\delta \to \infty$. That is, f symbolically corresponds to the right-hand side of the equation, while g_N^δ corresponds to the Neumann data.

Remark 3.9 We note that other ways of defining a nonlocal Neumann problem can also be used. For additional discussions about nonlocal Neumann problems, see, e.g., [4, 17, 23, 24, 28, 49, 54, 55, 63, 69]. In addition, we may also consider problems associated with the regional nonlocal and fractional Laplacian defined by (2.32) and(2.29) on the domain Ω. In particular, the bilinear form can be modified as

$$\mathcal{A}_{\delta,\Omega}(u, v) = (u, v)_{V_{c,\delta}(\Omega)} = \frac{1}{2}\int_{\Omega}\int_{\Omega} (v(\boldsymbol{y}) - v(\boldsymbol{x}))(u(\boldsymbol{y}) - u(\boldsymbol{x}))\gamma_\delta(\boldsymbol{x}, \boldsymbol{y})d\boldsymbol{y}d\boldsymbol{x} \tag{3.23}$$

for all $u, v \in V_{c,\delta}(\Omega)$.

Remark 3.10 Recalling the Robin-type constraint specified by (2.24) with (2.25), we can get similar well-posedness in such a case by modifying the bilinear form to

$$\mathcal{A}_\delta(u, v) = (u, v)_{V_{c,\delta}(\widehat{\Omega}_\delta)} + (\mathcal{D}_\delta\, u, v)_{\Omega_{\mathcal{I}_\delta}}, \quad \forall\, u, v \in V_\delta(\widehat{\Omega}_\delta). \tag{3.24}$$

For given $\delta > 0$, if $(\mathcal{D}_\delta\, u, u)_{\Omega_{\mathcal{I}_\delta}} = 0$ for $u \in V_\delta(\widehat{\Omega}_\delta)$ holds only if $u\,|_{\Omega_{\mathcal{I}_\delta}} = 0$, then the coercivity of the modified bilinear form can also be derived (e.g., by following similar discussions on nonlocal Poincaré inequality given in [54]).

To conclude this section, we note that one can establish the maximum principle for linear nonlocal diffusion problems as for the local case; see discussions in [33, 41].

4 Special Examples and Limiting Cases

In this section, we consider some special cases, including nonlocal models with an integrable kernel that are of particular interest in applications such as peridynamics. Then, we consider the local ($\delta \to 0$) and fractional ($\delta \to \infty$) limits of nonlocal diffusion models parameterized by $\delta > 0$. This again is to demonstrate the bridging role of the nonlocal models with a finite $\delta > 0$, i.e., as illustrated in Fig. 1, which generalizes some similar pictures in [75] for homogeneous Dirichlet problems.

4.1 Nonlocal Laplacian with Integrable Kernels

We consider the case with a kernel satisfying

$$\int_{\mathbb{R}^d} \gamma_\delta(\boldsymbol{x}, \boldsymbol{y}) d\boldsymbol{y} = \delta^d C_d(\delta) \int_{B_1(\boldsymbol{0})} \gamma(|\boldsymbol{z}|)\, d\boldsymbol{z} < \infty. \tag{4.1}$$

In this case, for any given $\delta > 0$, $\mathcal{L}_\delta$ and $\mathcal{N}_\delta$ become bounded operators on the L^2 spaces for functions defined on respective domains. One can see that the functions spaces are essentially the L^2 spaces on the corresponding domains [53]. Then for the problem,

$$\begin{cases} -\mathcal{L}_\delta\, u = f(\boldsymbol{x}), & \forall \boldsymbol{x} \in \Omega, \\ u = g_D^\delta(\boldsymbol{x}), & \forall \boldsymbol{x} \in \Omega_{\mathcal{I}_\delta}^D, \\ \mathcal{N}_\delta\, u = g_N^\delta(\boldsymbol{x}), & \forall \boldsymbol{x} \in \Omega_{\mathcal{I}_\delta}^N, \end{cases} \tag{4.2}$$

the basic existence of solution in $L^2(\widehat{\Omega}_\delta)$ can be obtained, similar to earlier discussion, for $f \in L^2(\Omega)$, $g_D^\delta \in L^2(\Omega_{\mathcal{I}_\delta}^D)$, and $g_N^\delta \in L^2(\Omega_{\mathcal{I}_\delta}^N)$.

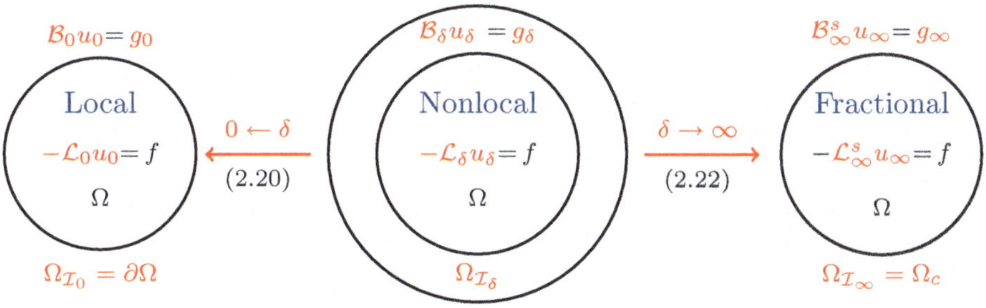

Fig. 1 Limits of nonlocal models with a finite interaction radius δ as $\delta \to 0$ (local limit) with a kernel given by (2.20) and as $\delta \to \infty$ (fractional limit) with a kernel given by (2.22)

We present this special case to highlight the fact that the solutions to the nonlocal diffusion problem may not share the smoothing properties enjoyed by the solutions to typical elliptic PDEs, that is, the solutions may not have a higher order of differentiability than the data.

4.2 Local Limit

We present some results concerning the local limits. We begin with a formal discussion. For example, Eq. (3.15) is a nonlocal analogue of the weak formulation corresponding to (2.3) with $\mathcal{B}_0\, u = u$ on $\partial\Omega^D$ and $\mathcal{B}_0\, u = \frac{\partial u}{\partial \mathbf{n}}$ on $\partial\Omega^N$ with the weak form given by

$$\int_\Omega \nabla u \cdot \nabla v d\boldsymbol{x} + \mathcal{Q}(u, v) = \int_\Omega f v d\boldsymbol{x} + \int_{\partial\Omega^N} g_N^0 v d\boldsymbol{x}. \tag{4.3}$$

For a pure linear Neumann problem with $\partial\Omega^N = \partial\Omega$, i.e., $\alpha = 0$ thus $\mathcal{Q}(u, v) = 0$ in (4.3), the functions f and g_N satisfy the conventional compatibility condition

$$\int_\Omega f d\boldsymbol{x} + \int_{\partial\Omega} g_N^0 d\boldsymbol{x} = 0.$$

Next, we present some rigorous results concerning the limit of $\delta \to 0$ of (2.12) for kernels satisfying (2.18) and (2.20). To properly discuss the local limit, we need to make several assumptions on the boundary domains and boundary data.

ASSUMPTION 4.1 *We first make the following assumptions on the domains:*

- *The domains $\Omega^D_{\mathcal{I}_\delta}$ and $\Omega^N_{\mathcal{I}_\delta}$ are monotone decreasing sets as $\delta \to 0$ and $\overline{\Omega^D_{\mathcal{I}_\delta}} \supset \partial\Omega^D$ and $\overline{\Omega^N_{\mathcal{I}_\delta}} \supset \partial\Omega^N$.*
- *Moreover, for any $\delta_1 > \delta_2 > 0$, we assume that $(\Omega^D_{\mathcal{I}_{\delta_1}} \backslash \Omega^D_{\mathcal{I}_{\delta_2}}) \cap \Omega_{\mathcal{I}_{\delta_2}} = \emptyset$.*

Next, we present the assumptions on the data:

- *For a constant $\delta_0 > 0$, we assume that for all $\delta < \delta_0$, the function w_D^δ in (3.13) and (3.14) is extended to $\widehat{\Omega}_{\delta_0}$ with the extension bounded in $V_{c,\delta_0}(\widehat{\Omega}_{\delta_0}) \cap \alpha L^m(\Omega)$ and w_D^δ converges in $L^2(\widehat{\Omega}_{\delta_0})$ to $w_D^0 \in H^1(\widehat{\Omega}_{\delta_0})$ as $\delta \to 0$, where $w_D^0|_{\partial\Omega^D} = g_D^0$.*
- *We also assume that the Neumann data g_N^δ, as a function defined on $\widehat{\Omega}_\delta$, are uniformly bounded in the dual space of the energy space. Moreover,*

$$\lim_{\delta \to 0} \int_{\Omega^N_{\mathcal{I}_\delta}} g_N^\delta(\boldsymbol{x}) v(\boldsymbol{x}) d\boldsymbol{x} = \int_{\partial\Omega^N} g_N^0(\boldsymbol{x}) v(\boldsymbol{x}) d\boldsymbol{x}\,, \quad \forall v \in C^\infty(\mathbb{R}^d). \tag{4.4}$$

Remark 4.2 Let us note that under the assumption on m and the Sobolev imbedding theorem, we have $\alpha w_D^0 \in \alpha L^m(\Omega)$.

Remark 4.3 Notice the special geometric assumptions on the domains $\Omega_{\mathcal{I}_\delta}^D$ and $\Omega_{\mathcal{I}_\delta}^N$ in the above discussions. In particular, we require $\{\overline{\Omega_{\mathcal{I}_\delta}^D}\}_{\delta>0}$ and $\{\overline{\Omega_{\mathcal{I}_\delta}^N}\}_{\delta>0}$ to be sequences of nested sets that decrease monotonically to $\partial\Omega^D$ and $\partial\Omega^N$, respectively, as $\delta \to 0$. These assumptions rule out the more complicated situations where $\Omega_{\mathcal{I}_\delta}^D$ or $\Omega_{\mathcal{I}_\delta}^N$ may not be attached to the boundary $\partial\Omega$. Some numerical experiments are provided in cases these geometric assumptions are violated.

With these assumptions, we present the convergence theorem for $\delta \to 0$.

Theorem 4.4 *Let u_δ and u_0 be the weak solutions to the nonlocal and local boundary value problems, respectively. Assume that $\|\widehat{f}\|_{V_\delta^*(\widehat{\Omega}_\delta)}$ and $\|g_D^\delta\|_{V_t(\Omega_{\mathcal{I}_\delta}^D)}$ are bounded uniformly in δ; then we have*

$$\|u_\delta - u_0\|_{L^2(\Omega)} \to 0,$$

as $\delta \to 0$.

Proof From the well-posedness theorem, it is easy to see that $\|u_\delta\|_{V_\delta(\widehat{\Omega}_\delta)}$, $\|u_\delta - w_D^\delta\|_{V_\delta(\widehat{\Omega}_\delta)}$, and $\mathcal{Q}(u_\delta, u_\delta)$ are uniformly bounded and $(u_\delta - w_D^\delta)|_{\Omega_{\mathcal{I}_\delta}^D} = 0$. Take $\delta_0 > 0$, and let $\Omega_1 := \overline{\Omega \cup \Omega_{\mathcal{I}_{\delta_0}}^D}$. By the assumptions on the boundary sets, for any $\delta < \delta_0$, we can always do a zero extension of the function $u_\delta - w_D^\delta$ onto Ω_1, and

$$\begin{aligned}&\int_{\Omega_1}\int_{\Omega_1} \gamma_\delta(\boldsymbol{x}, \boldsymbol{y}) \left(\left(u_\delta - w_D^\delta\right)(\boldsymbol{y}) - \left(u_\delta - w_D^\delta\right)(\boldsymbol{x})\right)^2 d\boldsymbol{y} d\boldsymbol{x}\\ &\quad\le \int_{(\Omega\cup\Omega_{\mathcal{I}_\delta}^D)^2\setminus(\Omega_{\mathcal{I}_\delta}^D)^2} \gamma_\delta(\boldsymbol{x}, \boldsymbol{y}) \left(\left(u_\delta - w_D^\delta\right)(\boldsymbol{y}) - \left(u_\delta - w_D^\delta\right)(\boldsymbol{x})\right)^2 d\boldsymbol{y} d\boldsymbol{x}.\end{aligned}$$

We can then use Lemma 3.1 to conclude that any $L^2(\Omega_1)$ limit (up to a subsequence) of $u_\delta - w_D^\delta$ as $\delta \to 0$ belongs to $H^1(\Omega_1)$. Let v^* denote the limit function and $u^* := v^* + w_D^0$. By the assumption on the Dirichlet data, u^* is the $L^2(\Omega_1)$ limit of u_δ. We want to show $u^* = u_0$. First, since $(u_\delta - w_D^\delta)|_{\Omega_{\mathcal{I}_{\delta_0}}^D} = 0$ for all $\delta < \delta_0$, we then have $v^*|_{\Omega_{\mathcal{I}_{\delta_0}}^D} = 0$. This implies that $u^* \in H^1(\Omega)$ and $u^*|_{\partial\Omega^D} = w_D^0|_{\partial\Omega^D} = g_D^0$. Now we only need to show that u^* satisfies the weak form (4.3) for any $v \in C^\infty(\overline{\Omega})$ with $v|_{\partial\Omega^D} = 0$. Assume that the extension of v is made such that $v \in C^\infty(\Omega \cup \Omega_{\mathcal{I}_{\delta_0}})$ and $v|_{\Omega_{\mathcal{I}_{\delta_0}}^D} = 0$. Since u_δ is the weak solution to the nonlocal problem, we have

$$\mathcal{A}_\delta(u_\delta, v) + \mathcal{Q}(u_\delta, v) = (f, v)_\Omega + \left(g_N^\delta, v\right)_{\Omega_{\mathcal{I}_\delta}^N}. \tag{4.5}$$

The right-hand side of (4.5) converges to the right-hand side of (4.3) from the assumption on the Neumann boundary data. Using the same argument in [69, Theorem 3.5], one can show that as $\delta \to 0$,

$$\int_\Omega \int_\Omega \gamma_\delta(\boldsymbol{x}, \boldsymbol{y})(u_\delta(\boldsymbol{y}) - u_\delta(\boldsymbol{x}))(v(\boldsymbol{y}) - v(\boldsymbol{x}))d\boldsymbol{y}d\boldsymbol{x} \to \int_\Omega \nabla u^* \cdot \nabla v d\boldsymbol{x} .$$

Note also that

$$\begin{aligned} &\left| \iint_{(\Omega_{\mathcal{I}_\delta} \times \Omega) \cup (\Omega \times \Omega_{\mathcal{I}_\delta})} \gamma_\delta(\boldsymbol{x}, \boldsymbol{y})(u_\delta(\boldsymbol{y}) - u_\delta(\boldsymbol{x}))(v(\boldsymbol{y}) - v(\boldsymbol{x}))d\boldsymbol{y}d\boldsymbol{x} \right| \\ &\le \|u_\delta\|^2_{V_\delta(\widehat{\Omega}_\delta)} \cdot \iint_{(\Omega_{\mathcal{I}_\delta} \times \Omega) \cup (\Omega \times \Omega_{\mathcal{I}_\delta})} \gamma_\delta(\boldsymbol{x}, \boldsymbol{y})(v(\boldsymbol{y}) - v(\boldsymbol{x}))^2 d\boldsymbol{y}d\boldsymbol{x} \\ &\le C \int_{\Omega_{\mathcal{I}_\delta}} |\nabla v|^2 d\boldsymbol{x} \to 0 , \end{aligned}$$

as $\delta \to 0$. We then have the convergence of $\mathcal{A}_\delta(u_\delta, v)$ to the integral of $\nabla u^* \cdot \nabla v$ on Ω. Meanwhile, by the assumption on m, we have $u_* \in L^m(\Omega)$. Moreover, we observe that there is a constant $C > 0$ only depends on Ω and m such that

$$\begin{aligned} \|u_\delta - u^*\|^{m-1}_{L^{m-1}(\Omega)} &\le \|u_\delta - u^*\|_{L^2(\Omega)} \|u_\delta - u^*\|^{m-2}_{L^{2(m-2)}(\Omega)} \\ &\le C\|u_\delta - u^*\|_{L^2(\Omega)} \|u_\delta - u^*\|^{m-2}_{L^m(\Omega)} \\ &\le C\|u_\delta - u^*\|_{L^2(\Omega)} (\|u_\delta\|_{L^m(\Omega)} + \|u^*\|_{L^m(\Omega)})^{m-2}, \end{aligned}$$

where we have used $2(m-2) \le m$ by the choice of m. Thus, we get the convergence of u_δ to u^* in $L^{m-1}(\Omega)$ (and consequently the convergence of $q(u_\delta)$ to $q(u^*)$ in $L^1(\Omega)$) from the convergence of u_δ to u^* in $L^2(\Omega)$. We thus have the convergence of $\mathcal{Q}(u_\delta, v)$ to $\mathcal{Q}(u_*, v)$. Therefore, u^* is the weak solution of the local boundary value problem, and this implies $u^* = u_0$. □

Remark 4.5 An example for data satisfying (4.4) for Ω being the unit ball and $\Omega^N_{\mathcal{I}_\delta} = \Omega_{\mathcal{I}_\delta}$ is the case of $g^\delta_N(\boldsymbol{x})$ being given by $\delta^{-1} g^0_N(\boldsymbol{x}/|\boldsymbol{x}|)$ for any $\boldsymbol{x} \in \Omega_{\mathcal{I}_\delta}$. Similar constant extensions along normal directions of the boundary $\partial\Omega$ can be constructed for smooth Ω. Other extensions can also be considered, see, for example, [23, 77]. One may also examine how the convergence order depends on the data and the solution; see, for example, [23, 41, 42]. It is no surprise that the constant extension may not be a desirable choice in order to achieve higher convergence order. Moreover, such a constant extension only makes sense if the resulting nonlocal data satisfy conditions specified in the assumption 4.1.

Remark 4.6 A direct connection between the fractional Laplacian and the local Laplacian is the case of $s \to 1^-$. It is known that, see, e.g., [6], the solution of integral fractional diffusion model (2.8) strongly converges to the solution of the local diffusion problem in $H^{1-\epsilon}(\Omega)$. Related discussions on Neumann boundary problems can be found in [44, 47].

Remark 4.7 We may also consider the limit of Robin-type problems considered in Remarks 2.7 and 3.10 assuming that (2.26) holds. Then one may also consider the local $\delta \to 0$ limit to the corresponding boundary value problem of the local Laplacian with the local Robin boundary condition.

Remark 4.8 The special example considered here corresponds to variational problems associated with convex energy so that the solutions are unique. One can consider extensions to problems associated with non-convex energy, see related discussions in [54] using Γ-convergence.

4.3 Fractional Limit

Here, we make more studies on treating fractional PDEs as either specialized nonlocal models or limiting cases.

We can let $\delta = \infty$ for kernels of the types in (2.22). We identify some of the spaces in this case:

$$
\begin{aligned}
V_\infty(\widehat{\Omega}_\infty) = H^s(\mathbb{R}^d), \quad V_{c,\infty}(\widehat{\Omega}_\infty) &= H^s_c(\mathbb{R}^d) \\
&:= \left\{ u \in H^s(\mathbb{R}^d) \colon u = 0 \ \forall \boldsymbol{x} \in \Omega^D_{\mathcal{I}_\infty} \right\}.
\end{aligned}
$$

Remark 4.9 We consider the case that $\Omega^D_{\mathcal{I}_\infty} = \Omega_{\mathcal{I}_\infty}$. For $s > 1/2$, then $H^s_c(\Omega)$ coincides with the space $H^s_0(\Omega)$ that is the closure of $C^\infty_0(\Omega)$ with respect to the $H^s(\Omega)$-norm, whereas for $s < 1/2$, $H^s_c(\Omega)$ is isomorphic to $H^s(\Omega)$ (i.e., functions in $H^s(\Omega)$ with zero extension to Ω_c). In the critical case $s = 1/2$, $H^s_c(\Omega) \subsetneq H^s_0(\Omega)$. See, e.g., [50, Chapter 3] for a detailed discussion.

Let us first consider an example of a pure Dirichlet-type problem. For $\Omega^D_{\mathcal{I}_\infty} = \Omega_{\mathcal{I}_\infty} = \Omega_c$, the following weak formulation has been given in [3]:

$$
\begin{aligned}
&\text{Find } u(\boldsymbol{x}) \in H^s(\mathbb{R}^d) \text{ such that } u(\boldsymbol{x}) = g^\infty_D(\boldsymbol{x}) \text{ for all } \boldsymbol{x} \in \Omega_c \text{ and} \\
&\mathcal{A}_s(u, v) = (f, v)_\Omega + (g^\infty_D, v)_{\Omega,\Omega_c} \quad \text{for all } v \in H^s_c(\mathbb{R}^d),
\end{aligned}
\tag{4.6}
$$

where

$$
\mathcal{A}_s(u, v) = \frac{1}{2} \int_{(\mathbb{R}^d \times \mathbb{R}^d) \setminus (\Omega_c \times \Omega_c)} (u(\boldsymbol{y}) - u(\boldsymbol{x}))(v(\boldsymbol{y}) - v(\boldsymbol{x}))\gamma_s(\boldsymbol{x}, \boldsymbol{y}) d\boldsymbol{y} d\boldsymbol{x}
$$

$$
\begin{aligned}
&= \underbrace{\frac{1}{2}\int_{\Omega}\int_{\Omega}(u(\boldsymbol{y})-u(\boldsymbol{x}))(v(\boldsymbol{y})-v(\boldsymbol{x}))\gamma_s(\boldsymbol{x},\boldsymbol{y})d\boldsymbol{y}d\boldsymbol{x}}_{\mathcal{A}_{\Omega,\Omega}(u,v)} \\
&+ \underbrace{\int_{\Omega}u(\boldsymbol{x})v(\boldsymbol{x})\int_{\Omega_c}\gamma_s(\boldsymbol{x},\boldsymbol{y})d\boldsymbol{y}d\boldsymbol{x}}_{\mathcal{A}_{\Omega,\Omega_c}(u,v)}, \qquad (4.7)
\end{aligned}
$$

and

$$
(g_D^\infty, v)_{\Omega,\Omega_c} = \int_{\Omega} v(\boldsymbol{x})\int_{\Omega_c} g_D^\infty(\boldsymbol{y})\gamma_s(\boldsymbol{x},\boldsymbol{y})d\boldsymbol{y}d\boldsymbol{x}. \qquad (4.8)
$$

It is easy to see that due to the Dirichlet constraint on $\Omega^D_{\mathcal{I}_\infty} = \Omega_{\mathcal{I}_\infty} = \Omega_c$, $u(\boldsymbol{x}) = g_D^\infty(\boldsymbol{x})$ and $v(\boldsymbol{x}) = 0$ for $\boldsymbol{x} \in \Omega_c$. Thus,

$$
\begin{aligned}
(g_D^\infty, v)_{\Omega,\Omega_c} &= \int_{\Omega} v(\boldsymbol{x})\int_{\Omega_c} g_D^\infty(\boldsymbol{y})\gamma_s(\boldsymbol{x},\boldsymbol{y})d\boldsymbol{y}d\boldsymbol{x} \\
&= \int_{\Omega} v(\boldsymbol{x})\int_{\Omega_c} u(\boldsymbol{y})\gamma_s(\boldsymbol{x},\boldsymbol{y})d\boldsymbol{y}d\boldsymbol{x}.
\end{aligned}
$$

Meanwhile, assuming that Fubini's theorem holds, then

$$
\begin{aligned}
&\mathcal{A}_{\Omega,\Omega_c}(u,v) - (g_D^\infty, v)_{\Omega,\Omega_c} \\
&= \int_{\Omega}\int_{\Omega_c} v(\boldsymbol{x})(u(\boldsymbol{x})-u(\boldsymbol{y}))\gamma_s(\boldsymbol{x},\boldsymbol{y})d\boldsymbol{y}d\boldsymbol{x} \\
&= \int_{\Omega}\int_{\Omega_c} (v(\boldsymbol{x})-v(\boldsymbol{y}))(u(\boldsymbol{x})-u(\boldsymbol{y}))\gamma_s(\boldsymbol{x},\boldsymbol{y})d\boldsymbol{y}d\boldsymbol{x} \\
&= \int_{\Omega_c}\int_{\Omega} (v(\boldsymbol{x})-v(\boldsymbol{y}))(u(\boldsymbol{x})-u(\boldsymbol{y}))\gamma_s(\boldsymbol{x},\boldsymbol{y})d\boldsymbol{y}d\boldsymbol{x} \\
&= \frac{1}{2}\int_{\Omega}\int_{\Omega_c} (v(\boldsymbol{x})-v(\boldsymbol{y}))(u(\boldsymbol{x})-u(\boldsymbol{y}))\gamma_s(\boldsymbol{x},\boldsymbol{y})d\boldsymbol{y}d\boldsymbol{x} \\
&+ \frac{1}{2}\int_{\Omega_c}\int_{\Omega} (v(\boldsymbol{x})-v(\boldsymbol{y}))(u(\boldsymbol{x})-u(\boldsymbol{y}))\gamma_s(\boldsymbol{x},\boldsymbol{y})d\boldsymbol{y}d\boldsymbol{x}.
\end{aligned}
$$

Thus, (4.3) is equivalent to that defined (3.15) for $\delta = \infty$ with $\mathcal{Q} = 0$.

Remark 4.10 One may also enforce the volume constraints via Lagrange multipliers, see, e.g., [2].

One can derive some rigorous results on the limit of (2.12) as $\delta \to \infty$ for kernels satisfying (2.22), particularly for the case of pure Dirichlet condition. For example, one can extend the result given [75] for the case of $g_D^\delta \equiv 0$ and $g_D^\infty \equiv 0$ (homogeneous Dirichlet problem) to inhomogeneous data.

Theorem 4.11 *For kernels satisfying* (2.22)*, let* u_δ *and* u_∞ *be the weak solutions to the nonlocal and fractional Dirichlet boundary value problems associated with an element* $w \in H^s(\mathbb{R}^d)$ *and* $g_D^\delta = w\,|_{\Omega_{\mathcal{I}_\delta}}$ *and* $g_D^\infty = w\,|_{\Omega_c}$*, respectively. Assume that* $\|\widehat{f}\|_{V_\delta^\star(\widehat{\Omega}_\delta)}$ *is bounded uniformly in* δ*, and then we have*

$$\|u_\delta - u_\infty\|_{H^s(\Omega)} \to 0,$$

as $\delta \to \infty$.

Indeed, one can get the above conclusion by considering the corresponding problems for $u_\delta - w$ and $u_\infty - w$ with homogeneous Dirichlet data and noticing that $\mathcal{A}_\delta(w, v) \to \mathcal{A}_s(w, v)$ as $\delta \to \infty$ for any $v \in H_c^s(\mathbb{R}^d)$. Then the convergence of $u_\delta \to u_\infty$ can be derived from a similar convergence result for the homogeneous Dirichlet problem, see [75] for more discussions on the latter.

Remark 4.12 In the more general case, further studies are needed to characterize the solution spaces and the weak form as $\delta \to \infty$ for kernels satisfying (2.22). For example, in [26], some weighted norm based on the Neumann data g is introduced, which is involved in the definition of weak solution.

5 Numerical Experiments and Discussion

In this section, we shall present some numerical examples to show the behaviors of the solution to the nonlocal model (2.12) with inhomogeneous boundary conditions. Since the Dirichlet boundary conditions have been intensively studied in existing works, we focus on the cases for the pure Neumann or mixed type boundary conditions here.

We consider a one-dimensional nonlocal problem in the unit interval $\Omega = (0, 1)$ with boundary region $\Omega_{\mathcal{I}_\delta} = (-\delta, 0) \cup (1, 1+\delta)$. Throughout the section, we choose the kernel functions in the fractional type:

$$\gamma_\delta(x, y) = C_{s,\delta}|x - y|^{-1-2s}\chi_{B_\delta(0)}(x - y).$$

Here the normalization constant $C_{s,\delta}(\delta)$ is defined by

$$C_{s,\delta} = \begin{cases} (2-2s)\delta^{-2+2s}, & \text{for } \delta \le \delta_s, \\ \dfrac{2^{2s} s\Gamma(s+1/2)}{\pi^{\frac{1}{2}}\Gamma(1-s)}, & \text{for } \delta \ge \delta_s, \end{cases} \quad \text{with} \quad \delta_s = \left(\frac{(2-2s)\pi^{\frac{1}{2}}\Gamma(1-s)}{2^{2s} s\Gamma(s+1/2)}\right)^{\frac{1}{2-2s}}.$$

In our computation, we use the standard Galerkin finite element method with uniform mesh size $h = 1/M$. Without loss of generality, we assume that $h \ll \delta$ and $r = \delta/h$ is an integer. Define the grid points $\{x_r\}_{i=-r}^{M+r}$ such that $x_r = rh$, and we then define X_h to be the set of functions in $V_{c,\delta}(\widehat{\Omega}_\delta)$ (see the definition in (3.5)) that are linear when restricted to the subintervals $[x_i, x_{i+1}]$, $i = -r, \ldots, M+r$. We fix a sufficiently small mesh size $h = 10^{-4}$.

Example 1 (Local Limit: Pure Neumann Boundary Conditions) To begin with, we examine the nonlocal model (2.12) with the inhomogeneous Neumann boundary conditions. We define the source term $f(x) = x(1-x)$ for all $x \in \Omega$ and the Neumann-type volumetric constraint

$$g_N^\delta(x) = \begin{cases} -\dfrac{1}{6\delta}, & x \in (-\delta, 0), \\ 0, & x \in (1, 1+\delta). \end{cases}$$

Note that the source term f and boundary data g_N^δ satisfy the compatibility condition (3.21), which implies the existence of the weak solution. We shall test two types of Neumann boundary conditions that correspond to the bilinear forms (3.16) and (3.23), respectively, and thus implying implicitly the use of the operators $\mathcal{N}_\delta$ and $\mathcal{N}_{\delta,\widehat{\Omega}_\delta}$, respectively, where

$$\mathcal{N}_\delta\, u(x) := -\int_\Omega (u(y) - u(x))\gamma_\delta(|x-y|)\, dy, \quad \forall x \in \Omega_{\mathcal{I}_\delta}, \tag{5.1}$$

and

$$\mathcal{N}_{\delta,\widehat{\Omega}_\delta}\, u(x) := -\int_{\widehat{\Omega}_\delta} (u(y) - u(x))\gamma_\delta(|x-y|)\, dy, \quad \forall x \in \Omega_{\mathcal{I}_\delta}. \tag{5.2}$$

For both cases, the local limit is the steady diffusion problem

$$-u''(x) = f(x) \text{ for } x \in \Omega, \qquad \text{with } u'(0) = 1/6 \text{ and } u'(1) = 0. \tag{5.3}$$

With the nonlocal Neumann boundary condition $\mathcal{N}_\delta\, u = g_N^\delta$, the finite element scheme for the nonlocal problem reads: find $u_h \in X_h$ such that for all $v_h \in X_h$,

$$\begin{aligned} \mathcal{A}_\delta(u_h, v_h) =& \frac{1}{2}\int_{\widetilde{\Omega}_\delta} (v_h(y) - v_h(x))(u_h(y) - u_h(x))\gamma_\delta(x-y)\, dy\, dx \\ =& \int_\Omega f(x) v_h(x)\, dx + \int_{\Omega_{\mathcal{I}_\delta}} g_N^\delta v_h(x)\, dx. \end{aligned} \tag{5.4}$$

Meanwhile, with the second kind of Neumann boundary condition $\mathcal{N}_{\delta,\widehat{\Omega}_\delta}\, u = g_N^\delta$, the corresponding finite element approximation reads: find $u_h \in X_h$ such that for all $v_h \in X_h$,

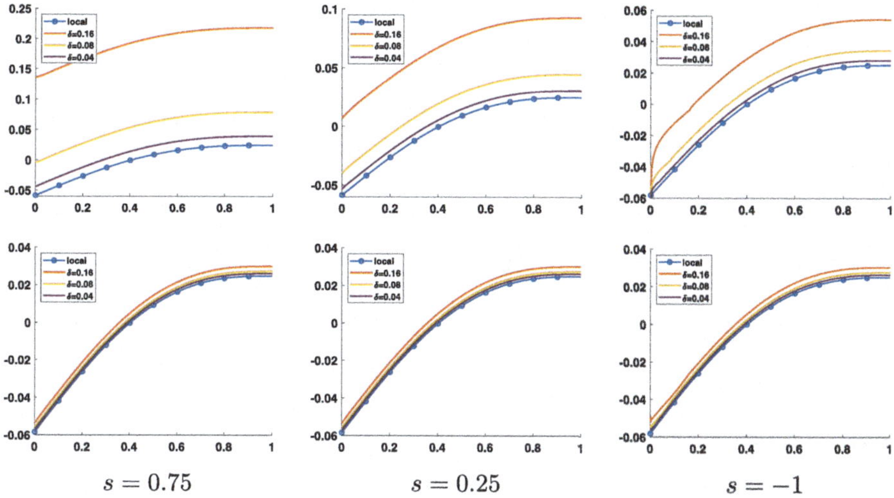

Fig. 2 Example 1. Solutions to the nonlocal diffusion problem with inhomogeneous Neumann-type boundary conditions, with operators $\mathcal{N}_\delta$ (upper) and $\mathcal{N}_{\delta,\widehat{\Omega}_\delta}$ (lower)

$$\begin{aligned}\mathcal{A}_{\delta,\widehat{\Omega}_\delta}(u_h, v_h) =& \frac{1}{2}\int_{\widehat{\Omega}_\delta\times\widehat{\Omega}_\delta}(v_h(y)-v_h(x))(u_h(y)-u_h(x))\gamma_\delta(x-y)\,dy\,dx\\ =& \int_\Omega f(x)v_h(x)\,dx + \int_{\Omega_{\mathcal{I}_\delta}} g_N^\delta v_h(x)\,dx.\end{aligned} \tag{5.5}$$

Both finite-dimensional problems (5.4) and (5.5) are well-posed by the Lax–Milgram theorem and the nonlocal version of Poincaré inequality in Lemma 3.2.

Figure 2 presents the numerical solutions in (0, 1) with various δ and s, while Fig. 3 zooms in on the solution profile near the boundary $x = 0$. In the case that $s = -1$, the solution is nonsmooth, and we observe discontinuity at $x = 0$, while the solution is continuous in $\widehat{\Omega}_\delta$ in cases that $s = 0.25$ and 0.75. As $\delta \to 0$, the numerical results clearly indicate that the solution to the nonlocal diffusion problem converges to u_0, the solution to the local problem. This fully supports our theoretical finding in Theorem 4.4. Meanwhile, with the relatively large nonlocal horizon δ, our empirical experiments also show that the solution of the boundary operator $\mathcal{N}_{\delta,\widehat{\Omega}_\delta}$ is closer to the local solution u_0. Finally, in Table 1, we present the convergence rate as $\delta \to 0$ for both cases. These interesting phenomena warrant further investigation in our future studies.

Example 2 (Local Limit: Mixed Boundary Conditions) In this example, we consider zero source term, i.e., $f(x) = 0$, and the following two types of boundary conditions:

$$\begin{aligned}&\mathcal{N}_\delta u(x) = -\frac{1}{3\delta} \text{ for } x \in (-\delta, -\delta/2), \quad u(x) = 1 \text{ for } x \in [-\delta/2, 0),\\ &u(x) = 1 \text{ for } x \in (1, 1+\delta), \qquad \textbf{Example 2(a)},\end{aligned}$$

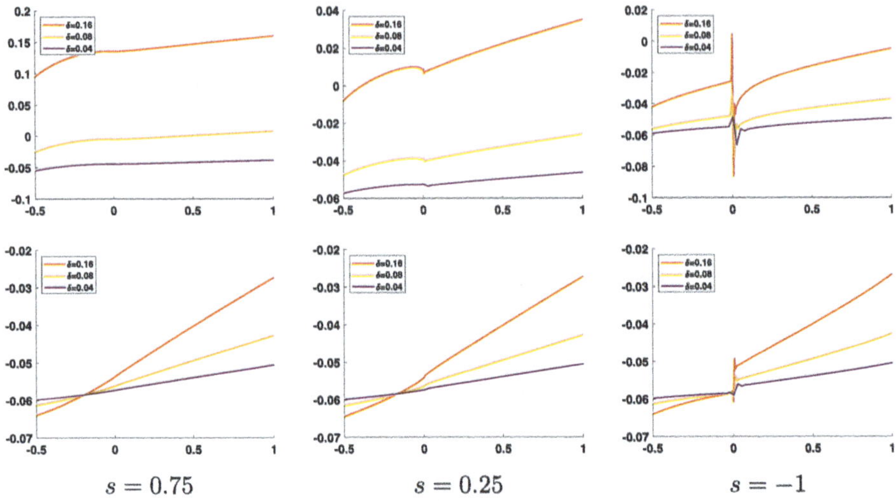

Fig. 3 Example 1. Profiles of $u_\delta(\delta y)$, $y \in (-1/2, 1)$, with various horizon parameters δ and fractional powers s, with operators $\mathcal{N}_\delta$ (upper) and $\mathcal{N}_{\delta,\widehat{\Omega}_\delta}$ (lower)

Table 1 Example 1. $\|u_\delta - u_0\|_{L^2(\Omega)}$ with $s = 0.75, 0.25$ and -1

	$s\backslash\delta$	0.08	0.04	0.02	0.01	0.005	Rate (δ)
$\mathcal{N}_\delta$	−1	9.33E-3	3.02E-3	1.14E-3	4.35E-4	1.97E-4	1.27
	0.25	2.15E-2	6.29E-3	1.92E-3	6.54E-4	2.48E-4	1.47
	0.75	5.97E-2	1.63E-2	4.44E-3	1.27E-3	3.95E-4	1.75
$\mathcal{N}_{\delta,\widehat{\Omega}_\delta}$	−1	2.71E-3	1.33E-3	6.63E-4	3.27E-4	1.60E-4	1.02
	0.25	2.65E-3	1.32E-3	6.60E-4	3.26E-4	1.60E-4	1.01
	0.75	2.62E-3	1.31E-3	6.57E-4	3.26E-4	1.60E-4	1.01

and

$$u(x) = 1 \ \text{ for } x \in (-\delta, -\delta/2), \quad \mathcal{N}_\delta u(x) = -\frac{1}{3\delta} \ \text{ for } x \in [-\delta/2, 0),$$
$$u(x) = 1 \ \text{ for } x \in (1, 1+\delta), \qquad \textbf{Example 2(b)}.$$

Noting that the main difference between Examples 2(a) and (b) is the order of Neumann and Dirichlet boundary conditions in the interval $(-\delta, 0)$. This leads to the case that a part of the data (Neumann data for the former and Dirichlet data for the latter, respectively) is defined in the interval disconnected from $(0, 1)$. It is interesting to observe that, in Fig. 4, both solutions converge to the same local limit, i.e., the solution to the steady diffusion problem: $-u''(x) = 0$ in $\Omega = (0, 1)$ with Dirichlet boundary conditions $u(0) = u(1) = 1$. The solution profiles near $x = 0$, i.e., $u_\delta(\delta y)$ with $y \in (-1, 2)$, are plotted in Fig. 5. We observe the discontinuity at $x = -\delta/2$, which might be due to the incompatibility between the nonlocal Neumann and Dirichlet boundary conditions. In Table 2, we present the difference between u_δ and u_0 in the L^2

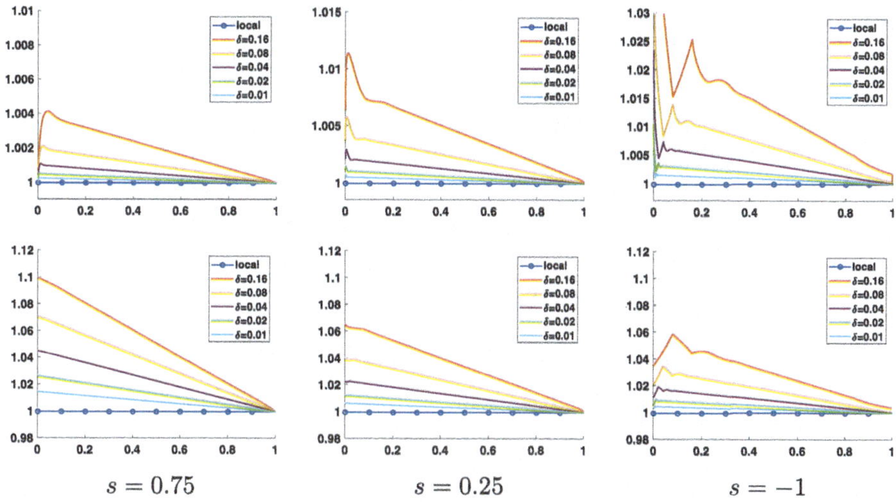

Fig. 4 Example 2 (**a**) (upper) and 2 (**b**) (lower). Solutions to the nonlocal diffusion problem with various horizon parameters δ and fractional powers s

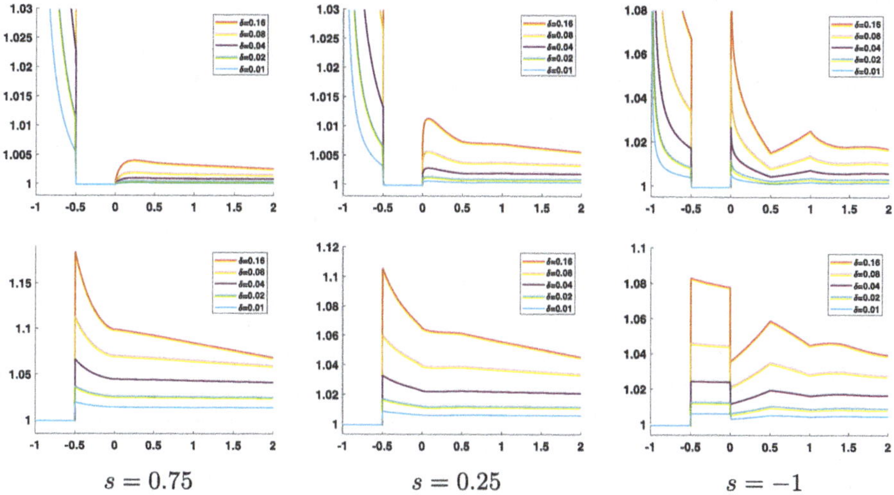

Fig. 5 Example 2(**a**)(upper) and 2(**b**)(lower). Profiles of $u_\delta(\delta y)$, $y \in (-1, 2)$, with various horizon parameters δ and fractional powers s

sense. Our numerical experiments show that the convergence of Example 2 (b) is slower than that of Example 2 (a). These observations deserve further theoretical analysis in the future.

Example 3 (Fractional Limit) In this example, we shall numerically check the limit of nonlocal diffusion models as $\delta \to \infty$. To this end, we consider the boundary value problem of steady fractional diffusion model

Table 2 Example 2 (a) and (b). $\|u_\delta - u_0\|_{L^2(\Omega)}$ with $s = 0.75, 0.25$, and -1

	$s\backslash\delta$	0.16	0.08	0.04	0.02	0.01	Rate (δ)
(a)	-1	1.67E-2	8.04E-3	3.95E-3	1.96E-3	9.84E-4	1.02
	0.25	5.11E-3	2.53E-3	1.27E-3	6.40E-4	3.27E-5	0.99
	0.75	2.32E-3	1.17E-3	5.92E-4	3.01E-4	1.55E-4	0.98
(b)	-1	3.33E-2	1.93E-2	1.06E-2	5.63E-3	2.96E-3	0.87
	0.25	3.90E-2	2.38E-2	1.35E-2	7.31E-3	3.90E-3	0.83
	0.75	5.87E-2	4.15E-2	2.63E-2	1.54E-2	8.56E-3	0.69

$$\begin{cases} -\mathcal{L}_\infty^s\, u_\infty(x) + u_\infty = f(x), & \forall x \in \Omega, \\ \mathcal{N}_\infty^s u_\infty(x) = g(x), & \forall x \in \Omega_c, \end{cases} \tag{5.6}$$

where the fractional operator $\mathcal{L}_\infty^s$ and the Neumann operator $\mathcal{N}_\infty^s$ are defined in (2.6) and (2.10), respectively. Since the closed form of the fractional model is unavailable, we assume that the solution is a Gaussian function:

$$u_\infty(x) = \exp\big(-(x-1/2)^2\big) \qquad \forall\ x \in \mathbb{R}. \tag{5.7}$$

Then it is easy to compute the source term f and the Neumann boundary data g numerically, since $u_\infty(x)$ decays double exponentially as $|x| \to \infty$. In our computation, we evaluate the numerical solution of the following boundary value problem of the steady nonlocal diffusion problem with a finite horizon:

$$\begin{cases} -\mathcal{L}_\delta u_\delta(x) + u_\delta = f(x), & \forall x \in \Omega, \\ \mathcal{N}_\delta u_\delta(x) = g(x), & \forall x \in \Omega_{\mathcal{I}_\delta}. \end{cases} \tag{5.8}$$

Note that, by adding lower order terms u_∞ and u_δ in (5.6) and (5.8), respectively, we will not need to be concerned with data incompatibility for the corresponding fractional and nonlocal problems.

In Fig. 6, we plot the numerical solutions u_δ with various δ and s. The numerical results clearly show the convergence of u_δ to u_∞ as $\delta \to \infty$. Besides, for smaller s, we observe a slower convergence, which might be due to the slower decay of the kernel function in the case of smaller s.

Similar to Example 1, we also consider the boundary value problem with another type of Neumann boundary condition:

$$\begin{cases} -\mathcal{L}_\infty^s\, u_\infty(x) + u_\infty = f(x), & \forall x \in \Omega, \\ \widetilde{\mathcal{N}}_\infty^s\, u_\infty(x) = \widetilde{g}(x), & \forall x \in \Omega_c, \end{cases} \tag{5.9}$$

where the fractional Neumann operator is defined by

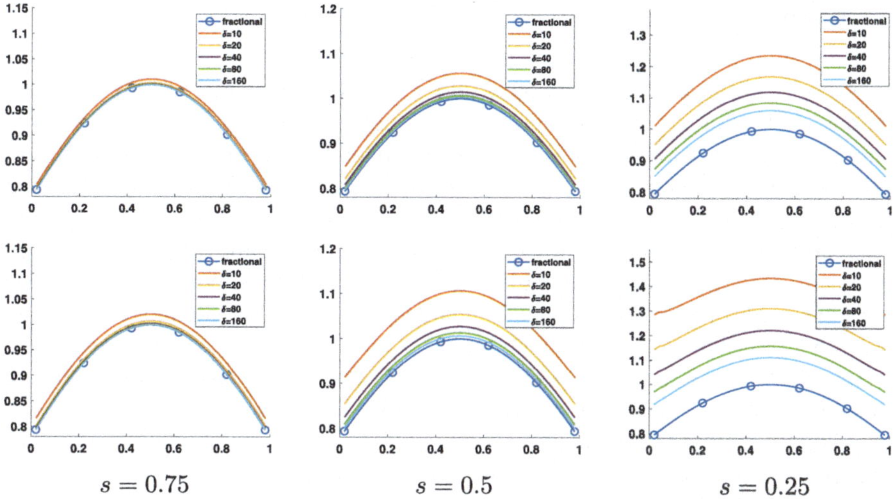

Fig. 6 Example 3. Solutions to the nonlocal diffusion problem with inhomogeneous Neumann-type boundary conditions, with operators $\mathcal{N}_\delta$ (upper) and $\mathcal{N}_{\delta,\widehat{\Omega}_\delta}$ (lower)

Table 3 Example 3. $\|u_\delta - u_\infty\|_{L^2(\Omega)}$ with $s = 0.25, 0.5$, and 0.75

	$s \backslash \delta$	10	20	40	80	160	Rate (δ)
$\mathcal{N}_\delta$	0.25	2.25E-1	1.61E-1	1.15E-1	8.12E-2	5.75E-2	0.49
	0.50	5.55E-2	2.84E-2	1.44E-2	7.22E-3	3.62E-3	0.98
	0.75	1.07E-2	3.93E-3	1.41E-3	5.03E-4	1.78E-4	1.48
$\mathcal{N}_{\delta,\widehat{\Omega}_\delta}$	0.25	4.40E-1	3.15E-1	2.24E-1	1.59E-1	1.12E-1	0.49
	0.50	1.01E-1	5.56E-2	2.81E-2	1.41E-2	7.08E-3	0.99
	0.75	2.11E-2	7.69E-3	2.76E-3	9.84E-3	3.48E-3	1.48

$$\widetilde{\mathcal{N}}^s_\infty u(x) = -C_{s,\infty} \int_{\mathbb{R}} \frac{u(y) - u(x)}{|x-y|^{1+2s}}\, dy, \qquad \forall x \in \Omega_c. \tag{5.10}$$

Let the solution to (5.9) be of the form (5.7). Then we are able to compute the new Neumann boundary data $\widetilde{g}(x) = \widetilde{\mathcal{N}}^s_\infty u_\infty(x)$. The corresponding nonlocal model reads

$$\begin{cases} -\mathcal{L}_\delta u_\delta(x) + u_\delta = f(x), & \forall x \in \Omega, \\ \mathcal{N}_{\delta,\widehat{\Omega}_\delta} u_\delta(x) = \widetilde{g}(x), & \forall x \in \Omega_{\mathcal{I}_\delta}. \end{cases} \tag{5.11}$$

Again, our numerical results indicate that the functions u_δ converge to u_∞ as $\delta \to \infty$, and the convergence rate deteriorates as $s \to 0$, see Fig. 6. For both cases, we observe that the convergence rate is around $O(\delta^{-2s})$(see Table 3), which still awaits theoretical justification.

6 Conclusion

The discussion here offers a glimpse into some of the recent studies of nonlocal models with a finite range of interactions defined on a bounded domain involving various types of nonlocal boundary conditions. In particular, we provide a consistent presentation of such models to classical PDEs, seen as the local limits, and fractional PDEs, seen as the limit with an infinite range of interactions. Obviously, there are many more known results in the literature and more interesting and open questions to be considered. For example, while the associated nonlocal boundary conditions are illustrated for problems defined in a bounded domain, the inhomogeneous data are only prescribed in abstract function spaces. It is interesting to give more precise characterizations of these spaces by applying recently developed trace and extension theorems for the nonlocal function spaces [38, 43]. Also, while we stress the consistent local and fractional limits, one can also further examine how the rates of convergence depend on the model and data in these limits [23, 41, 42]. Moreover, the formulations are only presented in the context of time-independent scalar equations, and one can consider the extensions to nonlinear systems and evolution equations, particularly those connected to peridynamics [28, 36, 40, 64]. Another interesting topic, as previously mentioned in the discussion, is on the connection of the nonlocal problems to the stochastic process and their limits, from the local diffusion equation and Brownian motion to fractional diffusion and Lévy jump processes [26, 39].

In addition, one may consider the numerical analysis of approximation methods and discuss, among many practical issues, the issue of asymptotic compatible approximation [72]. Further numerical experiments can also be carried out to provide more comparisons of the nonlocal models and their various limits.

References

1. N. Abatangelo and L. Dupaigne, *Nonhomogeneous boundary conditions for the spectral fractional Laplacian*, Ann. Inst. H. Poincaré Anal. Non Linéaire, 34 (2017), pp. 439–467.
2. G. Acosta, J.-P. Borthagaray, and N. Heuer, *Finite element approximations of the nonhomogeneous fractional Dirichlet problem*, IMA J. Numer. Anal., 39 (2019), pp. 1471–1501.
3. M. Ainsworth and C. Glusa, *Towards an efficient finite element method for the integral fractional Laplacian on polygonal domains*, in Contemporary Computational Mathematics—A Celebration of the 80th Birthday of Ian Sloan, Springer, 2018, pp. 17–57.
4. F. Andreu, J. Mazón, J. Rossi, and J. Toledo, *Nonlocal Diffusion Problems*, vol. 165 of Mathematical Surveys and Monographs, American Mathematical Society, 2010.
5. H. Antil, J. Pfefferer, and S. Rogovs, *Fractional operators with inhomogeneous boundary conditions: analysis, control, and discretization*, Commun. Math. Sci., 16 (2018), pp. 1395–1426.
6. U. Biccari and V. Hernández-Santamaría, *The Poisson equation from non-local to local*, Electron. J. Differential Equations, (2018), pp. Paper No. 145, 13.
7. K. Bogdan, K. Burdzy, and Z.-Q. Chen, *Censored stable processes*, Probab. Theory Related Fields, 127 (2003), pp. 89–152.

8. A. Bonito, J. P. Borthagaray, R. H. Nochetto, E. Otárola, and A. J. Salgado, *Numerical methods for fractional diffusion*, Computing and Visualization in Science, 19 (2018), pp. 19–46.
9. J. Bourgain, H. Brezis, and P. Mironescu, *Another look at Sobolev spaces*, IOS Press, 2001, pp. 439–455.
10. C. Bucur and E. Valdinoci, *Nonlocal diffusion and applications*, vol. 1, Springer, 2016.
11. D. Burago, S. Ivanov, and Y. Kurylev, *A graph discretization of the Laplace-Beltrami operator*, Journal of Spectral Theory, 4 (2014), pp. 675–714.
12. O. Burkovska and M. Gunzburger, *Regularity and approximation analyses of nonlocal variational equality and inequality problems*, J. Math. Anal. Appl., 478 (2019), pp. 1027–1048.
13. L. Caffarelli and L. Silvestre, *An extension problem related to the fractional Laplacian*, Communications in partial differential equations, 32 (2007), pp. 1245–1260.
14. Z.-Q. Chen and P. Kim, *Green function estimate for censored stable processes*, Probab. Theory Related Fields, 124 (2002), pp. 595–610.
15. R. R. Coifman and S. Lafon, *Diffusion maps*, Applied and Computational Harmonic Analysis, 21 (2006), pp. 5–30. Special Issue: Diffusion Maps and Wavelets.
16. R. R. Coifman, S. Lafon, A. B. Lee, M. Maggioni, F. Warner, and S. Zucker, *Geometric diffusions as a tool for harmonic analysis and structure definition of data: Diffusion maps*, in Proceedings of the National Academy of Sciences, 2005, pp. 7426–7431.
17. C. Cortazar, M. Elgueta, J. Rossi, and N. Wolanski, *How to approximate the heat equation with Neumann boundary conditions by nonlocal diffusion problems*, Archive for Rational Mechanics and Analysis, 187 (2008), pp. 137–156.
18. G.-H. Cottet and P. D. Koumoutsakos, *Vortex methods: theory and practice*, Cambridge University Press, 2000.
19. N. Cusimano, F. del Teso, L. Gerardo-Giorda, and G. Pagnini, *Discretizations of the spectral fractional Laplacian on general domains with Dirichlet, Neumann, and Robin boundary conditions*, SIAM J. Numer. Anal., 56 (2018), pp. 1243–1272.
20. M. D'Elia, Q. Du, C. Glusa, M. Gunzburger, X. Tian, and Z. Zhou, *Numerical methods for nonlocal and fractional models*, Acta Numerica, 29 (2020), pp. 1–124.
21. M. D'Elia, Q. Du, M. Gunzburger, and R. B. Lehoucq, *Nonlocal convection-diffusion problems on bounded domains and finite-range jump processes*, Comput. Meth. in Appl. Math., 17 (2017), pp. 707–722.
22. M. D'Elia and M. Gunzburger, *The fractional Laplacian operator on bounded domains as a special case of the nonlocal diffusion operator*, Comput. Math. Appl., 66 (2013), pp. 1245–1260.
23. M. D'Elia, X. Tian, and Y. Yu, *A physically consistent, flexible, and efficient strategy to convert local boundary conditions into nonlocal volume constraints*, SIAM Journal on Scientific Computing, 42 (2020), pp. A1935–A1949.
24. W. Deng, B. Li, W. Tian, and P. Zhang, *Boundary problems for the fractional and tempered fractional operators*, Multiscale Modeling & Simulation, 16 (2018), pp. 125–149.
25. E. Di Nezza, G. Palatucci, and E. Valdinoci, *Hitchhiker's guide to the fractional Sobolev spaces*, Bull. Sci. Math., 136 (2012), pp. 521–573.
26. S. Dipierro, X. Ros-Oton, and E. Valdinoci, *Nonlocal problems with Neumann boundary conditions*, Rev. Mat. Iberoam., 33 (2017), pp. 377–416.
27. Q. Du, *Local limits and asymptotically compatible discretizations*, in Handbook of peridynamic modeling, Adv. Appl. Math., CRC Press, Boca Raton, FL, 2017, pp. 87–108.
28. Q. Du, *Nonlocal Modeling, Analysis, and Computation*, vol. 94 of CBMS-NSF Conference Series in Applied Mathematics, SIAM, 2019.
29. Q. Du, B. Engquist, and X. Tian, *Multiscale modeling, homogenization and nonlocal effects: mathematical and computational issues*, in Contemporary Mathematics, Celebrating 75th Anniversary of Mathematics of Computation, vol. 754, AMS, 2020, pp. 115–140.
30. Q. Du, M. Gunzburger, R. Lehoucq, and K. Zhou, *Analysis and approximation of nonlocal diffusion problems with volume constraints*, SIAM Review, 54 (2012), pp. 667–696.

31. Q. Du, M. Gunzburger, R. Lehoucq, and K. Zhou, *A nonlocal vector calculus, nonlocal volume-constrained problems, and nonlocal balance laws*, Math. Mod. Meth. Appl. Sci., 23 (2013), pp. 493–540.
32. Q. Du, Z. Huang, and R. Lehoucq, *Nonlocal convection-diffusion volume constrained problems and jump processes*, Disc. Cont. Dyn. Syst. B, 19 (2014), pp. 373–389.
33. Q. Du, L. Ju, X. Li, and Z. Qiao, *Maximum bound principle for a class of semilinear parabolic equations and exponential time differencing schemes*, SIAM Review, 63 (2021), pp. 317–359.
34. Q. Du, T. Mengesha, and X. Tian, *L^p compactness criteria with an application to variational convergence of some nonlocal energy functionals*, Preprint, (2022).
35. Q. Du and Z. Shi, *A nonlocal stokes system with volume constraints*, Numerical Mathematics, Theory, Methods and Applications, under review (2022).
36. Q. Du, Y. Tao, and X. Tian, *A peridynamic model of fracture mechanics with bond-breaking*, Journal of Elasticity, 132 (2018), pp. 197–218.
37. Q. Du and X. Tian, *Mathematics of smoothed particle hydrodynamics via a nonlocal stokes equation*, Foundations of Computational Mathematics, 20 (2020), pp. 801–826.
38. Q. Du, X. Tian, C. Wright, and Y. Yu, *Nonlocal trace spaces and extension results for nonlocal calculus*, J. Functional Analysis, to appear, arXiv preprint arXiv:2107.00177 (2022).
39. Q. Du, L. Toniazzi, and Z. Zhou, *Stochastic representation of solution to nonlocal-in-time diffusion*, Stochastic Processes and their Applications, 130 (2020), pp. 2058–2085.
40. Q. Du, J. Yang, and Z. Zhou, *Analysis of a nonlocal-in-time parabolic equation*, Discrete Contin. Dyn. Syst. Ser. B, 22 (2017), pp. 339–368.
41. Q. Du and X. Yin, *A conforming DG method for linear nonlocal models with integrable kernels*, J. Scientific Computing, 80 (2019), pp. 1913–1935.
42. Q. Du, J. Zhang, and C. Zheng, *On uniform second order nonlocal approximations to linear two-point boundary value problems*, Communications in Mathematical Sciences, 17 (2019), pp. 1737–1755.
43. B. Dyda and M. Kassmann, *Function spaces and extension results for nonlocal Dirichlet problems*, Journal of Functional Analysis, 277 (2019), p. 108134.
44. G. Foghem and M. Kassmann, *A general framework for nonlocal Neumann problems*, arXiv preprint arXiv:2204.06793, (2022).
45. M. Foss, P. Radu, and Y. Yu, *Convergence analysis and numerical studies for linearly elastic peridynamics with Dirichlet-type boundary conditions*, arXiv preprint arXiv:2106.13878, (2021).
46. G. Gilboa and S. Osher, *Nonlocal operators with applications to image processing*, Multi. Model. Simul., 7 (2008), pp. 1005–1028.
47. G. F. F. Gounoue, *L^2-Theory for nonlocal operators on domains*, PhD thesis, Bielefeld University, 2020.
48. Z. Huang, *Nonlocal Models with Convection Effects*, PhD thesis, Pennsylvania State University, 2015.
49. Z. Li, Z. Shi, and J. Sun, *Point integral method for elliptic equations with variable coefficients on point cloud*, Communications in Computational Physics, 26 (2019), pp. 506–530.
50. W. McLean, *Strongly elliptic systems and boundary integral equations*, Cambridge University Press, Cambridge, 2000.
51. T. Mengesha, *Nonlocal Korn-type characterization of Sobolev vector fields*, Communications in Contemporary Mathematics, 14 (2012), p. 1250028.
52. T. Mengesha and Q. Du, *Analysis of a scalar nonlocal peridynamic model with sign changing kernel*, Discrete Contin. Dyn. Syst. Ser. B, 18 (2013), pp. 1415–1437.
53. T. Mengesha and Q. Du, *The bond-based peridynamic system with Dirichlet-type volume constraint*, Proceedings of the Royal Society of Edinburgh A: Mathematics, 144 (2014), pp. 161–186.
54. T. Mengesha and Q. Du, *On the variational limit of a class of nonlocal functionals related to peridynamics*, Nonlinearity, 28 (2015), pp. 3999–4035.

55. T. Mengesha and Q. Du, *Characterization of function spaces of vector fields and an application in nonlinear peridynamics*, Nonlinear Anal., 140 (2016), pp. 82–111.
56. R. Metzler and J. Klafter, *The random walk's guide to anomalous diffusion: a fractional dynamics approach*, Phys. Rep., 339 (2000), p. 77.
57. R. Metzler and J. Klafter, *The restaurant at the end of the random walk: recent developments in the description of anomalous transport by fractional dynamics*, Journal of Physics A: Mathematical and General, 37 (2004), p. R161.
58. J. Monaghan, *Smoothed particle hydrodynamics*, Reports on progress in physics, 68 (2005), p. 1703.
59. R. Nochetto, E. Otárola, and A. J. Salgado, *A PDE approach to fractional diffusion in general domains: a priori error analysis*, Found. Comput. Math., 15 (2015), pp. 733–791.
60. G. Pang, M. D'Elia, M. Parks, and G. E. Karniadakis, *nPINNs: nonlocal physics-informed neural networks for a parametrized nonlocal universal Laplacian operator. Algorithms and applications*, Journal of Computational Physics, 422 (2020), p. 109760.
61. A. C. Ponce, *An estimate in the spirit of Poincaré's inequality*, J. Eur. Math. Soc, 6 (2004), pp. 1–15.
62. R. Servadei and E. Valdinoci, *On the spectrum of two different fractional operators*, Proc. Roy. Soc. Edinburgh Sect. A, 144 (2014), pp. 831–855.
63. Z. Shi and J. Sun, *Convergence of the point integral method for Laplace–Beltrami equation on point cloud*, Research in the Mathematical Sciences, 4 (2017), pp. 1–39.
64. S. Silling, *Reformulation of elasticity theory for discontinuities and long-range forces*, Journal of the Mechanics and Physics of Solids, 48 (2000), pp. 175–209.
65. S. Silling, *Introduction to Peridynamics*, in Handbook of Peridynamic Modeling, Adv. Appl. Math., CRC Press, Boca Raton, FL, 2017, pp. 25–59.
66. I. Sokolov, J. Klafter, and A. Blumen, *Fractional kinetics*, Physics Today, 55 (2002), pp. 48–54.
67. P. R. Stinga, *Fractional powers of second order partial differential operators: extension problem and regularity theory*, PhD thesis, UAM, 2010.
68. P. R. Stinga, *User's guide to the fractional Laplacian and the method of semigroups*, in Handbook of fractional calculus with applications. Vol. 2, De Gruyter, Berlin, 2019, pp. 235–265.
69. Y. Tao, X. Tian, and Q. Du, *Nonlocal diffusion and peridynamic models with Neumann type constraints and their numerical approximations*, Appl. Math. Comput., 305 (2017), pp. 282–298.
70. H. Tian, L. Ju, and Q. Du, *Nonlocal convection-diffusion problems and finite element approximations*, Comput. Methods Appl. Mech. Engrg., 289 (2015), pp. 60–78.
71. H. Tian, L. Ju, and Q. Du, *A conservative nonlocal convection-diffusion model and asymptotically compatible finite difference discretization*, Comput. Methods Appl. Mech. Engrg., 320 (2017), pp. 46–67.
72. X. Tian and Q. Du, *Asymptotically compatible schemes and applications to robust discretization of nonlocal models*, SIAM J. Numer. Anal., 52 (2014), pp. 1641–1665.
73. X. Tian and Q. Du, *Nonconforming discontinuous Galerkin methods for nonlocal variational problems*, SIAM J. Numer. Anal., 53 (2015), pp. 762–781.
74. X. Tian and Q. Du, *Asymptotically compatible schemes for robust discretization of parametrized problems with applications to nonlocal models*, SIAM Review, 62 (2020).
75. X. Tian, Q. Du, and M. Gunzburger, *Asymptotically compatible schemes for the approximation of fractional Laplacian and related nonlocal diffusion problems on bounded domains*, Adv. Comput. Math., 42 (2016), pp. 1363–1380.
76. E. Valdinoci, *From the long jump random walk to the fractional Laplacian*, Bol. Soc. Esp. Mat. Apl. SeMA, 49 (2009), pp. 33–44.
77. H. You, X. Lu, N. Task, and Y. Yu, *An asymptotically compatible approach for Neumann-type boundary condition on nonlocal problems*, ESAIM: Mathematical Modelling and Numerical Analysis, 54 (2020), pp. 1373–1413.
78. Y. Zhang and Z. Shi, *A second-order nonlocal approximation for surface Poisson model with Dirichlet boundary*, arXiv preprint arXiv:2101.01016, (2021).

A One-Dimensional Symmetric-Force-Based Blending Method for Atomistic-to-Continuum Coupling

Elaine Gorom-Alexander and Xingjie Helen Li

Abstract Inspired by the blending method developed by [P. Seleson, S. Beneddine, and S. Prudhome, *A Force-Based Coupling Scheme for Peridynamics and Classical Elasticity*, (2013)] for the nonlocal-to-local coupling, we create a symmetric and consistent blended force-based atomistic-to-continuum (a/c) scheme for the atomistic chain in one-dimensional space. The conditions for the well-posedness of the underlying model are established by analyzing an optimal blending size and blending type to ensure the H^1 semi-norm stability for the blended force-based operator. We present several numerical experiments to test and confirm the theoretical findings.

Keywords Atomistic-to-continuum coupling · Symmetric-force-based blending · Stability and blending size analysis

1 Introduction

Many important materials from airplane wings to computer chips can be improved by a better understanding of failure modes such as fracture and fatigue. Thus, one of the most important goals of computational materials science is to efficiently and reliably predict phenomena such as crack growth and to facilitate the design of new materials better able to resist failure. Scientists and engineers have proposed several multi-scale methods (e.g., [1–10]) to overcome the computational challenges in fidelity and efficiency.

The two main strategies in multi-scale modeling are: (1) bottom-up atomistic-to-continuum: coarse-graining of microscopic descriptions (e.g., atomistic models) of material behavior [2, 3, 11–13]; (2) top-down local-to-nonlocal: informing

E. Gorom-Alexander · X. H. Li (✉)
Department of Mathematics and Statistics, University of North Carolina at Charlotte, Charlotte, NC, USA
e-mail: egorom@uncc.edu; xli47@uncc.edu

T. Mengesha, A. J. Salgado (eds.), *A^3N^2M: Approximation, Applications, and Analysis of Nonlocal, Nonlinear Models*, The IMA Volumes in Mathematics and its Applications 165, https://doi.org/10.1007/978-3-031-34089-5_6

macroscopic models (e.g., continuum equations) with physics gleaned from the microscopic scales [5, 7–10, 14–19]. The former provides a "closer" comparison with macroscopic experiments, and the latter predicts the materials' microscopic properties [6]. Meanwhile, the two approaches have many interconnections, and their development and understanding often inspire each other.

In this chapter, we employ a symmetric blending strategy developed by P. Seleson et al. for the nonlocal-to-local coupling [8, 9] and develop a new force-based atomistic-to-continuum model for a 1D atomistic chain. We then study the stability property of the new coupling scheme in terms of the blending function and its blending size using similar mathematical tools from [20]. We investigate the optimal number of atoms within the blending region to ensure the positive definiteness of the resulting force blending operator under the discrete H^1 semi-norm. The results admit a very narrow blending region to maintain the coercivity and efficiency when the number of atoms is large. In addition, the stability analysis developed in this chapter is crucial for the convergence for several popular iterative methods for solving large-force equilibrium systems.

We will arrange the paper as follows. In Sect. 2, we introduce the force-based symmetric blending method for a 1D atomistic chain. We construct an atomistic, linearized force equation and a continuum, linearized force equation from the atomistic energy equation. The consistency between these methods is discussed in Proposition 2.2. A blending function is then introduced to symmetrically combine these two force-based equations.

In Sect. 3, we establish the optimal conditions on the size of the blending region for the blending force operator with respect to the H^1 stability. Theorem 3.1 establishes these conditions.

In Sect. 4, a uniform stretch is applied to compute the critical strain errors for various types of blending functions with different blending sizes. It is found that the cubic blending function is optimal. We find that a larger polynomial size on the blending region is suggested from the numerical trials when the number of atoms in the chain is just moderately large.

Also in Sect. 4, we test a sine and Gaussian external force to the system to model the displacement with our force-based blending method. The displacements produced by the blending methods with sufficient blending size agree with those of fully atomistic models. In addition, we compare the impact of interaction range of the atomistic model and observe that an interaction range potential greater than 2 neighbors does not change the displacement significantly.

2 Derivation of the Symmetric- and Consistent-Force-Based Scheme

In this section, we will introduce notations, introduce the reference atomistic model, and then derive the continuum approximation. After this, we introduce a blending equation to symmetrically combine these two models. Utilizing this blending equation, a coupling scheme for the blended atomistic and continuum forces is created.

2.1 Notations

We consider a 1D atomistic chain with finite interaction range up to the N-th nearest neighbor and a total number of $2M$ atoms within the domain Ω. We denote the scaled reference lattice $x_\ell = a\ell$ for $\ell \in \mathbb{Z}$ with fixed reference lattice spacing constant $a := \frac{1}{M}$ such that we can select a reference domain that is fixed to be $\Omega = (-1, 1]$. Throughout, the interaction range N will be fixed. The chain is deformed to a current configuration $y_\ell = x_\ell + u_\ell$.

The displacement field $u = (u_\ell)_{\ell\in\mathbb{Z}} : \mathbb{Z} \to \mathbb{R}$ is assumed to be $2M$ periodic discrete function, and $\mathcal{U}$ denotes the space of all $2M$ periodic displacement functions

$$\mathcal{U} := \{u : u_{\ell+2M} = u_\ell | \ell \in \mathbb{Z}\}.$$

Accordingly, we set the deformation space by

$$\mathcal{Y} := \{y : y_\ell = x_\ell + u_\ell | u \in \mathcal{U}, \ell \in \mathbb{Z}\}.$$

We also define the discrete differentiation operator for simplicity, $u^{'}$, on periodic displacements by

$$u_\ell^{'} := \frac{u_{\ell+1} - u_\ell}{a}.$$

Then we may define the higher-order discrete differentiation $u^{''}$, $u^{(3)}$, and $u^{(4)}$ for ℓ by

$$\begin{cases} u_\ell^{''} := \frac{u_\ell' - u_{\ell-1}'}{a}, \\ u_\ell^{(3)} := \frac{u_{\ell+1}'' - u_\ell''}{a}, \\ u_\ell^{(4)} := \frac{u_\ell^{(3)} - u_{\ell-1}^{(3)}}{a}. \end{cases} \tag{2.1}$$

For a displacement $u \in \mathcal{U}$ and its discrete derivatives, we employ the discrete ℓ^2 and ℓ^∞ norms by

$$||u||_{\ell^2}^2 := \sum_{\ell=-M+1}^{M} |u_\ell|^2 a, \quad \text{and} \quad ||u||_{\ell^\infty} := \max_{-M+1 \le \ell \le M} |u_\ell|. \tag{2.2}$$

In particular, the associated inner product for ℓ_2 is

$$\langle u, w\rangle := \sum_{\ell=-M+1}^{M} u_\ell w_\ell \, a.$$

We also employ the discrete H^1 semi-norm, $|u|_{H^1}^2 = \left\| u' \right\|_{\ell_2}^2$, in the stability analysis.

Meanwhile, we proceed with $\widetilde{u} : \mathbb{R} \to \mathbb{R}$ as a quintic spline interpolation of u such that

$$\begin{aligned} \widetilde{u}(a\ell) &= u_\ell, \\ \widetilde{u}(-Ma) &= \widetilde{u}(Ma), \\ \lim_{t\to(a\ell)^-} \frac{d^\omega \widetilde{u}}{dx^\omega}(t) &= \lim_{t\to(a\ell)^+} \frac{d^\omega \widetilde{u}}{dx^\omega}(t), \quad \omega = 1, \ldots, 4 \end{aligned} \quad . \tag{2.3}$$

As $\widetilde{u}$ is a continuous function, we can introduce notations for its derivatives, for instance, $\widetilde{u}_x$ as its first derivative at $(a\ell)$, and $\widetilde{u}_{xx}$ as its second derivative at $(a\ell)$, etc.

We can compare the derivatives of $\widetilde{u}(x)$ with the differencing of u_ℓ. Clearly, we have

$$\begin{aligned} u'_\ell &= \widetilde{u}_x(a\ell) + \frac{a}{2}\widetilde{u}_{xx}(\xi) \\ u''_\ell &= \widetilde{u}_{xx}(a\ell) + \frac{a^2}{12}\widetilde{u}_{xxxx}(\tilde{\xi}) \end{aligned} \quad . \tag{2.4}$$

Note that throughout, subscript ℓ is used to denote when discrete differentiation is employed, whereas subscript x is used to denote when considering the true derivatives.

As in [20], we will frequently use the following discrete summation by parts identity:

Lemma 2.1 *Suppose* $\{u_\ell\}_{\ell=a+1}^b$ *and* $\{v_\ell\}_{\ell=a+1}^b$ *are two sequences; then we have*

$$\sum_{\ell=a+1}^{b} u_\ell(v_\ell - v_{\ell-1}) = [u_b v_b - u_a v_a] - \sum_{\ell=a+1}^{b} (u_\ell - u_{\ell-1}) v_{\ell-1}.$$

Furthermore, when both $\{u_\ell\}_{\ell=a+1}^b$ *and* $\{v_\ell\}_{\ell=a+1}^b$ *are periodic sequences with* $u(a) = u(b)$ *and* $v(a) = v(b)$*, we have*

$$\sum_{\ell=a+1}^{b} u_\ell(v_\ell - v_{\ell-1}) = -\sum_{\ell=a+1}^{b} (u_\ell - u_{\ell-1}) v_{\ell-1}.$$

We use this lemma to find conditions on coupling to ensure the positive definiteness of the bilinear form of the symmetric, blended force-based operator.

2.2 1D Atomistic and Continuum Models

Energy Formulations and Consistency Analysis

We now consider a one-dimensional atomistic periodic chain deformed into configuration $y \in \mathcal{Y}$. Recall that the atomistic periodicity is fixed to $2M$, and the interaction range is fixed to N-th neighbors. The total atomistic energy for this periodic chain is given by

$$E^{a,tot}(y) := \sum_{\ell=-M+1}^{M} \sum_{\substack{k=-N,\\ k\neq 0}}^{N} \frac{a}{2}\phi\left(\frac{y_{\ell+k}-y_\ell}{a}\right), \tag{2.5}$$

where $\phi(\cdot) : \mathbb{R} \to \mathbb{R}$ is a Lennard–Jones-type potential. It can also be viewed as the energy density per unit volume for pairwise interactions. We assume that $\phi(\cdot)$ has the following properties:

- $\phi(r) = \phi(|r|)$.
- ϕ is at least four times differentiable.
- $\phi_{xx}(1) > 0$ and $\phi_{xx}(k) \leq 0$ for $k \geq 2$.

In the numerical simulation, we employ the Morse potential, and a graphical illustration can be found in Fig. 1.

Next, we derive the continuum model by only using the first neighbor distance, that is, we only use the differencing u'_ℓ as we approximate the argument for ϕ. For $k = 2, \ldots, N$, we have

$$\begin{aligned}\frac{y_{\ell+k}-y_\ell}{a} &= \frac{x_{\ell+k}-x_\ell+u_{\ell+k}-u_\ell}{a}\\ &= k + u'_\ell + \sum_{j=1}^{k-1} u'_{\ell+j}.\end{aligned}$$

Using the error estimates listed in (2.4), we can replace $u'_{\ell+j}$ by u'_ℓ and estimate the discrepancy

$$\begin{aligned}\frac{y_{\ell+k}-y_\ell}{a} &= k + u'_\ell + \sum_{j=1}^{k-1} \widetilde{u}_x\left(a(\ell+j)\right) + c_1 a\\ &= k + u'_\ell + \sum_{j=1}^{k-1} \widetilde{u}_x\left((a\ell)\right) + c_2 a\\ &= k + ku'_\ell + c_3 a,\end{aligned} \tag{2.6}$$

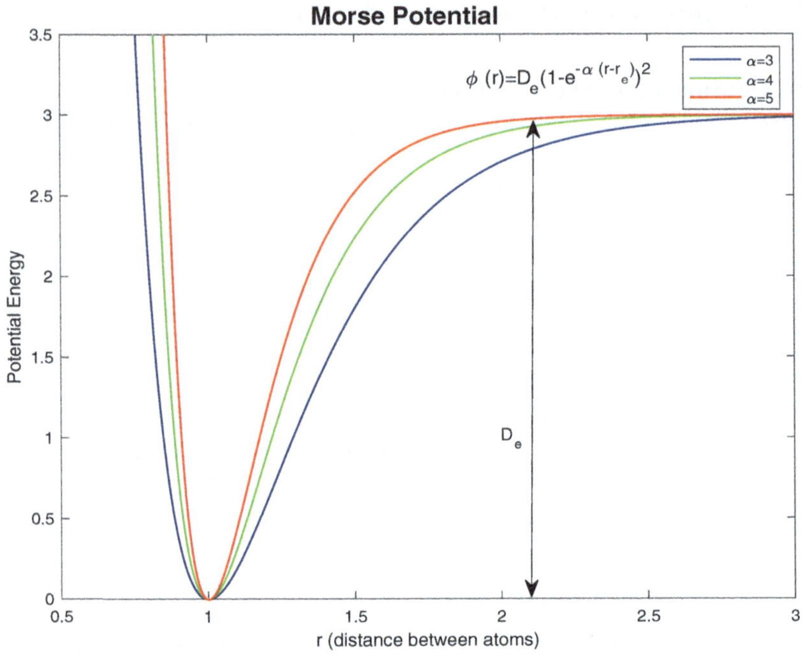

Fig. 1 Graphical illustration of the Lennard–Jones-type interaction potential (Morse potential) used in numerical experiments. Notice that the local minimum is achieved at the nearest neighbor distance $r = r_e$

where c_1, c_2, and c_3 are constants depending on k and regularity of $\widetilde{u}$. For $k = -N, \ldots, -2$, we can obtain similar consistency estimates.

The atomistic energy equation is rooted in discrete, nonlocal energy descriptions. Assuming the finest mesh with each atom regarded as a node and substituting the previous approximation into the atomistic energy (2.5), we thus defined the continuum energy as

$$E^c(u) := \sum_{\ell=-M+1}^{M} \sum_{\substack{k=-N,\\ k\neq 0}}^{N} \frac{a}{2}\phi\left(k + ku_\ell^{'}\right). \tag{2.7}$$

So far, both the atomistic energy and the continuum energy are non-linearly dependent on the displacement field $\{u_\ell\}_{\ell=-M+1}^{M}$, and we would like to apply further simplifications to obtain linear models. To linearize the total atomistic energy, we follow a similar argument as deriving the continuum energy,

$$\begin{aligned}\frac{y_{\ell+k} - y_\ell}{a} &= \frac{x_{\ell+k} - x_\ell + u_{\ell+k} - u_\ell}{a} \\ &= k + \frac{u_{\ell+k} - u_\ell}{a}.\end{aligned}$$

Therefore, the total atomistic energy can be written as

$$\begin{aligned} E^{a,tot}(y) &:= \sum_{\ell=-M+1}^{M} \sum_{\substack{k=-N,\\ k\neq 0}}^{N} \frac{a}{2}\phi\left(\frac{y_{\ell+k}-y_{\ell}}{a}\right) \\ &= \sum_{\ell=-M+1}^{M} \sum_{\substack{k=-N,\\ k\neq 0}}^{N} \frac{a}{2}\phi\left(k+\frac{u_{\ell+k}-u_{\ell}}{a}\right) \end{aligned} . \tag{2.8}$$

Next, we utilize Taylor expansion to $\phi(k+\frac{u_{\ell+k}-u_{\ell}}{a})$ at the reference configuration to linearize the expression.

$$\begin{aligned} \phi\left(k+\frac{u_{\ell+k}-u_{\ell}}{a}\right) &= \phi(k)+\frac{u_{\ell+k}-u_{\ell}}{a}\phi_x(k) \\ &\quad +\frac{1}{2}\left(\frac{u_{\ell+k}-u_{\ell}}{a}\right)^2\phi_{xx}(k)+O\left(\left(\frac{u_{\ell+k}-u_{\ell}}{a}\right)^3\right). \end{aligned}$$

Inserting the Taylor approximation into the atomistic energy equation (2.8), we obtain

$$\begin{aligned} &\sum_{\ell=-M+1}^{M} \sum_{\substack{k=-N,\\ k\neq 0}}^{N} \frac{a}{2}\left[\phi(k)+\frac{u_{\ell+k}-u_{\ell}}{a}\phi_x(k)\right. \\ &\quad \left. +\frac{1}{2}\left(\frac{u_{\ell+k}-u_{\ell}}{a}\right)^2\phi_{xx}(k)+O\left(\left(\frac{u_{\ell+k}-u_{\ell}}{a}\right)^3\right)\right]. \end{aligned} \tag{2.9}$$

Then, without loss of generality, we assume $\sum_{\substack{k=-N,\\ k\neq 0}}^{N} \phi(k)=0$ since this term will not contribute to force. Also, since the reference configuration is a local minimizer and the potential is symmetric, we have

$$\sum_{\substack{k=-N,\\ k\neq 0}}^{N} \frac{u_{\ell+k}-u_{\ell}}{a}\phi_x(k)=0.$$

Thus, the linearized atomistic energy is

$$E^{a,lin}(u) := \sum_{\ell=-M+1}^{M} \sum_{\substack{k=-N,\\ k\neq 0}}^{N} \frac{a}{2}\left(\frac{1}{2}\left(\frac{u_{\ell+k}-u_{\ell}}{a}\right)^2\phi_{xx}(k)\right). \tag{2.10}$$

Following the linearization of the atomistic energy, for the continuum energy, we utilize Taylor expansion to $\phi(k+ku_\ell')$ at the reference configuration to linearize the expression

$$\phi\left(k+ku_\ell'\right)=\phi(k)+ku_\ell'\phi_x(k)+\frac{1}{2}\left(ku_\ell'\right)^2\phi_{xx}(k)+O\left(\left(ku_\ell'\right)^3\right).$$

Inserting the Taylor approximation into the continuum energy equation (2.7), we obtain

$$\sum_{\ell=-M+1}^{M}\sum_{\substack{k=-N,\\k\neq 0}}^{N}\frac{a}{2}\left(\phi(k)+ku_\ell'\phi_x(k)+\frac{1}{2}\left(ku_\ell'\right)^2\phi_{xx}(k)+O\left(\left(ku_\ell'\right)^3\right)\right). \tag{2.11}$$

Following the same justification as above regarding the terms of the Taylor expansion, the linearized continuum energy associated with the finest mesh is defined as

$$\begin{aligned} E^{c,lin}(u) &:= \sum_{\ell=-M+1}^{M}\sum_{\substack{k=-N,\\k\neq 0}}^{N}\frac{a}{2}\left(\frac{k^2}{2}\left(u_\ell'\right)^2\phi_{xx}(k)\right) \\ &= \sum_{\ell=-M+1}^{M}\left(\sum_{k=1}^{N}\frac{k^2}{2}\phi_{xx}(k)\right)\left(u_\ell'\right)^2 a. \end{aligned} \tag{2.12}$$

Remark 2.1 Notice that the classical continuum mechanics for interaction range up to the N-th nearest neighbor has the following form:

$$\widetilde{E}^{c,lin}(\widetilde{u}) = \int W\left(\frac{d\widetilde{u}}{dx}\right)^2 dx, \tag{2.13}$$

where $\frac{d\widetilde{u}}{dx}$ is the deformation field and W represents the strain energy density

$$W := \sum_{k=1}^{N}\frac{k^2}{2}\phi_{xx}(k). \tag{2.14}$$

Notice that if we use a fine mesh and applying a Riemann sum to approximate the integral of (2.13), we can convert (2.13) into (2.12). Therefore, the linearized continuum energy for a discrete lattice system is consistent with the theory of classical continuum mechanics. For the consistency between the classical continuum mechanics and the linearized continuum energy, we refer to Sect. 6 as it follows closely with the consistency for the linearized continuum energy equation and the atomistic energy description.

Since we obtain the atomistic and continuum energies, in the next proposition, we will summarize the truncation errors.

Proposition 2.1 (Consistency Analysis of Linearized Energy Formulations) *Given a fully refined continuum mesh on the 1D atomistic chain, we derive the linearized continuum energy equation* (2.12),

$$E^{c,lin}(u) := \sum_{\ell=-M+1}^{M} \sum_{\substack{k=-N,\\ k\neq 0}}^{N} \frac{a}{2}\left(\frac{k^2}{2}\left(u_\ell^{'}\right)^2 \phi_{xx}(k)\right)$$

from the atomistic energy description, (2.5),

$$E^{a,lin}(u) := \sum_{\ell=-M+1}^{M} \sum_{\substack{k=-N,\\ k\neq 0}}^{N} \frac{a}{2}\left(\frac{1}{2}\left(\frac{u_{\ell+k}-u_\ell}{a}\right)^2 \phi_{xx}(k)\right),$$

with the deformed configuration and the displacement field linked by $y_\ell = x_\ell + u_\ell$. *Then, the consistency between the linearized continuum energy equation and the atomistic energy equation is* $O(a^2)$.

Proof Comparing $E^{c,lin}$ and $E^{a,lin}$, we have

$$E^{c,lin} - E^{a,lin} = \sum_{\ell=-M+1}^{M} \sum_{\substack{k=-N,\\ k\neq 0}}^{N} \frac{a}{2}\left(\frac{k^2}{2}(u_\ell^{'})^2 \phi_{xx}(k)\right)$$
$$- \sum_{\ell=-M+1}^{M} \sum_{\substack{k=-N,\\ k\neq 0}}^{N} \frac{a}{2}\left(\frac{1}{2}\left(\frac{u_{\ell+k}-u_\ell}{a}\right)^2 \phi_{xx}(k)\right).$$

For any $k = 2, ..., N$, we compare $u_{\ell+k}$ around ℓ. Recalling the quintic spline interpolation $\widetilde{u}$ defined in (2.3), we have

$$u_{\ell+k} = u_\ell + ka\widetilde{u}_x + O(a^2).$$

Thus, for all k,

$$u_{\ell+k} - u_\ell = ka\widetilde{u}_x + O(a^2).$$

Then, the consistency analysis of energies yields

$$
\begin{aligned}
E^{c,lin} &- E^{a,lin}\\
&= \sum_{\ell=-M+1}^{M} \sum_{\substack{k=-N,\\ k\neq 0}}^{N} \frac{a}{2}\left(\frac{k^2}{2}(u_\ell^{'})^2\phi_{xx}(k)\right)\\
&\quad - \sum_{\ell=-M+1}^{M} \sum_{\substack{k=-N,\\ k\neq 0}}^{N} \frac{a}{2}\left(\frac{1}{2}\left(\frac{u_{\ell+k}-u_\ell}{a}\right)^2\phi_{xx}(k)\right)\\
&= \sum_{\ell=-M+1}^{M} \sum_{\substack{k=-N,\\ k\neq 0}}^{N} \frac{a}{2}\left(\frac{k^2}{2}\left(\frac{u_{\ell+1}-u_\ell}{a}\right)^2\phi_{xx}(k)\right)\\
&\quad - \sum_{\ell=-M+1}^{M} \sum_{\substack{k=-N,\\ k\neq 0}}^{N} \frac{a}{2}\left(\frac{1}{2}\left(\frac{u_{\ell+k}-u_\ell}{a}\right)^2\phi_{xx}(k)\right)\\
&= \sum_{\ell=-M+1}^{M} \sum_{\substack{k=-N,\\ k\neq 0}}^{N} \frac{a}{2}\left(\frac{k^2}{2}\left(\frac{a\widetilde{u}_x+O(a^2)}{a}\right)^2\phi_{xx}(k)\right)\\
&\quad - \sum_{\ell=-M+1}^{M} \sum_{\substack{k=-N,\\ k\neq 0}}^{N} \frac{a}{2}\left(\frac{1}{2}\left(\frac{ka\widetilde{u}_x+O(a^2)}{a}\right)^2\phi_{xx}(k)\right)\\
&= \sum_{\ell=-M+1}^{M} \sum_{\substack{k=-N,\\ k\neq 0}}^{N} \frac{a}{2}\frac{k^2}{2}\phi_{xx}(k)+O(a^2) - \sum_{\ell=-M+1}^{M} \sum_{\substack{k=-N,\\ k\neq 0}}^{N} \frac{a}{2}\frac{k^2}{2}\phi_{xx}(k)+O(a^2)\\
&= O(a^2).
\end{aligned}
$$

□

Next, we derive the formulae of forces for both linear atomistic and continuum models. Because the mesh is fully refined, the linearized continuum force of atom ℓ can be obtained from taking the first-order variation of the linearized continuum energy (2.12) with respect to u_ℓ, and we thus get

$$
\begin{aligned}
F_{\ell,}^{c,lin}(u) &:= \frac{1}{a}\frac{\delta E^{c,lin}(u)}{\delta u_\ell} \\
&= \frac{\delta\left[\sum_{j=-M+1}^{M}\left(\sum_{k=1}^{N}\frac{k^2}{2}\phi_{xx}(k)\right)\left(u_j^{'}\right)^2\right]}{\delta u_\ell} \\
&= -\left(\left(\sum_{k=1}^{N}\frac{k^2}{2}\phi_{xx}(k)\right)\frac{2(u_{\ell+1}-u_\ell)}{a^2} - \left(\sum_{k=1}^{N}\frac{k^2}{2}\phi_{xx}(k)\right)\frac{2(u_\ell-u_{\ell-1})}{a^2}\right) \\
&= -\left(\sum_{k=1}^{N}k^2\phi_{xx}(k)\right)u_\ell^{''},
\end{aligned}
\tag{2.15}
$$

where we recall the shorthand notation $u_\ell^{''}$ as

$$
u_\ell^{''} := \frac{u_{\ell+1} - 2u_\ell + u_{\ell-1}}{a^2}.
$$

For the atomistic forces, we recall $y_\ell = x_\ell + u_\ell$, take the first-order variation of the total atomistic energy (2.5) at atom ℓ, and notice we employ the forward finite differencing; hence, we obtain

$$
\begin{aligned}
F_{\ell,}^{a}(u) &:= \frac{\delta E^{a,tot}}{\delta u_\ell} = \frac{\delta}{\delta u_\ell}\sum_{j=-M+1}^{M}\sum_{\substack{k=-N,\\ k\neq 0}}^{N}\frac{1}{2}\phi\left(\frac{y_{j+k}-y_j}{a}\right) \\
&= -\sum_{\substack{k=-N,\\ k\neq 0}}^{N}\frac{1}{2a}\left(\phi_x\left(k+\frac{u_{\ell+k}-u_\ell}{a}\right) - \phi_x\left(k+\frac{u_\ell-u_{\ell-k}}{a}\right)\right).
\end{aligned}
\tag{2.16}
$$

Next, we linearize the forces around the reference configuration by applying Taylor expansion to $\phi_x(\cdot)$, so we obtain the linearized atomistic forces

$$
\begin{aligned}
F_{\ell,}^{a,lin}(u) &:= -\sum_{\substack{k=-N,\\ k\neq 0}}^{N}\frac{1}{2}\phi_{xx}(k)\left(\frac{u_{\ell+k}-2u_\ell+u_{\ell-k}}{a^2}\right) \\
&= -\sum_{k=1}^{N}\phi_{xx}(k)\left(\frac{u_{\ell+k}-2u_\ell+u_{\ell-k}}{a^2}\right).
\end{aligned}
\tag{2.17}
$$

In the next proposition, we summarize the consistency errors between the linearized atomistic and linearized continuum forces.

Proposition 2.2 (Consistency Analysis of Force) *Given a fully refined continuum mesh on the 1D atomistic chain, the linearized atomistic force equation* (2.17) *for atom* ℓ *is*

$$F^{a,lin}_{\ell,}(u) := -\sum_{k=1}^{N} \phi_{xx}(k) \left(\frac{u_{\ell+k} - 2u_\ell + u_{\ell-k}}{a^2} \right),$$

and the linearized continuum force equation (2.15) *for node* ℓ *is*

$$F^{c,lin}_{\ell,}(u) = -\left(\sum_{k=1}^{N} k^2 \phi_{xx}(k) \right) u^{''}_{\ell}.$$

Thus, the consistency error between (2.17) *and* (2.15) *is* $O(a^2)$.

Proof Comparing $F^{c,lin}$ and $F^{a,lin}$, we have

$$F^{c,lin}_{\ell,}(u) -_{\ell,} F^{a,lin}(u) = -\sum_{k=1}^{N} \phi_{xx}(k) \left(k^2 u^{''}_{\ell} \right) + \sum_{k=1}^{N} \phi_{xx}(k) \left(\frac{u_{\ell+k} - 2u_\ell + u_{\ell-k}}{a^2} \right).$$

For any $k = 2, \ldots, N$, we compare $u_{\ell+k}$ and $u_{\ell-k}$ around ℓ. Recalling the quintic spline interpolation $\widetilde{u}$ defined in (2.3), we have

$$\begin{aligned}
u_{\ell+k} &= \widetilde{u}\left(a(\ell+k)\right) - \widetilde{u}\,(a\ell) + \widetilde{u}\,(a\ell) \\
&= u_\ell + ka\widetilde{u}_x + \frac{1}{2}(ka)^2 \widetilde{u}_{xx} + \frac{1}{6}(ka)^3 \widetilde{u}_{xxx} + O(a^4), \\
u_{\ell-k} &= \widetilde{u}\left(a(\ell-k)\right) - \widetilde{u}\,(a\ell) + \widetilde{u}\,(a\ell) \\
&= u_\ell - ka\widetilde{u}_x + \frac{1}{2}(ka)^2 \widetilde{u}_{xx} - \frac{1}{6}(ka)^3 \widetilde{u}_{xxx} + O(a^4).
\end{aligned}$$

Utilizing this Taylor expansion for the atomistic and continuous linear force equations, the consistency analysis yields

$$\begin{aligned}
&F^{c,lin}_{\ell,}(u) - F^{a,lin}_{\ell,}(u) \\
&= -\sum_{k=1}^{N} \phi_{xx}(k) \left(k^2 u^{''}_{\ell} \right) + \sum_{k=1}^{N} \phi_{xx}(k) \left(\frac{u_{\ell+k} - 2u_\ell + u_{\ell-k}}{a^2} \right) \\
&= -\sum_{k=1}^{N} \phi_{xx}(k) \left(k^2 \frac{u_{\ell+1} - 2u_\ell + u_{\ell-1}}{a^2} \right) + \sum_{k=1}^{N} \phi_{xx}(k) \left(k^2 \widetilde{u}_{xx} + O(a^2) \right)
\end{aligned}$$

$$= -\sum_{k=1}^{N} \phi_{xx}(k) \left(k^2 \widetilde{u}_{xx} + O(a^2)\right) + \sum_{k=1}^{N} \phi_{xx}(k) \left(k^2 \widetilde{u}_{xx} + O(a^2)\right)$$
$$= O(a^2).$$

A more thorough proof can be seen in Sect. 6.1. □

Remark 2.2 Notice that a is chosen to be $\frac{1}{M}$ with M being large; hence, the consistency error becomes small when the number of atoms within Ω is sufficiently large.

2.3 Derivation of a Symmetric Blending Model for the AtC Coupling

In this section, we will derive a symmetric- and consistent-force-based atomistic-to-continuum scheme for the 1D atomistic chain.

We first divide the domain of interest into three distinct sub-domains: Ω^a: the domain described by the atomistic force; Ω^c: the domain described by the continuum force; and Ω^b: the blending region where the atomistic and continuum force models are both used.

We now introduce a smooth blending function β that can be defined as such:

Definition 2.1 (Definition of Blending Function) We may define a smooth blending function β_ℓ such that:

$$\beta_\ell = \begin{cases} 1, & \ell \in \Omega^a \\ 0, & \ell \in \Omega^c \\ \in (0, 1), & \ell \in \Omega^b. \end{cases} \tag{2.18}$$

This blending function can take many forms, and we will employ linear spline, cubic spline, and quintic spline blending functions in the numerical experiments in Sect. 4.

Notice that creating a linearized force equation will give way to easier analysis in studying the stability of the scheme and providing insights on the coupling conditions of more general cases, so we focus on blending the linearized atomistic and continuum forces. In order for consistent symmetry, we start from $k = -N, \ldots, N$, and $k \neq 0$. Consequently, we start from the linearized, atomistic force equation in (2.17) and incorporate the blending function β_ℓ as follows:

$$F_{\ell,}^{a,lin} := -\sum_{k=1}^{N} \phi_{xx}(k) \frac{u_{\ell+k} - 2u_\ell + u_{\ell-k}}{a^2} = -\sum_{\substack{k=-N,\\ k\neq 0}}^{N} \frac{1}{2}\phi_{xx}(k) \frac{u_{\ell+k} - 2u_\ell + u_{\ell-k}}{a^2}$$

$$= - \sum_{\substack{k=-N,\\ k\neq 0}}^{N} \left(\frac{\beta_\ell + \beta_{\ell+k}}{2}\right) \frac{1}{2}\phi_{xx}(k) \frac{u_{\ell+k} - 2u_\ell + u_{\ell-k}}{a^2}$$

$$- \sum_{\substack{k=-N,\\ k\neq 0}}^{N} \left(1 - \frac{\beta_\ell + \beta_{\ell+k}}{2}\right) \frac{1}{2}\phi_{xx}(k) \frac{u_{\ell+k} - 2u_\ell + u_{\ell-k}}{a^2},$$

such that the term $\frac{u_{\ell+|k|} - 2u_\ell + u_{\ell-|k|}}{a^2}$ is multiplied by the pair $\left(\frac{\beta_\ell + \beta_{\ell+|k|}}{2}\right)$ and $\left(\frac{\beta_\ell + \beta_{\ell-|k|}}{2}\right)$, respectively. Next, we further simplify and get

$$F_{\ell,}^{a,lin} = - \sum_{k=1}^{N} \left(\frac{\beta_{\ell-k} + 2\beta_\ell + \beta_{\ell+k}}{4}\right) \phi_{xx}(k) \frac{u_{\ell+k} - 2u_\ell + u_{\ell-k}}{a^2}$$

$$- \sum_{k=1}^{N} \left(1 - \frac{\beta_{\ell-k} + 2\beta_\ell + \beta_{\ell+k}}{4}\right) \phi_{xx}(k) \frac{u_{\ell+k} - 2u_\ell + u_{\ell-k}}{a^2}. \tag{2.19}$$

Then, we approximate the second term of the equation by using the linearized continuum portion. Therefore, we get the blended force function that is defined as

$$F_{\ell,}^{bqcf,lin}(u) := - \sum_{k=1}^{N} \left(\frac{\beta_{\ell-k} + 2\beta_\ell + \beta_{\ell+k}}{4}\right) \phi_{xx}(k) \frac{u_{\ell+k} - 2u_\ell + u_{\ell-k}}{a^2}$$

$$- \sum_{k=1}^{N} \left(1 - \frac{\beta_{\ell-k} + 2\beta_\ell + \beta_{\ell+k}}{4}\right) \phi_{xx}(k) k^2 \frac{u_{\ell+1} - 2u_\ell + u_{\ell-1}}{a^2} \tag{2.20}$$

$$= - \sum_{k=1}^{N} \left(\frac{\beta_{\ell-k} + 2\beta_\ell + \beta_{\ell+k}}{4}\right) \phi_{xx}(k) \frac{u_{\ell+k} - 2u_\ell + u_{\ell-k}}{a^2}$$

$$- \sum_{k=1}^{N} \left(1 - \frac{\beta_{\ell-k} + 2\beta_\ell + \beta_{\ell+k}}{4}\right) \phi_{xx}(k) k^2 u_\ell''.$$

Remark 2.3 As $\ell - k$ and $\ell + k$ are both employed in the term of the blending function, this blending operator is symmetric as in [8].

We also see that the first term recovers (2.17) for $\beta \equiv 0$ and the second term recovers (2.15) for $\beta \equiv 1$. Thus, the consistency error between the linearized $F_{\ell,}^{bqcf,lin}$ and $F_{\ell,}^{a,lin}$ is also of $O(a^2)$.

3 Stability Analysis for the Linearized Blending Model

3.1 *Bilinear Form of Linearized Blending Model*

In this section, we study the size of the blending region with respect to the H^1 stability of the blending operator. This is achieved by obtaining the optimal conditions in which the linearized coupling operator is positive definite under the discrete H^1 semi-norm.

From (2.20), we consider the bilinear form

$$\left\langle F^{bqcf,lin}_{,}(u), v\right\rangle = \left\langle F^{bqcf,lin}_{,1}(u), v\right\rangle + \sum_{k=2}^{N}\left\langle F^{bqcf,lin}_{,k}(u), v\right\rangle, \quad \forall v \in \mathcal{U}, \tag{3.1}$$

where the first neighbor interaction—the first term—is set apart due to its simplicity in analysis. The second term accounts for the next-nearest neighbor to the N-th nearest neighbor.

To observe the stability of the operator, we first look at the nearest and next-nearest-neighbor interaction, for simplicity as well as for the coercivity assumption on $\phi_{xx}(1) > 0$ and $\phi_{xx}(k) \le 0$ with $k \ge 2$, to find the constraints on the size of blending region. Then, we discuss how the next-nearest-neighbor analysis can be extended to the general N-th-neighbor interaction.

The discrete stability analysis is inspired by and similar to the analogous continuous analysis for the force-based operator that can be seen in Appendix 6. We proceed with the analysis for the discrete case.

3.2 *Stability Analysis for Next-Nearest-Neighbor Interaction Range: N = 2*

Lemma 3.1 *For any displacements $u = (u_\ell)_{\ell=-M+1}^{M}$ from the deformed configuration $y_\ell = x_\ell + u_\ell$, the bilinear forms of nearest neighbor and the next-nearest-neighbor interaction operator can be written as*

$$\left\langle F^{bqcf,lin}_{,1}(u), u\right\rangle = \phi_{xx}(1)\left\| u' \right\|^2_{\ell_2}$$

$$\left\langle F^{bqcf,lin}_{,2}(u), u\right\rangle$$

$$= 2\left\{2\phi_{xx}(2)\left\| u' \right\|^2_{\ell_2} - \frac{\phi_{xx}(2)}{2}a^2\left\| \sqrt{\beta} u'' \right\|^2_{\ell_2} + \frac{\phi_{xx}(2)}{2}a^2\left\| \sqrt{|\beta''|}u' \right\|^2_{\ell_2} + R + S\right\}, \tag{3.2}$$

where

$$R := \sum_{\ell=-M+1}^{M} \frac{\phi_{xx}(2)}{2}\left(u'_{\ell}\beta^{(3)}_{\ell-1}(u_{\ell-1}) - u''_{\ell}\beta''_{\ell}u'_{\ell}a\right)a^3$$

and

$$S := \sum_{\ell=-M+1}^{M} \frac{\phi_{xx}(2)}{2}\left(u'_{\ell}\right)a^2\left(\beta'_{\ell}u'_{\ell} - \beta'_{\ell-2}u'_{\ell-2}\right).$$

Proof For the nearest-neighbor interaction, as the $F^{a,lin}$ and $F^{c,lin}$ coincide, we have

$$\begin{aligned}\left\langle F^{bqcf,lin}_{,1}(u), u\right\rangle &= -\sum_{\ell=-M+1}^{M}\left[\phi_{xx}(1)a\left(\frac{u_{\ell+1} - 2u_{\ell} + u_{\ell-1}}{a^2}\right)\right]u_{\ell} \\ &= -\sum_{\ell=-M+1}^{M} \phi_{xx}(1)u''_{\ell}u_{\ell}a.\end{aligned}$$

Then, using the discrete derivative and summation by parts formula from Lemma 2.1, we conclude that

$$\left\langle F^{bqcf,lin}_{,1}(u), u\right\rangle = -\sum_{\ell=-M+1}^{M} \phi_{xx}(1)u''_{\ell}u_{\ell}a = \phi_{xx}(1)\left\|u'\right\|^2_{\ell_2}.$$

For the next-nearest-neighbor interaction, recall from (2.20) for the linearized blending operator $F^{bqcf,lin}_{\ell,2}$

$$\begin{aligned}&F^{bqcf,lin}_{\ell,2}(u) \\ &= -\Bigg[\left(\frac{\beta_{\ell+k} + 2\beta_{\ell} + \beta_{\ell-k}}{4}\right)\phi_{xx}(k)\left(\frac{u_{\ell+k} - 2u_{\ell} + u_{\ell-k}}{a^2}\right) \\ &\quad + \left(1 - \frac{\beta_{\ell+k} + 2\beta_{\ell} + \beta_{\ell-k}}{4}\right)\phi_{xx}(k)\left(\frac{u_{\ell+1} - 2u_{\ell} + u_{\ell-1}}{a^2}\right)k^2\Bigg]\Bigg|_{k=2},\end{aligned}$$

so the bilinear form with test function u is

$$
\begin{aligned}
&\left\langle F_{,2}^{bqcf,lin}(u), u\right\rangle \\
&\quad = \sum_{\ell=-M+1}^{M} F_{\ell,2}^{bqcf,lin}(u)\cdot u_\ell a \\
&\quad = \sum_{\ell=-M+1}^{M} \left\{ -\left(\frac{\beta_{\ell+2}+2\beta_\ell+\beta_{\ell-2}}{4}\right)\phi_{xx}(2)\left(\frac{u_{\ell+2}-2u_\ell+u_{\ell-2}}{a^2}\right) a \right. \\
&\qquad \left. -\left(1-\frac{\beta_{\ell+2}+2\beta_\ell+\beta_{\ell-2}}{4}\right)\phi_{xx}(2)\left(\frac{u_{\ell+1}-2u_\ell+u_{\ell-1}}{a^2}\right)4a\right\}\cdot u_\ell .
\end{aligned}
\tag{3.3}
$$

We particularly focus on terms contributed to β_ℓ as the other terms could be similarly treated. Hence, we divide the constant "1" in (3.3) by $1 = \frac{1}{4}+\frac{2}{4}+\frac{1}{4}$, collect terms contributed to β_ℓ, and recall the finite difference defined in (2.1); then we have

$$
\begin{aligned}
T &:= \frac{-\phi_{xx}(2)}{2}\left[\sum_{\ell=-M+1}^{M}\left(\frac{u_{\ell+2}-2u_\ell+u_{\ell-2}}{a}\right)(\beta_\ell u_\ell)\right. \\
&\qquad \left. +(1-\beta_\ell)\left(\frac{u_{\ell+1}-2u_\ell+u_{\ell-1}}{a}\right)4u_\ell\right] \\
&= \frac{-\phi_{xx}(2)}{2}\left[\sum_{\ell=-M+1}^{M}\left(u'_{\ell+1}+u'_\ell-u'_{\ell-1}-u'_{\ell-2}\right)(\beta_\ell u_\ell)\right. \\
&\qquad \left. +(1-\beta_\ell)\left(u'_\ell-u'_{\ell-1}\right)4u_\ell\right] \\
&= \frac{-\phi_{xx}(2)}{2}\left[\sum_{\ell=-M+1}^{M}\left(u'_\ell-u'_{\ell-1}\right)4u_\ell\right]+\frac{-\phi_{xx}(2)}{2} \\
&\qquad \times \sum_{\ell=-M+1}^{M}\left[\left(u'_{\ell+1}-u'_\ell\right)-2\left(u'_\ell-u'_{\ell-1}\right)+\left(u'_{\ell-1}-u'_{\ell-2}\right)\right]\cdot(\beta_\ell u_\ell) \\
&=: T_1+T_2 .
\end{aligned}
\tag{3.4}
$$

We consider T_1 first by using Lemma 2.1

$$T_1 = \sum_{\ell=-M+1}^{M} \frac{-\phi_{xx}(2)}{2}\left(u_\ell^{'} - u_{\ell-1}^{'}\right)4u_\ell = \sum_{\ell=-M+1}^{M} \frac{\phi_{xx}(2)}{2}u_{\ell-1}^{'}4\left(u_\ell - u_{\ell-1}\right)$$
$$= \sum_{\ell=-M+1}^{M} \frac{\phi_{xx}(2)}{2}u_{\ell-1}^{'}4\left(u_{\ell-1}^{'}\right)a = 2\phi_{xx}(2)\|u^{'}\|_{\ell_2}^2. \tag{3.5}$$

Treating T_2 is more tedious and is still mainly based on Lemma 2.1. We have

$$T_2 = \sum_{\ell=-M+1}^{M} \frac{-\phi_{xx}(2)}{2}\Big[\Big(\big(u_{\ell+1}^{'} - u_\ell^{'}\big) - \big(u_\ell^{'} - u_{\ell-1}^{'}\big)\Big)$$
$$-\Big(\big(u_\ell^{'} - u_{\ell-1}^{'}\big) - \big(u_{\ell-1}^{'} - u_{\ell-2}^{'}\big)\Big)\Big]\cdot(\beta_\ell u_\ell)$$
$$= \sum_{\ell=-M+1}^{M} \frac{\phi_{xx}(2)}{2}\Big[\Big(\big(u_{\ell+1}^{'} - u_\ell^{'}\big) - \big(u_\ell^{'} - u_{\ell-1}^{'}\big)\Big)\Big]\cdot(\beta_{\ell+1}u_{\ell+1} - \beta_\ell u_\ell)$$
$$= \sum_{\ell=-M+1}^{M} \frac{\phi_{xx}(2)}{2}(u_\ell^{''} - u_{\ell-1}^{''})a\cdot(\beta_{\ell+1}u_{\ell+1} - \beta_\ell u_\ell)$$
$$= \sum_{\ell=-M+1}^{M} \frac{-\phi_{xx}(2)}{2}(u_\ell^{''})a\cdot\Big[(\beta_{\ell+1}u_{\ell+1} - \beta_\ell u_\ell) - (\beta_\ell u_\ell - \beta_{\ell-1}u_{\ell-1})\Big]$$
$$= \sum_{\ell=-M+1}^{M} \frac{-\phi_{xx}(2)}{2}(u_\ell^{''})a\cdot\Big[(\beta_{\ell+1}u_{\ell+1} - \beta_\ell u_{\ell+1})$$
$$+(\beta_\ell u_{\ell+1} - 2\beta_\ell u_\ell + \beta_\ell u_{\ell-1}) - (\beta_\ell u_{\ell-1} - \beta_{\ell-1}u_{\ell-1})\Big]$$
$$= \sum_{\ell=-M+1}^{M} \frac{-\phi_{xx}(2)}{2}(u_\ell^{''})\cdot\beta_\ell u_\ell^{''}a^3 + \frac{-\phi_{xx}(2)}{2}(u_\ell^{''})\cdot\Big[\beta_\ell^{'}u_{\ell+1} - \beta_{\ell-1}^{'}u_{\ell-1}\Big]a^2$$
$$=: -\frac{\phi_{xx}(2)}{2}a^2\|\sqrt{\beta}u^{''}\|_{\ell_2}^2 + T_{22}.$$

Now we mainly focus on T_{22} term, which can be treated as

$$T_{22} = \sum_{\ell=-M+1}^{M} \frac{-\phi_{xx}(2)}{2}(u_\ell^{'} - u_{\ell-1}^{'})\cdot\Big[\beta_\ell^{'}u_{\ell+1} - \beta_{\ell-1}^{'}u_{\ell-1}\Big]a$$
$$= \sum_{\ell=-M+1}^{M} \frac{\phi_{xx}(2)}{2}(u_\ell^{'})\cdot\Big[\Big(\beta_\ell^{'}u_{\ell+1} - \beta_{\ell-1}^{'}u_{\ell-1}\Big) - \Big(\beta_{\ell-1}^{'}u_\ell - \beta_{\ell-2}^{'}u_{\ell-2}\Big)\Big]a.$$

We work on

$$
\begin{aligned}
&\left(\beta_{\ell}' u_{\ell+1} - \beta_{\ell-1}' u_{\ell-1}\right) - \left(\beta_{\ell-1}' u_{\ell} - \beta_{\ell-2}' u_{\ell-2}\right) \\
&\quad = \beta_{\ell}'(u_{\ell+1} - u_{\ell-1}) + u_{\ell-1}\left(\beta_{\ell}' - 2\beta_{\ell-1}' + \beta_{\ell-2}'\right) \\
&\qquad + \beta_{\ell-1}'(u_{\ell-1} - u_{\ell}) + \beta_{\ell-2}'(u_{\ell-2} - u_{\ell-1}) \\
&\quad = \beta_{\ell}'(u_{\ell}' + u_{\ell-1}')a + u_{\ell-1}\beta_{\ell-1}^{(3)}a^2 - \beta_{\ell-1}' u_{\ell-1}' a - \beta_{\ell-2}' u_{\ell-2}' a.
\end{aligned}
$$

Then we plug into T_{22} to get

$$
\begin{aligned}
T_{22} = &\sum_{\ell=-M+1}^{M} \frac{\phi_{xx}(2)}{2}(u_{\ell}') \\
&\cdot \left[\beta_{\ell}'(u_{\ell}' + u_{\ell-1}')a + u_{\ell-1}\beta_{\ell-1}^{(3)}a^2 - \beta_{\ell-1}' u_{\ell-1}' a - \beta_{\ell-2}' u_{\ell-2}' a\right] a \\
= &\frac{\phi_{xx}(2)}{2} \sum_{\ell=-M+1}^{M} (u_{\ell}') \left(\left[u_{\ell-1}\beta_{\ell-1}^{(3)}a^2 + (\beta_{\ell}' u_{\ell-1}' - \beta_{\ell-1}' u_{\ell-1}')\, a\right] a \right. \\
&\left. + \left[\beta_{\ell}' u_{\ell}' - \beta_{\ell-2}' u_{\ell-2}'\right] a^2 \right) \\
= &\frac{\phi_{xx}(2)}{2} \sum_{\ell=-M+1}^{M} \left((u_{\ell}')\beta_{\ell-1}^{(3)}(u_{\ell-1}) + (u_{\ell}')^2 \beta_{\ell}'' - u_{\ell}'' \beta_{\ell}'' u_{\ell}' a\right) a^3 \\
&+ \frac{\phi_{xx}(2)}{2} \sum_{\ell=-M+1}^{M} \left(\beta_{\ell}' u_{\ell}' - \beta_{\ell-2}' u_{\ell-2}'\right) u_{\ell}' a^2.
\end{aligned}
$$

Summarizing all terms T_1, T_2, T_{22} for terms belong to β_{ℓ} in (3.3), and treat those for $\beta_{\ell-2}$ and $\beta_{\ell+2}$ in a similar way, we get (3.2) and R, S terms. □

Remark 3.1 The terms R and S from above are viewed as "residual term." Thus, we will estimate their bounds controlled by the support of β_{ℓ}', the size of blending region, in pursuit of the positive definiteness of the bilinear form.

Lemma 3.2 *Let R, S be defined as above; then we have the following estimates:*

$$
\begin{aligned}
|R| &\le \frac{(-\phi_{xx}(2))}{2}\left(c_3 L^{-\frac{5}{2}} a^{-\frac{1}{2}} + 2c_2(L)^{-2}\right) \left\|u'\right\|_{\ell_2}^2, \\
|S| &\le (-\phi_{xx}(2)) c_1 (L)^{-1} \|u'\|_{\ell_2}^2,
\end{aligned}
\tag{3.6}
$$

where L is the number of atoms within the blending region Ω_b, $a = \frac{1}{M}$ being the lattice spacing, and $2M$ being the total number of atoms within the periodic domain $\Omega = (-1, 1]$, and the constants c_1, c_2, and c_3 depend on $\beta^{(j)}$ with $j = 1, \ldots, 3$, respectively.

Proof Recall that

$$R = \sum_{\ell=-M+1}^{M} \frac{\phi_{xx}(2)}{2} \left(u'_\ell \beta^{(3)}_{\ell-1}(u_{\ell-1}) - u''_\ell \beta''_\ell u'_\ell a \right) a^3.$$

Also notice that the finite differences of β are nonzero only on Ω_b, so utilizing the fact that $\phi_{xx}(2) \le 0$ and Hölder's inequality, we get

$$\begin{aligned} |R| &\le \left| \sum_{\ell\in\Omega^b} \frac{\phi_{xx}(2)}{2} (u'_\ell) \left(\beta^{(3)}_\ell (u_{\ell-1}) \right) a^3 \right| + \left| \sum_{\ell\in\Omega^b} \frac{\phi_{xx}(2)}{2} u''_\ell \beta''_\ell u'_\ell a^4 \right| \\ &\le \frac{-\phi_{xx}(2)}{2} \left\| \beta^{(3)} \right\|_{\ell^\infty} \left\| u' \right\|_{\ell_2(\Omega^b)} \|u\|_{\ell_2(\Omega^b)} a^2 \\ &\quad + \frac{-\phi_{xx}(2)}{2} \|\beta^{(2)}\|_{\ell^\infty} \|u''\|_{\ell_2(\Omega^b)} \|u'\|_{\ell_2(\Omega^b)} a^3. \end{aligned}$$

Also notice that $|\Omega_b| = La$, and by the discrete Poincaré inequality, we have $\|u\|_{\ell_{2(\Omega^b)}} \le (La)^{\frac{1}{2}} \left\| u' \right\|_{\ell_2}$. In addition, we use the fact that $\left\| \beta^{(j)} \right\|_{\ell^\infty} \le c\,(La)^{-j}$. Then

$$\begin{aligned} |R| &\le \frac{-\phi_{xx}(2)}{2} \left\| \beta^{(3)} \right\|_{\ell^\infty} \left\| u' \right\|_{\ell_2(\Omega^b)} \|u\|_{\ell_2(\Omega^b)} a^2 \\ &\quad + \frac{-\phi_{xx}(2)}{2} \|\beta^{(2)}\|_{\ell^\infty} \|u''\|_{\ell_2(\Omega^b)} \|u'\|_{\ell_2(\Omega^b)} a^3 \\ &\le \frac{-\phi_{xx}(2)}{2} c_3 (La)^{-3} (La)^{\frac{1}{2}} \left\| u' \right\|_{\ell_2}^2 a^2 + \frac{-\phi_{xx}(2)}{2} c_2 (La)^{-2} \frac{2}{a} \|u'\|_{\ell_2}^2 a^3 \\ &\le \frac{-\phi_{xx}(2)}{2} \left(c_3 L^{-\frac{5}{2}} a^{-\frac{1}{2}} + 2c_2 (L)^{-2} \right) \left\| u' \right\|_{\ell_2}^2. \end{aligned}$$

Next, we bound S. Recall that

$$S := \sum_{\ell=-M+1}^{M} \frac{\phi_{xx}(2)}{2} (u'_\ell) a^2 \left(\beta'_\ell u'_\ell - \beta'_{\ell-2} u'_{\ell-2} \right).$$

Utilizing similar inequalities, we obtain

$$|S| \leq \frac{-\phi_{xx}(2)}{2}\|\beta'\|_{\ell^\infty} \cdot 2\|u'\|_{\ell_2}^2 a \leq \frac{-\phi_{xx}(2)}{2} c_1 (La)^{-1} \cdot 2\|u'\|_{\ell_2}^2 a$$
$$= \big(-\phi_{xx}(2)\big)c_1(L)^{-1}\|u'\|_{\ell_2}^2.$$

Hence, we prove the lemma. □

Theorem 3.1 *Suppose that the number of atoms M within the chain model is sufficiently large, or equivalently, the lattice spacing $a = \frac{1}{M}$ is sufficiently small; also suppose that the fully atomistic model with next-nearest-neighbor interaction $N = 2$ is stable so that $[\phi_{xx}(1) + 4\phi_{xx}(2)] > 0$. Let L denote the number of atoms within the blending region and let the blending function β be sufficiently smooth such that $\left\|\beta^{(j)}\right\|_\infty \leq (La)^{-j}$. Then there exists a positive constant $\widetilde{C}$ such that*

$$\sum_{k=1}^{2}\sum_{\ell=-M+1}^{M}\left\langle F_{\ell,k}^{bqcf,lin}u, u\right\rangle \geq \widetilde{C}\left\|u'\right\|_{\ell_2}^2. \tag{3.7}$$

$\widetilde{C}$ is strictly bounded above zero as $a = \frac{1}{M} \to 0$.

Proof For $N = 2$ and from Lemma 3.1, the blended force-based operator satisfies

$$\begin{aligned}\sum_{\ell=-M+1}^{M}\left\langle F_{\ell,2}^{bqcf,lin}(u), u\right\rangle = \big(\phi_{xx}(1) + 4\phi_{xx}(2)\big)\left\|u'\right\|_{\ell_2}^2 \\ -\phi_{xx}(2)a^2\left\|\sqrt{\beta}u''\right\|_{\ell_2}^2 + \phi_{xx}(2)a^2\left\|\sqrt{\beta''}u'\right\|_{\ell_2}^2 \\ +2R + 2S.\end{aligned} \tag{3.8}$$

From Lemma 3.2, we have

$$\begin{aligned}|2R + 2S| &\leq -\frac{\phi_{xx}(2)}{2}\left(2c_3L^{-\frac{5}{2}}a^{-\frac{1}{2}} + 4c_2L^{-2} + c_1L^{-1}\right)\left\|u'\right\|_{\ell_2}^2 \\ &\leq -\frac{\phi_{xx}(2)}{2}\left(c_4L^{-\frac{5}{2}}a^{-\frac{1}{2}}\right)\left\|u'\right\|_{\ell_2}^2;\end{aligned} \tag{3.9}$$

hence, we have

$$2R + 2S \geq \frac{\phi_{xx}(2)}{2}\left(c_4L^{-\frac{5}{2}}a^{-\frac{1}{2}}\right)\left\|u'\right\|_{\ell_2}^2$$

with the latter inequality following from $L^{-2} \leq L^{-1} \leq L^{-\frac{5}{2}}a^{-\frac{1}{2}}$ as $L^{-\frac{5}{2}}a^{-\frac{1}{2}}$ dominates the latter two terms when a is sufficiently small.

From (3.8), we want to observe the terms that do not favor the coercivity in the terms of the H^1 semi-norm. For the second term, since $\phi_{xx}(2) \leq 0$, we have

$$- \phi_{xx}(2)a^2 \left\| \sqrt{\beta} u'' \right\|_{\ell_2}^2 \geq 0,$$

and thus it does not negatively contribute. For the third term, we observe that

$$\phi_{xx}(2)a^2 \left\| \sqrt{\beta''} u' \right\|_{\ell_2}^2 \geq \phi_{xx}(2) c_5 L^{-2} \left\| u' \right\|_{\ell_2}^2$$

because of $\phi_{xx}(2) \leq 0$. For up to $N = 2$ neighbor interaction, we have from the coercivity of atomistic model that $\phi''(1) + 4\phi''(2) > 0$; hence, for the linearized force-blended model, we have

$$\begin{aligned}\left\langle F_{,1}^{bqcf,lin} + F_{,2}^{bqcf,lin}, u \right\rangle &\geq \left[\phi_{xx}(1) + \phi_{xx}(2) \left(4 + c_4 L^{-\frac{5}{2}} a^{-\frac{1}{2}} + c_5 L^{-2} \right) \right] \left\| u' \right\|_{\ell_2}^2 \\ &\geq \widetilde{C} \left\| u' \right\|_{\ell_2}^2\end{aligned}$$

for $\widetilde{C} = \left(\phi_{xx}(1) + 2\phi_{xx}(2) \right) > 0$ strictly positive, and independent of $a \to 0$. Thus, we can conclude that the necessary and optimal blending size for coercivity is $L^{-\frac{5}{2}} a^{-\frac{1}{2}} \lesssim 1/c_4$, that is, $L = \widehat{C} a^{-\frac{1}{5}} = \widehat{C} M^{\frac{1}{5}}$ for some $\widehat{C} > 0$. □

3.3 Stability Analysis for General N-th-Nearest-Neighbor Interaction Range

To observe the general form of the N-th-neighbor interaction range, we notice that for $k = 3, \ldots, N$

$$\begin{aligned}&\left\langle F_{,k}^{bqcf,lin}(u), u \right\rangle \\ &= -\phi_{xx}(k) \sum_{\ell=-M+1}^{M} \left\{ \left(\frac{\beta_{\ell+k} + 2\beta_\ell + \beta_{\ell-k}}{4} \right) \left(\frac{u_{\ell+k} - 2u_\ell + u_{\ell-k}}{a^2} \right) \right. \\ &\quad \left. + \left(1 - \frac{\beta_{\ell+k} + 2\beta_\ell + \beta_{\ell-k}}{4} \right) k^2 \left(\frac{u_{\ell+1} - 2u_\ell + u_{\ell-1}}{a^2} \right) \right\} u_\ell a.\end{aligned} \tag{3.10}$$

The k-th-neighbor interaction differs by having $\beta_{\ell \pm k}$ terms that are treated similarly to β_ℓ terms. $\phi_{xx}(k)$ is a non-positive constant term for all $k \geq 2$.

Similarly to the previous subsection, we fix an interaction range k and at the moment only consider all terms that contribute to β_ℓ; thus we get

$$
\begin{aligned}
T :=& -\frac{\phi_{xx}(k)}{2} \sum_{\ell=-M+1}^{M} \left(k^2 \frac{u_{\ell+1} - 2u_\ell + u_{\ell-1}}{a}\right)(u_\ell) \\
&- \frac{\phi_{xx}(k)}{2} \sum_{\ell=-M+1}^{M} \left[\left(\frac{u_{\ell+k} - 2u_\ell + u_{\ell-k}}{a} - k^2 \frac{u_{\ell+1} - 2u_\ell + u_{\ell-1}}{a}\right)(\beta_\ell u_\ell)\right] \\
=& \, T_1 + T_2.
\end{aligned}
$$

For T_1, we similarly have

$$
\begin{aligned}
T_1 &= -\frac{\phi_{xx}(k)}{2} \sum_{\ell=-M+1}^{M} \left(k^2 \frac{u_{\ell+1} - 2u_\ell + u_{\ell-1}}{a}\right)(u_\ell) \\
&= -\frac{\phi_{xx}(k)}{2} \sum_{\ell=-M+1}^{M} k^2 \left(u'_\ell - u'_{\ell-1}\right) u_\ell = \frac{k^2}{2} \phi_{xx}(k) \|u'\|^2_{\ell_2}.
\end{aligned} \tag{3.11}
$$

For T_2, the simplification is more difficult; we have

$$
\begin{aligned}
T_2 &= -\frac{\phi_{xx}(k)}{2} \sum_{\ell=-M+1}^{M} \left[\left(\frac{u_{\ell+k} - 2u_\ell + u_{\ell-k}}{a} - k^2 \frac{u_{\ell+1} - 2u_\ell + u_{\ell-1}}{a}\right)(\beta_\ell u_\ell)\right] \\
&= -\frac{\phi_{xx}(k)}{2} \sum_{\ell=-M+1}^{M} \left[\left(\sum_{j=1}^{k} \sum_{s=1}^{k} u''_{\ell-j+s}\right) - k^2 u''_\ell\right] (a\beta_\ell u_\ell) \\
&= -\frac{\phi_{xx}(k)}{2} \sum_{\ell=-M+1}^{M} \left[\sum_{j=1}^{k} \sum_{s=1}^{k} \left(u''_{\ell-j+s} - u''_\ell\right)\right] (a\beta_\ell u_\ell).
\end{aligned}
$$

Due to the exact symmetry of j and s, we have

$$
\sum_{j=1}^{k} \sum_{s=1}^{k} u''_{\ell-j+s} = \sum_{j=1}^{k} \sum_{s=1}^{k} u''_{\ell-s+j}.
$$

Hence, T_2 can be converted into a symmetrical form

$$
\begin{aligned}
T_2 &= -\frac{\phi_{xx}(k)}{2} \sum_{\ell=-M+1}^{M} \left[\sum_{j=1}^{k} \sum_{s=1}^{k} \left(u''_{\ell-j+s} - u''_{\ell} \right) \right] (a\beta_\ell u_\ell) \\
&= -\frac{\phi_{xx}(k)}{4} \sum_{\ell=-M+1}^{M} \left[\sum_{j=1}^{k} \sum_{s=1}^{k} \left(u''_{\ell-j+s} - 2u''_{\ell} + u''_{\ell+j-s} \right) \right] (a\beta_\ell u_\ell) \\
&= -\frac{\phi_{xx}(k)}{4} \sum_{\ell=-M+1}^{M} \left[\sum_{j=1}^{k} \sum_{s=1}^{k} (u''_{\ell-j+s} - u''_{\ell}) - (u''_{\ell} - u''_{\ell+j-s}) \right] (a\beta_\ell u_\ell).
\end{aligned}
\tag{3.12}
$$

Therefore, by the discrete summation by parts formula, we have

$$
\begin{aligned}
T_2 &= \frac{\phi_{xx}(k)}{4} \sum_{\ell=-M+1}^{M} \left[\sum_{j=1}^{k} \sum_{s=1}^{k} (u''_{\ell-j+s} - u''_{\ell}) \right] a(\beta_{\ell-j+s} u_{\ell-j+s} - \beta_\ell u_\ell) \\
&= \frac{-\phi_{xx}(k)}{4} \sum_{\ell=-M+1}^{M} \left[\sum_{j=1}^{k} \sum_{s=1}^{k} u''_{\ell} a \right] \\
&\quad \times \left(\left(\beta_{\ell-j+s} u_{\ell-j+s} - \beta_\ell u_\ell \right) - \left(\beta_\ell u_\ell - \beta_{\ell+j-s} u_{\ell+j-s} \right) \right) \\
&= \frac{-\phi_{xx}(k)}{4} \sum_{\ell=-M+1}^{M} \left[\sum_{j=1}^{k} \sum_{s=1}^{k} u''_{\ell} a \right] \left(\beta_{\ell-j+s} u_{\ell-j+s} - 2\beta_\ell u_\ell + \beta_{\ell+j-s} u_{\ell+j-s} \right).
\end{aligned}
\tag{3.13}
$$

We can carefully work on the symmetrical term

$$
\begin{aligned}
&\left(\beta_{\ell-j+s} u_{\ell-j+s} - 2\beta_\ell u_\ell + \beta_{\ell+j-s} u_{\ell+j-s} \right) \\
&\overset{r:=j-s}{=} \left(\beta_{\ell-r} u_{\ell-r} - 2\beta_\ell u_\ell + \beta_{\ell+r} u_{\ell+r} \right) \\
&= \beta_{\ell-r} u_{\ell-r} - \beta_\ell u_{\ell+1} + \left(\beta_\ell u_{\ell+1} - 2\beta_\ell u_\ell + \beta_\ell u_{\ell-1} \right) - \beta_\ell u_{\ell-1} + \beta_{\ell+r} u_{\ell+r} \\
&= \beta_\ell u''_\ell a^2 + \beta_{\ell-r} u_{\ell-r} - \beta_\ell u_{\ell+1} + \beta_{\ell+r} u_{\ell+r} - \beta_\ell u_{\ell-1}.
\end{aligned}
$$

Without loss of generality, we assume $r > 0$, and then

$$
\begin{aligned}
&\Big(\beta_{\ell-r}u_{\ell-r} - 2\beta_\ell u_\ell + \beta_{\ell+r}u_{\ell+r}\Big) \\
&\quad = \beta_\ell u_\ell'' a^2 + \beta_{\ell-r}u_{\ell-r} - \beta_\ell u_{\ell+1} + \beta_{\ell+r}u_{\ell+r} - \beta_\ell u_{\ell-1} \\
&\quad = \beta_\ell u_\ell'' a^2 + \big(\beta_\ell u_{\ell-r} - \beta_\ell u_{\ell-1}\big) + \big(\beta_\ell u_{\ell+r} - \beta_\ell u_{\ell+1}\big) \\
&\quad\quad + u_{\ell+r}\sum_{t=0}^{r-1}\beta'_{\ell+t}a - u_{\ell-r}\sum_{t=0}^{r-1}\beta'_{\ell-r+t}a.
\end{aligned}
\tag{3.14}
$$

Therefore, we can handle T_2 for general k-th-neighbor interaction range in a similar way to that for the case of $k = 2$ and obtain that

$$
|T_2| \le (-\phi_{xx}(k))\big(|\widetilde{c}_4|L^{-\frac{5}{2}}a^{-\frac{1}{2}} + |\widetilde{c}_5|L^{-2}\big)\|u'\|_{\ell^2}^2,
$$

which suggests that for $k \ge 2$

$$
T_2 \ge (\phi_{xx}(k))\big(\widetilde{c}_4 L^{-\frac{5}{2}}a^{-\frac{1}{2}} + \widetilde{c}_5 L^{-2}\big)\|u'\|_{\ell^2}^2.
$$

Combining with T_1 estimate (3.11), we have for any $k \ge 2$

$$
\langle F_{,k}^{bqcf,lin}(u), u\rangle \ge \phi_{xx}(k)\left(k^2 + \widehat{C}_k L^{-\frac{5}{2}}a^{-\frac{1}{2}} + \widehat{C}_k L^{-2}\right)\|u'\|_{\ell^2}^2 \tag{3.15}
$$

with $L = \widetilde{C}a^{-\frac{1}{5}}$ when $a = \frac{1}{M}$ being sufficiently small.

Thus, collecting all interactions up to the N-th-nearest-neighbor interaction range, we have

$$
\begin{aligned}
\Big\langle F_{,}^{bqcf,lin}(u), u\Big\rangle &= \Big\langle F_{,1}^{bqcf,lin}(u), u\Big\rangle + \sum_{k=2}^{N}\Big\langle F_{,k}^{bqcf,lin}(u), u\Big\rangle \\
&\ge \left[\phi_{xx}(1) + \sum_{k=2}^{N}\phi_{xx}(k)\big(k^2 + \widehat{C}_k L^{-\frac{5}{2}}a^{-\frac{1}{2}}\big)\right]\|u'\|_{\ell^2}^2.
\end{aligned}
$$

Hence, similar to Theorem 3.1, we summarize the stability conditions on the blending size for a general atomistic chain with N-th-nearest-neighbor interaction in the following theorem.

Theorem 3.2 *Suppose that the number of atoms M is sufficiently large, which is equivalent to $a = \frac{1}{M}$ being sufficiently small, and the blending function β is sufficiently smooth. Also, we assume that the fully atomistic model is stable so that*

$$
\left[\phi_{xx}(1) + \sum_{k=1}^{N}k^2\phi_{xx}(k)\right] > 0.
$$

If the blending size L satisfies $L = \widetilde{C}a^{-\frac{1}{5}}$, *then the linear B-QCF operator* $F^{bqcf,lin}$ *is positive definite in terms of the* H^1 *semi-norm*

$$\sum_{k=1}^{N}\sum_{\ell=-M+1}^{M}\left\langle F_{\ell,k}^{bqcf,lin}u,u\right\rangle \geq \widetilde{C}\left\|u'\right\|_{\ell_2}^2, \tag{3.16}$$

where $\widetilde{C}$ *is strictly bounded above zero as* $a = \frac{1}{M} \to 0$.

Meanwhile, we can see that the bounds are dependent on the smoothness of β. Therefore, we aim to find the optimal types of blending function to preserve the bounds in the coming numerical session.

Remark 3.2 Throughout, the term "optimal" is used to describe the conclusion that $L = \widetilde{C}a^{-\frac{1}{5}}$ is the optimal blending size for keeping coercivity of the force operator when $a = \frac{1}{M} \to 0$. Optimality in these instances describes the smallest asymptotic order that could be expected to ensure stability. However, the analysis of consistency conditions is beyond the purview of this chapter. Note that $\widehat{C}$ is dependent on the blending function, β, and the choice for potential energy, ϕ; and it is not on the lattice spacing constant, a.

Also notice from the inequality (3.15) that $\widehat{C}_k L^{-\frac{5}{2}}a^{-\frac{1}{2}}$ dominates the other terms only when the lattice spacing $a = \frac{1}{M}$ is very small. As a result, the asymptotically rate $L \sim a^{-\frac{1}{5}} = M^{\frac{1}{5}}$ might be not observed when the size of atomistic chain is only moderately large. In this case, we suggest to take $L \sim a^{-\frac{1}{3}} = M^{\frac{1}{3}}$, which is observed in the numerical test.

4 Numerical Experiments

We conduct numerical experiments to verify the theoretical findings from the stability analysis.

4.1 Blending Size and the Stability Constant

We consider a periodic chain with atom indices from $-M+1$ to M. For the following numerical simulations, we set $M = 2000$. The optimal blending size as was found analytically in the previous section is $M^{1/5} \approx 5$. We will test to see if the numerical experiments coincide with this value.

First, we apply the uniform stretch to the atomistic chain and compute the critical strain when $F^{a,lin}$ and $F^{bqcf,lin}$ lose the coercivity. We compare the critical strain values between the atomistic model and blending models with different blending sizes and various types of blending functions. By comparing these values to the atomistic critical strain error, we obtain the optimal blending function and attempt to verify the optimal blending size of $M^{\frac{1}{3}}$ found before.

In numerical experiments, the Morse potential

$$\phi(r) = D_e \times [1 - e^{-\alpha(r-r_e)}]^2$$

was utilized for the interaction potential due to its popularity in applications. We use the values $D_e = 3$ and $\alpha = 3, 4,$ and 5, respectively. Recall from Fig. 1, the local minimal value is set to be $\phi(1)$ and the local height of the potential is D_e. Also, as α grows larger, the more narrow the potential becomes.

We will consider our computational domain as $\Omega = (-1, 1]$ with periodic boundary conditions. The computational domain will be decomposed into several sub-domains following [8]. An interaction range is introduced to serve as a buffer region to simplify the treatment of periodic boundary conditions. The lattice spacing constant that was used had a value of $a = \frac{1}{M}$, which helped to start the blending region.

For the numerical experiments, we denote the blending region, $\Omega^b = (b_1, b_2)$, for some $b_1, b_2 \in \Omega$. The numerical blending size, L, will be defined as $L = b_2 - b_1$. We can compare the numerical blending size to that which was found in the stability analysis.

Recall from Definition (2.18),

$$\beta(x) = \begin{cases} 1, & x \in \Omega^a \\ 0, & x \in \Omega^c \\ \in (0, 1), & x \in \Omega^b = (b_1, b_2). \end{cases}$$

We conduct numerical experiments using a piecewise linear spline, piecewise cubic spline, and piecewise quintic spline blending function that are defined as follows:

$$\beta^{linear}(x) := \begin{cases} 1, & x \in \Omega^a, \\ 0, & x \in \Omega^c, \\ 1 - \frac{x-b_1}{L}, & x \in \Omega^b, \end{cases}$$

and

$$\beta^{cubic}(x) := \begin{cases} 1, & x \in \Omega^a, \\ 0, & x \in \Omega^c, \\ 1 + 2(\frac{x-b_1}{L})^3 - 3(\frac{x-b_1}{L})^2, & x \in \Omega^b, \end{cases}$$

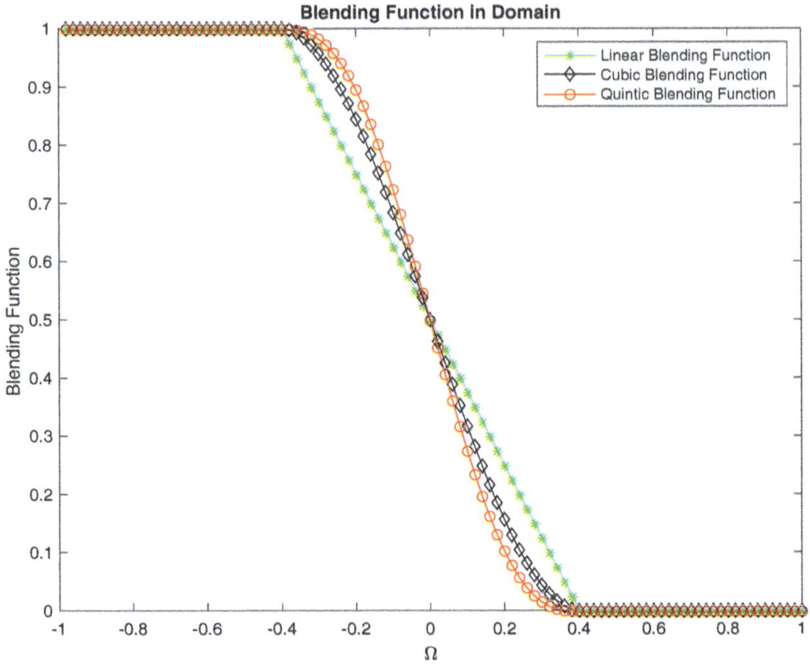

Fig. 2 Pictorial representation of the blending functions used in numerical experiments. Recall, at $\beta = 1$, the purely atomistic model is obtained and at $\beta = 0$, the purely continuum model is obtained

and

$$\beta^{quintic}(x) := \begin{cases} 1, & x \in \Omega^a, \\ 0, & x \in \Omega^c, \\ 1 - 6(\frac{x-b_1}{L})^5 + 15(\frac{x-b_1}{L})^4 - 10(\frac{x-b_1}{L})^3 & x \in \Omega^b. \end{cases}$$

Graphical demonstrations of these various blending functions can be found in Fig. 2.

We apply a uniform stretch to the atomistic chain. From this, we numerically compute the critical strains of the atomistic model and compare this to the coupling model with different blending sizes in pursuit of the critical stretch value that makes the atomistic chain unstable. The step size for increasing γ is $\Delta\gamma = 10^{-5}$. We also model the different values of α in the Morse potential using the cubic blending function, and the results are plotted in Fig. 3a. As can be seen in Table 1, the cubic blending function reaches the atomistic critical stretch value quicker than the other two blending functions (Fig. 3b).

The results from Table 1 suggest the blending size to be $L \approx M^{\frac{1}{3}}$, and this might be due to the other terms in the inequality (3.15) when observing only a moderately large atomistic chain. For further clarification, see Remark 4.1.

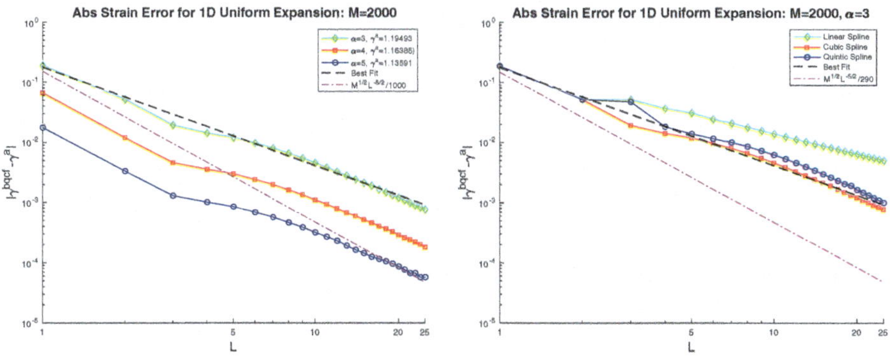

Fig. 3 The absolute critical strain errors are plotted for the 1D uniform stretching. We set $M = 2000$ and γ^a and γ^{bqcf} are the critical strains for the atomistic and B-QCF models, respectively. (**a**) Models the cubic spline blending for various values of α; and (**b**) models the critical strain errors of linear, cubic, and quintic blending functions with $M = 2000$ and $\alpha = 3$

Table 1 Shown are the critical stretching values for linear, cubic, and quintic blending models for a blending size from a to $10a$. The critical value for the purely atomistic model was found to be $\gamma^a = 1.195$. The numerical increment for $\Delta\gamma$ is 10^{-5}

Blend size	Linear	Cubic	Quintic
1	1	1	1
2	1.1400	1.1409	1.1409
3	1.1269	1.1747	1.1456
4	1.1562	1.1801	1.1759
5	1.1624	1.1824	1.1804
6	1.1692	1.1848	1.1828
7	1.1735	1.1866	1.1847
10	1.1811	1.1950	1.1950

The results from the numerical experiments find the cubic blending function as that which converges quickest toward the atomistic strain value and is thus the optimal blending function from those we tested.

Remark 4.1 Recall from (3.9), the assumption that

$$\left(2c_3L^{-\frac{5}{2}}a^{-\frac{1}{2}} + 4c_2L^{-2} + c_1L^{-1}\right) \leq \left(c_4L^{-\frac{5}{2}}a^{-\frac{1}{2}}\right).$$

The latter two terms on the left side of the inequality would not necessarily be negligible if the number of atoms M were not large enough. This accounts for the difference observed in the blending size between the analysis and the numerical simulation.

It must be noted that the analysis conducted in the previous section only applies to the cubic or quintic blending function used in these simulations. Due to less regularities near the boundaries of the blending region, the analysis does not encompass the linear blending function. In these experiments, we can also see that the linear blending leads to the most discrepancies.

4.2 Simulation of Deformed Configuration

Now that we found that the cubic blending function as the optimal blending function, we utilize this for the remaining numerical tests. We use a blend size $L = 5$ since $2000^{\frac{1}{5}} \approx 4.57$ as well as $\alpha = 3$ for both numerical experiments. Next, we test two functions with periodic boundary conditions as the external force of the system to ensure the blended coupling scheme performs as imagined:

- **First, we use a sinusoidal external force**

$$F_{\ell,}^{ext} = 0.01a \times \sin(-x_\ell \times \pi).$$

  We use $a = \frac{1}{5}$ numerically and incorporate -1 into sine because of our left domain boundary.
  We obtain the expected force plot for our domain as can be seen in Fig. 4. The displacement is also observed for an interaction range potential from first neighbor up to third neighbor. We observe that the difference between an interaction range potential of $N = 2$ versus an interaction range potential of $N = 3$ is much smaller than the difference between the change in displacement for the interaction range potential for $N = 1$ and $N = 2$. Recall the features of the Morse potential are such that $\phi_{xx}(1) > 0$ and $\phi_{xx}(k) \leq 0$ for $k \geq 2$. Thus, after the next-nearest neighbor, $k = 2$, the change in displacement will not differ by as much.
- **Next, we test a Gaussian external force:**

$$F_{\ell,}^{ext} = 0.01 \times a \times e^{\frac{-(x_\ell - \mu)^2}{2\sigma^2}},$$

  where $\mu = 4a, \sigma = 50a$, and $a = 1/M$ with $M = 2000$ was used.

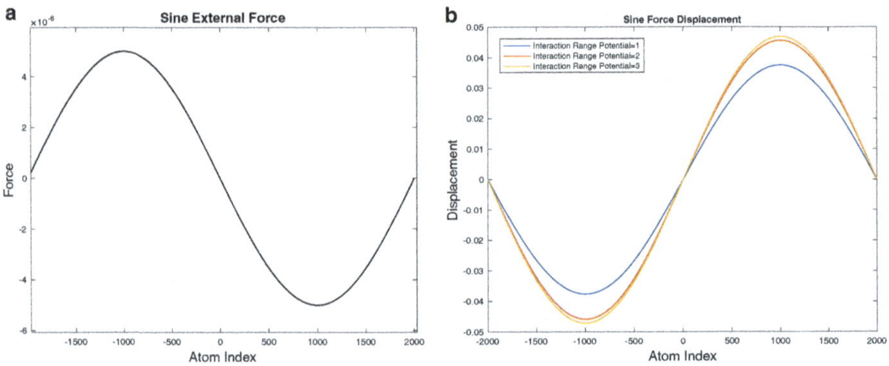

Fig. 4 (**a**) A sinusoidal external force is shown. (**b**) The various displacements for this external force are displayed within the domain

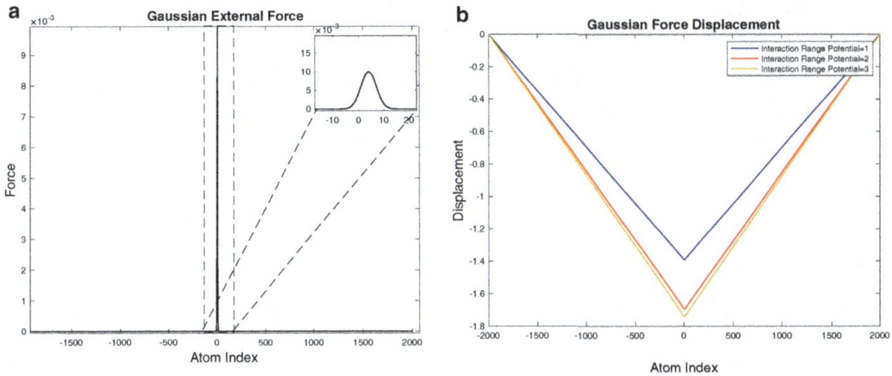

Fig. 5 (**a**) A Gaussian external force is shown. (**b**) The various displacement for this external force is displayed within the domain

Again, we show the force output for our domain and show the various displacements for three interaction range potentials. Similarly to the sinusoidal external force, once the interaction range reaches a value of $N = 2$, the change in displacement becomes less significant (Fig. 5).

5 Conclusion

In this chapter, inspired by the force-based coupling of the peridynamics model of [8], we have formulated a similar symmetric and consistent blended force-based atomistic-to-continuum coupling scheme in one-dimensional space. We were able to identify the optimal asymptotic conditions on the width of the blending region, $L \approx M^{\frac{1}{5}}$ to ensure the H^1 stability of the linearized force blending operator when chain size is huge.

We have verified the theoretical findings with numerical experiments on the blending function and the blending region. From these numerical experiments, we find that the cubic blending provides the best results compared to the critical stretch of the fully atomistic model. We also find that the optimal blending width from these numerical experiments is $L \approx M^{\frac{1}{3}}$ due to non-negligible terms when *not working with* a large enough atomistic chain.

In the future, extension of this scheme to two-dimensional atomistic-to-continuum coupling with a triangular crystal lattice in regard to the neighbors will be pursued.

6 Appendix

6.1 A More Rigorous Proof for Proposition 2.2, A Consistency Analysis of Force

Proof Comparing $F^{c,lin}$ and $F^{a,lin}$, we have

$$F^{c,lin}_{\ell,}(u) - F^{a,lin}_{\ell,}(u) = -\sum_{\substack{k=-N,\\ k\neq 0}}^{N} \frac{1}{2}\phi_{xx}(k)\left(k^2 u_\ell''\right) + \sum_{\substack{k=-N,\\ k\neq 0}}^{N} \frac{1}{2}\phi_{xx}(k)\left(\frac{u_{\ell+k} - 2u_\ell + u_{\ell-k}}{a^2}\right).$$

Let $N \in \mathbb{N}$ be fixed. For any $k = -N, \ldots, N$ and $k \neq 0$, we apply the Taylor expansion to $\widetilde{u}_{\ell+k} = u_{\ell+k}$ and $\widetilde{u}_{\ell-k} = u_{\ell-k}$ around ℓ. We compare the differencing notation used to define the discrete displacement field and with choosing a smooth spline interpolation. We proceed with defining $\widetilde{u}_x$ as this smooth interpolation of the discrete displacement field u in order to compute this approximation.

$$u_{\ell+k} = u_\ell + ka\widetilde{u}_x + \frac{1}{2}(ka)^2\widetilde{u}_{xx} + \frac{1}{6}(ka)^3\widetilde{u}_{xxx} + O(a^4),$$
$$u_{\ell-k} = u_\ell - ka\widetilde{u}_x + \frac{1}{2}(ka)^2\widetilde{u}_{xx} - \frac{1}{6}(ka)^3\widetilde{u}_{xxx} + O(a^4).$$

Thus, for all k,

$$\begin{aligned} u_{\ell+k} - 2u_\ell + u_{\ell-k} &= u_\ell + ka\widetilde{u}_x + \frac{1}{2}(ka)^2\widetilde{u}_{xx} + \frac{1}{6}(ka)^3\widetilde{u}_{xxx} + O(a^4) - 2u_\ell \\ &\quad + u_\ell - ka\widetilde{u}_x + \frac{1}{2}(ka)^2\widetilde{u}_{xx} - \frac{1}{6}(ka)^3\widetilde{u}_{xxx} + O(a^4) \\ &= (ka)^2\widetilde{u}_{xx} + O(a^4). \end{aligned} \tag{6.1}$$

Utilizing (6.1) for the atomistic and continuous force equations, the consistency analysis yields

$$\begin{aligned} &F^{c,lin}_{\ell,}(u) - F^{a,lin}_{\ell,}(u) \\ &= -\sum_{\substack{k=-N,\\ k\neq 0}}^{N} \frac{1}{2}\phi_{xx}(k)\left(k^2 u_\ell''\right) + \sum_{\substack{k=-N,\\ k\neq 0}}^{N} \frac{1}{2}\phi_{xx}(k)\left(\frac{u_{\ell+k} - 2u_\ell + u_{\ell-k}}{a^2}\right) \end{aligned}$$

$$
\begin{aligned}
&= -\sum_{\substack{k=-N,\\k\neq 0}}^{N} \frac{1}{2}k^2\phi_{xx}(k)\left(\frac{u_{\ell+1}-2u_\ell+u_{\ell-1}}{a^2}\right)\\
&\quad+\sum_{\substack{k=-N,\\k\neq 0}}^{N} \frac{1}{2}\phi_{xx}(k)\left(\frac{u_{\ell+k}-2u_\ell+u_{\ell-k}}{a^2}\right)\\
&= -\sum_{\substack{k=-N,\\k\neq 0}}^{N} \frac{1}{2}k^2\phi_{xx}(k)\frac{a^2\widetilde{u}_{xx}+O(a^4)}{a^2}+\sum_{\substack{k=-N,\\k\neq 0}}^{N} \frac{1}{2}k^2\phi_{xx}(k)\frac{(ka)^2\widetilde{u}_{xx}+O(a^4)}{a^2}\\
&= -\sum_{\substack{k=-N,\\k\neq 0}}^{N} \frac{1}{2}k^2\phi_{xx}(k)\widetilde{u}_{xx}+\sum_{\substack{k=-N,\\k\neq 0}}^{N} \frac{1}{2}k^2\phi_{xx}(k)\widetilde{u}_{xx}+O(a^2)\\
&= O(a^2).
\end{aligned}
$$

Thus, the consistency error between the linearized atomistic force equation and the continuum force equation is $O(a^2)$. □

6.2 Analysis in the Continuous Setting

Before finding the nearest neighbor and the next-nearest-neighbor interaction for the discrete case, the continuous case was observed. The continuous case was meant to shed light on the nature of the discrete case as it would be easier to find.

From (2.20), we look at the next-nearest-neighbor interaction. Also, we will approximate $\frac{\beta_{\ell+k}+2\beta_\ell+\beta_{\ell-k}}{4} \approx \beta(x_\ell)$. Thus, (2.20) becomes for the force-based operator:

$$
\begin{aligned}
F_{\ell,}^{bqcf,lin} &= \left(\frac{\beta_{\ell-1}+2\beta_\ell+\beta_{\ell+1}}{4}\right)\phi_{xx}(1)\left(\frac{u_{\ell+1}-2u_\ell+u_{\ell-1}}{a^2}\right)\\
&\quad+\left(1-\frac{\beta_{\ell-1}+2\beta_\ell+\beta_{\ell+1}}{4}\right)\phi_{xx}(1)u_\ell''\\
&\quad+\left(\frac{\beta_{\ell-2}+2\beta_\ell+\beta_{\ell+2}}{4}\right)\phi_{xx}(2)\left(\frac{u_{\ell+2}-2u_\ell+u_{\ell-2}}{a^2}\right)\\
&\quad+\left(1-\frac{\beta_{\ell-2}+2\beta_\ell+\beta_{\ell+2}}{4}\right)\phi_{xx}(2)4u_\ell''
\end{aligned}
$$

$$\approx \beta(x_\ell)\left(\phi_{xx}(1)u_\ell'' + \phi_{xx}(2)\left(\frac{u_{\ell+2} - 2u_\ell + u_{\ell-2}}{a^2}\right)\right)$$
$$+ (1 - \beta(x_\ell))\left(\phi_{xx}(1)u_\ell'' + 4\phi_{xx}(2)u_\ell''\right).$$

Using a Taylor approximation on $u_{\ell+2}$ and $u_{\ell-2}$, the next-nearest-neighbor operator becomes

$$F_{\ell,}^{bqcf,lin} = \phi_{xx}(1)u_\ell'' + \beta(x_\ell)\phi_{xx}(2)\left(4u_\ell'' + \frac{4}{3}u_\ell^{(4)}a^2\right) + (1 - \beta(x_\ell))\, 4\phi_{xx}(2)u_\ell''. \tag{6.2}$$

As the nearest-neighbor interaction is not difficult to find, we drop it from the continuous case to find the approximation for the next-nearest neighbor as well as utilize the fact that ϕ is a Lennard–Jones-type potential. Also, we denote $\beta(x_\ell)$ by β when there is no ambiguity. Therefore, the continuous next-nearest operator becomes

$$\begin{aligned} Fu_x &= \beta(-u_{xx} + a^2Au_{xxxx}) + (1 - \beta)(-u_{xx}) \\ &= \beta a^2 Au_{xxxx} - u_{xx}, \end{aligned} \tag{6.3}$$

where $A = \frac{4}{3}$.

Lemma 6.1 *For any displacements $u = (u_\ell)_{\ell\in\mathbb{Z}}$ from y_ℓ, the nearest neighbor and up to the next-nearest-neighbor interaction operator can be written in the form*

$$\begin{aligned} \left\langle F_{,1}^{bqcf,lin}, u\right\rangle &= \|u_x\|^2 \\ \left\langle F_{,2}^{bqcf,lin}, u\right\rangle &= 4\|u_x\|^2 + a^2 16A\left\|\sqrt{\beta}u_{xx}\right\|^2 + R + S, \end{aligned} \tag{6.4}$$

where R and S are given by

$$R = -a^2A\int \beta_{xx}(u_x)^2dx, \quad S = a^2A\int \beta_{xx}(u_{xx}u)dx. \tag{6.5}$$

Proof Since the proof of the first identity of Lemma 6.1 is straightforward, the proof for the second-neighbor interaction operator will be given. The main tool used is integration by parts based on the periodic boundary conditions.

$$\begin{aligned} \left\langle F_{,2}^{bqcf}u, u\right\rangle &= \int (\beta a^2 Bu_{xxxx} - u_{xx})udx \\ &= \int -u_{xx}udx + \int \beta a^2 Au_{xxxx}udx \end{aligned}$$

$$
\begin{aligned}
&= \int (u_x)^2 dx + a^2 A \int \beta u u_{xxxx} dx \\
&= \|u_x\|^2 - a^2 A \int (\beta u)_x u_{xxx} dx \\
&= \|u_x\|^2 + a^2 A \left\| \sqrt{\beta} u_{xx} \right\|^2 \\
&\quad + a^2 A \int \beta_x (u_x)_x^2 dx - a^2 A \int \beta_x (u\, u_{xx})_x dx \\
&= \|u_x\|^2 + a^2 A \left\| \sqrt{\beta} u_{xx} \right\|^2 \\
&\quad - a^2 A \int \beta_{xx} (u_x)^2 dx + a^2 A \int \beta_{xx} (u_{xx} u) dx \\
&= \|u_x\|^2 + a^2 A \left\| \sqrt{\beta} u_{xx} \right\|^2 + R + S.
\end{aligned}
\tag{6.6}
$$

□

Using the continuous analysis as a roadmap, we thus derive the discrete analysis in Sect. 3.

6.3 Symbols and Notation

Superscript e.g., $'$, $''$	First-order backward finite difference, second-order central finite difference, etc., for the discrete case
Subscript e.g., $_x$, $_{xx}$	First derivative, second derivative, etc., for the continuous case
$\frac{\delta}{\delta u_\ell}$	First-order variation evaluated at u_ℓ
$\frac{d}{dx}$	Derivative evaluated at x
$\|\cdot\|_{\ell_2}$	ℓ_2-norm
$\|\cdot\|_{\ell_\infty}$	∞-norm
$\|\cdot\|_{\ell_2(\Omega^b)}$	2-norm evaluated on the blending region Ω_b
$\lvert\cdot\rvert_{H^1}$	Discrete H^1 semi-norm
$<\cdot,\cdot>$	Inner product
ϕ	Atomistic interaction potential per unit cell
$(\cdot)^c$	Continuum equation
$(\cdot)^a$	Atomistic equation
$(\cdot)^{lin}$	Linearized equation
$(\cdot)^{bqcf}$	Force-based blended equation
Ω	Whole domain
Ω^a	Atomistic domain

Superscript e.g., $'$, $''$	First-order backward finite difference, second-order central finite difference, etc., for the discrete case
Ω^c	Continuum domain
Ω^b	Blending domain
β	Blending function
L	The number of atoms within the blending region Ω_b
a	Lattice spacing constant
y_ℓ	Deformation at ℓ
u_ℓ	Displacement at ℓ
E	Energy equation
$F_{\ell,k}$	Force equation at atom ℓ with kth-neighbor interaction
$F_{\ell,}$	Force equation at atom ℓ with the summation of all neighbor interaction
$F_{,k}$	Force equation with kth-neighbor interaction with the summation of all atoms
α	Parameter in the Morse potential
N	The number of neighbors within the interaction range
M	A half number of atoms in the atomistic chain
F^{ext}.	External force

Acknowledgments Elaine Gorom-Alexander and Dr. X. Li are supported by NSF CAREER award: DMS-1847770 and the University of North Carolina at Charlotte Faculty Research Grant.

We would like to thank the helpful discussions from Dr. Pablo Seleson and Dr. Christoph Ortner. We also want to thank the opportunity provided by the 50th John H. Barrett Memorial Lectures organizing committee.

References

1. R. Phillips, Crystals, defects and microstructures: modeling across scales, Cambridge University Press, 2001.
2. E. Tadmor, R. Miller, Modeling materials continuum, atomistic and multiscale techniques, 1st Edition, Cambridge University Press, 2011.
3. W. E, Principles of multiscale modeling, Cambridge University Press, 2011.
4. I. G. Graham, T. Y. Hou, O. Lakkis, R. Scheichl, Numerical analysis of multiscale problems, Springer, 2012.
5. Q. Du, M. Gunzburger, R. Lehoucq, K. Zhou, Analysis and approximation of nonlocal diffusion problems with volume constraints, SIAM Review 56 (2012) 676–696.
6. I. Kondov, G. Sutmann, Multiscale modeling methods for applications in materials science, Forschungszentrum Julich, 2013.
7. S. Silling, Reformulation of elasticity theory for discontinuities and long-range forces, Journal of the Mechanics and Physics of Solids 48 (2000) 175–209.
8. P. Seleson, S. Beneddine, S. Prudhomme, A force based coupling scheme for peridynamics and classical elasticity, Computational Materials Science 66 (2013) 34–49.
9. P. Seleson, Y. D. Ha, S. Beneddine, Concurrent coupling of bond-based peridynamics and the Navier equation of classical elasticity by blending, Journal for Multiscale Computational Engineering 13 (2015) 91–113.

10. K. Weinberg, A. Pandolfi, Innovative numerical approaches for multi-field and multi-scale problems: in honor of Michael Ortiz's 60th birthday, Springer, 2016.
11. R. Miller, E. Tadmor, A unified framework and performance benchmark of fourteen multiscale atomistic/continuum coupling methods, Modelling and Simulation in Materials Science and Engineering 17(5) (2009) 053001.
12. A. V. Shapeev, Consistent energy-based atomistic/continuum coupling for two-body potentials in one and two dimensions, SIAM J. Multiscale Model. Simul. 9 (3) (2011) 905–932.
13. M. Luskin, C. Ortner, Atomistic-to-continuum-coupling, Acta Numerica 22(4) (2013) 397–508.
14. Q. Du, R. Lipton, Peridynamics, fracture, and nonlocal continuum models, SIAM News 47 (3).
15. M. D'Elia, M. Gunzburger, Optimal distributed control of nonlocal steady diffusion problems, SIAM Journal on Control and Optimization 52 (2014) 243–273.
16. M. D'Elia, M. Perego, P. Bochev, D. Littlewood, A coupling strategy for nonlocal and local diffusion models with mixed volume constraints and boundary conditions, Computers and Mathematics with Applications 71(11) (2015) 2218–2230.
17. M. D'Elia, X. Li, P. Seleson, X. Tian, Y. Yu, A review of local-to-nonlocal coupling methods in nonlocal diffusion and nonlocal mechanics, To appear on Journal of Peridynamics and Nonlocal Modeling.
18. N. Trask, H. You, Y. Yu, M. L. Parks, An asymptotically compatible meshfree quadrature rule for nonlocal problems with applications to peridynamics, Computer Methods in Applied Mechanics and Engineering 343 (2019) 151–165.
19. H. You, Y. Yu, D. Kamensky, An asymptotically compatible formulation for local-to-nonlocal coupling problems without overlapping regions, Computer Methods in Applied Mechanics and Engineering 366.
20. X. Li, M. Luskin, C. Ortner, Positive definiteness of the blended force-based quasicontinuum method, Multiscale Modeling & Simulation 10 (3) (2012) 1023–1045.

A Note on Estimates of Level Sets and Their Role in Demonstrating Regularity of Solutions to Nonlocal Double-Phase Equations

Tadele Mengesha and James M. Scott

Abstract In this note, we prove an estimate on the level sets of a function with (p, q) growth that depends on the difference quotient of a bounded weak solution to a nonlocal double-phase equation. This estimate is related to a self-improving property of these solutions.

Keywords Nonlocal double phase · Fractional Gehring's lemma · Self-improving property · Sobolev regularity · Dual pairs

1 Introduction and Main Results

This note is a companion to the article [6], in which the authors study regularity properties of weak solutions u to

$$\mathcal{L}u(x) = f(x)\,, \tag{1.1}$$

where for measurable functions $u : \mathbb{R}^n \to \mathbb{R}$ and for $x \in \mathbb{R}^n$ the nonlocal *double-phase* operator $\mathcal{L}$ is defined as

The research of T. Mengesha was supported by NSF grants DMS-1910180 and DMS-2206252. J. Scott acknowledges support from NSF DMS-1937254 and DMS-2012562.

T. Mengesha (✉)
Department of Mathematics, University of Tennessee, Knoxville, Knoxville, TN, USA
e-mail: mengesha@utk.edu

J. M. Scott
Department of Applied Physics and Applied Mathematics, Columbia University, New York, NY, USA

T. Mengesha, A. J. Salgado (eds.), *A³N²M: Approximation, Applications, and Analysis of Nonlocal, Nonlinear Models*, The IMA Volumes in Mathematics and its Applications 165, https://doi.org/10.1007/978-3-031-34089-5_7

$$\mathcal{L}u(x) := \text{P.V.} \int_{\mathbb{R}^n} \frac{|u(x)-u(y)|^{p-2}}{|x-y|^{n+sp}}(u(x)-u(y))$$

$$+ a(x,y)\frac{|u(x)-u(y)|^{q-2}}{|x-y|^{n+tq}}(u(x)-u(y))\,dy\,.$$

Throughout, we assume $n \geq 2$ and the integrability indices p, q belong to $(1, \infty)$ with $p \leq q$ and differentiability indices s, t belong to $(0, 1)$. The abbreviation P.V. stands for principal value.

The operator $\mathcal{L}$ is the archetype of a class of nonlocal double-phase operators first introduced in [2], in which the Hölder continuity of bounded viscosity solutions to $\mathcal{L}u = f$ with bounded data f was obtained. In the work [6], the authors demonstrate regularity of solutions on a different scale; that under suitable assumptions on the data f, the modulating coefficient $a(\cdot, \cdot)$, and a certain ratio of integrability and differentiability exponents solutions u to $\mathcal{L}u = f$ exhibits a self-improvement property. Precisely, distributional solutions u belonging to the fractional Sobolev space $W^{s,p}(\mathbb{R}^n)$ in fact belong to a Sobolev space with higher exponents of integrability and differentiability.

For ease of reference, we summarize the relevant definitions, assumptions, properties, and results found in [6]. We assume that the modulating coefficient a is measurable and satisfies

$$a(x,y) \in L^\infty(\mathbb{R}^{2n}), \qquad 0 \leq a(x,y) \leq M, \qquad a(x,y) = a(y,x). \tag{A1}$$

We also require that

$$p \leq q, \quad t \leq s, \qquad \frac{1}{p'} \leq \frac{tq}{sp} \leq 1, \tag{A2}$$

where p' is the Hölder conjugate of p: $\frac{1}{p} + \frac{1}{p'} = 1$. Additionally, we will restrict ourselves to the case

$$sp < n. \tag{A3}$$

The higher differentiability and integrability described above and in [6] apply to bounded solutions $u \in W^{s,p}(\mathbb{R}^n)$ of a weak formulation of Eq. (1.1), that is,

$$\mathcal{E}(u,\varphi) = \int_{\mathbb{R}^n} f(x)\varphi(x)\,dx\,, \qquad \text{for any } \varphi \in C_c^\infty(\mathbb{R}^n)\,, \tag{1.2}$$

where the form $\mathcal{E}(u, \varphi)$ is defined as

$$\mathcal{E}(u,\varphi) := \int_{\mathbb{R}^n}\int_{\mathbb{R}^n} \frac{|u(x)-u(y)|^{p-2}}{|x-y|^{n+sp}}(u(x)-u(y))(\varphi(x)-\varphi(y))$$
$$+ a(x,y)\frac{|u(x)-u(y)|^{q-2}}{|x-y|^{n+tq}}(u(x)-u(y))(\varphi(x)-\varphi(y))\,\mathrm{d}y\,\mathrm{d}x\,. \tag{1.3}$$

We assume the data f belong to a Lebesgue space with sufficiently high exponent. Precisely, for a given $\delta_0 > 0$,

$$f \in L^{p_{*s}+\delta_0}_{loc}(\mathbb{R}^n)\,,$$

where we are using standard notation for Hölder and Sobolev exponents; that is, for any $r \in (1,\infty)$ and any $\sigma \in (0,1)$, we write

$$r' = \frac{r}{r-1}\,, \qquad r^* = r^{*\sigma} = \frac{nr}{n-\sigma r}\,, \qquad r_* = r_{*\sigma} = \frac{nr'}{n+\sigma r'} = (r^*)'\,.$$

(The dependence of the embedding exponents on σ will be suppressed whenever it is clear from context.)

Weak solutions u are assumed to be *a priori* bounded, a point clarified by the following definition:

Definition 1.1 A function $u \in W^{s,p}(\mathbb{R}^n) \cap L^\infty(\mathbb{R}^n)$ is a *bounded weak solution* to (1.1) with data f if the nonlocal double-phase energy $\mathcal{E}(u,u) < \infty$ and if u satisfies (1.2).

If we denote the integrand of $\mathcal{E}(u,u)$ by $P(x,y,u)$ so that

$$\mathcal{E}(u,u) = \int_{\mathbb{R}^n}\int_{\mathbb{R}^n} P(x,y,u)\,\mathrm{d}y\,\mathrm{d}x\,,$$

then by definition of u as a bounded weak solution the function $P(\cdot,\cdot,u)$ belongs to $L^1(\mathbb{R}^{2n})$. The following theorem concerning P constitutes the main result of [6]:

Theorem 1.2 *Let $p, q \geq 2$, and $s, t \in (0,1)$ satisfy* (A2)–(A3) *and let $a(x,y)$ satisfy* (A1). *Fix $\delta_0 > 0$, and let $f \in L^{p_{*s}+\delta_0}_{loc}(\mathbb{R}^n)$. Let $u \in W^{s,p}(\mathbb{R}^n) \cap L^\infty(\mathbb{R}^n)$ be any bounded weak solution to* (1.1) *with data f. Then there exists $\varepsilon_0 \in (0,1)$ depending only on n, p, q, s, t, M, δ_0, and $\|u\|_{L^\infty(\mathbb{R}^n)}$ such that for every $\tau \in (0,\varepsilon_0)$*

$$P(\cdot,\cdot,u) \in L^{1+\tau}_{loc}(\mathbb{R}^{2n})\,.$$

In particular, there exist positive constants ε_1 and ε_2 such that $u \in W^{s+\varepsilon_1,p+\varepsilon_2}_{loc}(\mathbb{R}^n)$, and if $(s+\varepsilon_1)(p+\varepsilon_2) > n$, then u is locally Hölder continuous.

To prove Theorem 1.2, we use an argument developed by Kuusi, Mingione and Sire announced in [3] and presented in [4] that builds a nonlocal fractional Gehring lemma in order to prove a self-improvement result for solutions to a class of monotone operators with quadratic growth related to the fractional Laplacian. The arguments in [6] and in this note are heavily based on the work and presentation done for the case $p = 2$ in [4]. While it is apparent from a careful reading of that work that their methods apply to functionals with more general p-growth, the precise treatment of such classes of operators does not appear in the literature. Since we are further working with operators of mixed (p, q) growth, in [6] and in this note, we have written the arguments of [4] for a general exponent p so that the robustness of their technique and results can be clearly seen as applicable in a wealth of contexts. One such instance is an extension of these arguments to vector-valued solutions of nonlocal systems. A specific example is the strongly coupled system of nonlinear equations studied in [5]. We would also like to mention a sequence of recent papers in [9, 10] where the same argument is applied to obtain higher Sobolev regularity estimates for nonlocal equations with VMO coefficients corresponding to the case $p = 2$.

The fractional Gehring lemma relies on a level set estimate of a quantity related to the solution, and its proof is the main contribution of this note. If a reader of [6] accepts the level set estimate as true, then the rest of the proof in [6] of the Gehring lemma follows in a straightforward way. However, because the proof of the estimate itself is quite technical and very closely resembles the argument from [4], we have written it here instead of in [6].

Due to its technical nature, we do not write a statement of the level set estimate precisely until the beginning of Sect. 4. To ensure there is no interruption in the thread of reasoning between this work and [6], the prerequisite results are given in Sect. 3 as a summarized version of their counterparts in [6].

Following the structure introduced in [4], we define *dual pairs* of measures and functions (U, ν). For small $\varepsilon \in (0, 1/p)$, we define the locally finite doubling Borel measure in $\mathbb{R}^{2n}$

$$\nu(A) := \int_A \frac{\mathrm{d}x\,\mathrm{d}y}{|x-y|^{n-\varepsilon p}}, \qquad A \subset \mathbb{R}^{2n} \text{ measurable}, \tag{1.4}$$

and we define the function

$$U(x,y) := \frac{|u(x)-u(y)|}{|x-y|^{s+\varepsilon}}. \tag{1.5}$$

It is then clear that

$$u \in W^{s,p}(\mathbb{R}^n) \qquad \text{if and only if} \qquad U \in L^p(\mathbb{R}^{2n};\nu).$$

The integrand $P(x, y, u)$ of the energy $\mathcal{E}(u, u)$ can be expressed in terms of U as

$$[U^p + A(x, y)U^q]|x - y|^{-n+\epsilon p}, \quad \text{where } A(x, y) := a(x, y)|x - y|^{(s-t)q+\varepsilon(q-p)}. \tag{1.6}$$

We can therefore write the double-phase energy $\mathcal{E}(u.u)$ in terms of the dual pair as

$$\mathcal{E}(u, u) = \int_{\mathbb{R}^{2n}} (U^p + A(x, y)U^q)\, \mathrm{d}\nu =: \int_{\mathbb{R}^{2n}} G(x, y, U)\, \mathrm{d}\nu \tag{1.7}$$

where the integrand $G(x, y, U) := U^p + A(x, y)U^q$. Then it now becomes clear that

$$P(\cdot, \cdot, u) \in L^1(\mathbb{R}^{2n}) \quad \text{if and only if} \quad G(\cdot, \cdot, U) \in L^1(\mathbb{R}^{2n}; \nu).$$

Theorem 1.3 (Higher Regularity Result) *With all the assumptions of Theorem 1.2, there exists $\varepsilon_0 > 0$ depending only on* `data` *such that for every $\delta \in (0, \varepsilon_0)$ we have*

$$G(x, y, U) \in L^{1+\delta}_{loc}(\mathbb{R}^{2d}; \nu), \tag{1.8}$$

where `data` *represents n, p, q, s, t, M, and $\|u\|_{L^\infty}$.*

Theorem 1.2 is a simple consequence of the above theorem. In [6], we show (1.8) directly by way of a fractional Gehring lemma applied to the dual pair of function and measure (G, ν). This fractional Gehring lemma in turn relies on a kind of fractional reverse Hölder inequality; for the exact statement, see Theorem 3.3 below. This inequality holds only for diagonal sets of the type $B \times B \subset \mathbb{R}^{2n}$, and it is insufficient to apply tools traditionally used to prove Gehring's lemma such as the maximal function. Nevertheless, Kuusi, Mingione, and Sire in [4] used a novel localization technique to show that the fractional reverse Hölder-type inequality over diagonal balls is sufficient to prove a special fractional version of Gehring's lemma that is applicable for dual pairs of the above type. A key ingredient of this localization technique is a level set estimate

$$\frac{1}{\lambda^2}\int_{\mathcal{B}(x_0,\beta)\cap\{U>\lambda\}} U^2\, \mathrm{d}\nu \precsim \frac{1}{\lambda^r}\int_{\mathcal{B}(x_0,\alpha)\cap\{U>\lambda\}} U^r\, \mathrm{d}\nu + \text{ terms involving level sets of } f, g$$

for some fixed $r < 2$ and for any $\lambda \geq \lambda_0$, where λ_0 is a finite constant depending on the solution. Here $\mathcal{B} = B \times B$, and $B \subset \mathbb{R}^n$ is a ball. We adapt the statement and proof of this level set estimate to our setting; see Proposition 4.1. Key steps of the proof are described in Sect. 4, and we additionally refer to the original discussions and summaries of the technique in [3, 4].

We finally remark that the fractional Gehring lemma incorporates the level set estimate, and the proof of the fractional Gehring lemma itself is contained in [6].

This note is organized as follows: In the next section, we identify notation and conventions. The reverse Hölder inequality is proved in [6] and for reference is stated in Sect. 3. Section 4 contains the proof of the level set estimate in its entirety.

2 Preliminaries

Throughout, we denote positive constants by c, C, etc., and they may change from line to line. We list the dependencies in parentheses after the constant when we wish to make them explicit, i.e., if a constant C depends only on n, p, and s, we write $C = C(n, p, s)$. We will abbreviate the following set of parameters as

$$\texttt{data} \equiv (n, p, q, s, t, M, \|u\|_{L^\infty}).$$

In $\mathbb{R}^n$, denote the open ball of radius R centered at x_0 by

$$B(x_0, R) = B_R(x_0) := \{x \in \mathbb{R}^n : |x - x_0| < R\}.$$

We will sometimes denote the ball $B \equiv B_R \equiv B_R(x_0)$ whenever the center and/or radius is clear from context. If B is a ball centered at x_0 with radius R, then σB is the ball centered at x_0 with radius σR. Given any measure μ, denote the average of a μ-measurable function h over a set $\mathcal{A}$ by

$$(h)_{\mathcal{A}} := \fint_{\mathcal{A}} h \, \mathrm{d}\mu = \frac{1}{\mu(\mathcal{A})} \int_{\mathcal{A}} h(x) \, \mathrm{d}\mu.$$

In dealing with functions defined on $\mathbb{R}^{2n}$ such as U, we consider the norm on $\mathbb{R}^{2n}$ defined by

$$\|(x, y)\| := \max\{|x|, |y|\},$$

where $|\cdot|$ denotes the Euclidean norm on $\mathbb{R}^n$. Denote the balls defined by this norm as

$$\begin{aligned} \mathcal{B}(x_0, y_0, R) &:= \{(x, y) \in \mathbb{R}^n \times \mathbb{R}^n : \|(x, y) - (x_0, y_0)\| < R\} \\ &= B(x_0, R) \times B(y_0, R). \end{aligned}$$

If we denote

$$B_{\mathbb{R}^{2n}}(x_0, y_0, R) := \{(x, y) \in \mathbb{R}^n \times \mathbb{R}^n : \sqrt{|x - x_0|^2 + |y - y_0|^2} < R\},$$

then clearly

$$B_{\mathbb{R}^{2n}}(x_0, y_0, R) \subset \mathcal{B}(x_0, y_0, R) \subset B_{\mathbb{R}^{2n}}(x_0, y_0, 2R).$$

Often we will need to consider balls in $\mathbb{R}^{2n}$ centered at a point on the "diagonal," that is, a point of the form (x_0, x_0) for $x_0 \in \mathbb{R}^n$. In this case, we abbreviate $\mathcal{B}(x_0, x_0, R) \equiv \mathcal{B}(x_0, R)$. We will also use the abbreviations $\mathcal{B}(x_0, R) \equiv \mathcal{B}_R(x_0) \equiv$

$\mathcal{B}_R \equiv \mathcal{B}$ whenever the center and/or radius is clear from context. Whenever there is no ambiguity, we write $\mathcal{B}(x_0, \sigma R) = \sigma \mathcal{B}$. We also denote

$$\text{Diag} := \{(x, x) : x \in \mathbb{R}^n\}.$$

We will use the elementary inequality

$$2^{kr} \sum_{j=k-1}^{\infty} 2^{-jr} \leq \frac{4^r}{r \ln(2)}, \qquad \text{for } k \geq 1 \quad \text{and} \quad r \in (0, \infty). \tag{2.1}$$

The cardinality of a finite set $\mathcal{A}$ is denoted by $\#\mathcal{A}$. The set of nonnegative integers $\{0, 1, 2, \ldots\}$ is designated by $\mathbb{Z}_+$.

For any domain $\Omega \subset \mathbb{R}^n$, $0 < \sigma < 1$, and $r \in [1, \infty)$, the fractional Sobolev spaces are defined by the Gagliardo seminorm

$$W^{\sigma,r}(\Omega) := \left\{ u \in L^r(\Omega) : [u]_{W^{\sigma,r}(\Omega)} := \int_\Omega \int_\Omega \frac{|u(x) - u(y)|^r}{|x - y|^{n+\sigma r}} \, \mathrm{d}y \, \mathrm{d}x < \infty \right\}$$

with norm $\|\cdot\|^r_{W^{\sigma,r}(\Omega)} := \|\cdot\|^r_{L^r(\Omega)} + [\cdot]^r_{W^{\sigma,r}(\Omega)}$.

We will also use the following fractional Poincaré–Sobolev-type inequalities throughout the note. A proof of the first can be found in several places; see, for instance, [1, 8]. The second can be found in [7, 11].

Theorem 2.1 (Fractional Poincaré–Sobolev Inequality) *Let $r \in [1, \infty)$, $0 < \sigma < 1$. Let $B = B_R(x_0)$ for some $R > 0$, $x_0 \in \mathbb{R}^n$. Then there exists $C = C(n, r, \sigma) > 0$ such that*

$$\left(\fint_B \left| \frac{v(x) - (v)_B}{R^\sigma} \right|^{r^{*\sigma}} \mathrm{d}x \right)^{1/r^{*\sigma}} \leq C \left(\int_B \fint_B \frac{|v(x) - v(y)|^r}{|x - y|^{n+\sigma r}} \, \mathrm{d}y \, \mathrm{d}x \right)^{1/r}$$

for every $v \in W^{\sigma,r}(B)$.

Theorem 2.2 (Fractional Poincaré Inequality) *Let $r \in [1, \infty)$, $0 < \sigma < 1$. Let $B = B_R(x_0)$ for some $R > 0$, $x_0 \in \mathbb{R}^n$. Then there exists $C = C(n, r) > 0$ such that*

$$\left(\fint_B \left| \frac{v(x) - (v)_B}{R^\sigma} \right|^{r} \mathrm{d}x \right)^{1/r} \leq C \left(\int_B \fint_B \frac{|v(x) - v(y)|^r}{|x - y|^{n+\sigma r}} \, \mathrm{d}y \, \mathrm{d}x \right)^{1/r}$$

for every $v \in W^{\sigma,r}(B)$.

3 Sobolev Inequality for Dual Pairs and Reverse Hölder Inequality

3.1 The Dual Pair Measure

We summarize some basic properties of the measure ν defined in (1.4). These properties are natural extensions of those established in [4, Proposition 4.1]; their proof is sketched in [6].

Theorem 3.1 *For any $\varepsilon \in (0, 1/p)$, the measure ν defined as*

$$\nu(\mathcal{A}) := \int_{\mathcal{A}} \frac{1}{|x-y|^{n-\varepsilon p}} \, dy \, dx \,, \qquad \mathcal{A} \subset \mathbb{R}^{2n} \,,$$

is absolutely continuous with respect to Lebesgue measure on $\mathbb{R}^{2n}$. Additionally:

- *For $\mathcal{B} = B_R(x_0) \times B_R(x_0)$,*

$$\nu(\mathcal{B}) = \frac{c(n, p, \varepsilon) R^{n+\varepsilon p}}{\varepsilon} \,, \tag{3.1}$$

  where $c(n, p, \varepsilon)$ is a constant depending only on n, p, and ε that satisfies $1/\widetilde{c}(n, p) \leq c(n, p, \varepsilon) \leq \widetilde{c}(n, p)$, where $\widetilde{c}$ is another constant depending only on n and p.
- *For every $x \in \mathbb{R}^n$ and for $R \geq r > 0$,*

$$\frac{\nu(\mathcal{B}(x, R))}{\nu(\mathcal{B}(x, r))} = \left(\frac{R}{r}\right)^{n+\varepsilon p} \,. \tag{3.2}$$

- *For every $a \leq 1$, $R > 0$, and $x \in \mathbb{R}^n$, there exists a constant $C_d = C_d(n, p)$ such that*

$$\frac{\nu(\mathcal{B}(x, R))}{\nu(K_1 \times K_2)} \leq \frac{C_d}{a^{2n} \varepsilon} \tag{3.3}$$

  for any two cubes $K_1, K_2 \subset B_R(x)$ with sides parallel to the coordinate axes and such that $|K_1| = |K_2| = (aR)^n$.

3.2 Reverse Hölder Inequality

Recall that

$$U(x, y) = \frac{|u(x) - u(y)|}{|x-y|^{s+\varepsilon}} \,, \qquad \text{and define } F(x, y) := |f(x)| \,. \tag{3.4}$$

Then $F \in L^{p_*+\delta}_{loc}(\mathbb{R}^{2n})$ for every $\delta \in (0, \delta_0)$, as a direct calculation using the properties of measure ν.

We now report the compatibility of the Sobolev–Poincaré inequality with the definition of U. Given $B = B_R(x_0)$, define $\tau \in (0,1)$, and $\eta \in (1,\infty)$ to be differentiability and integrability constants, respectively, that have yet to be fixed. Letting $\varepsilon \in (0, \min\{\frac{s}{p}, 1-s\})$ and using (3.1),

$$\fint_B \int_B \frac{|u(x)-u(y)|^\eta}{|x-y|^{n+\tau\eta}}\,dy\,dx = \frac{CR^{\varepsilon p}}{\varepsilon} \fint_{\mathcal{B}} U^\eta\,d\nu$$

so long as

$$\tau + \frac{\varepsilon p}{\eta} = s + \varepsilon\,.$$

Since $\varepsilon \in (0, \frac{s}{p})$ and $\varepsilon < 1-s$, the exponent τ remains in $(0,1)$ for every $\eta \in (1,\infty)$. With this choice of τ, by the fractional Sobolev inequality, Theorem 2.1,

$$\left(\fint_B \left|\frac{u(x)-(u)_B}{R^\tau}\right|^m dx\right)^{1/m} \le C\left(\fint_B \int_B \frac{|u(x)-u(y)|^\eta}{|x-y|^{n+\tau\eta}}\,dy\,dx\right)^{1/\eta}$$

for every $m \in [1, \eta^{*_\tau}]$ with $\eta \in (1,\infty)$. We choose η to satisfy the relation

$$p = \eta^{*_\tau} = \frac{n\eta}{n-\tau\eta} = \frac{n\eta}{n-\eta(s+\varepsilon-\frac{\varepsilon p}{\eta})} \iff \eta = \frac{np+\varepsilon p^2}{n+sp+\varepsilon p}. \tag{3.5}$$

This choice of η is a valid Lebesgue exponent; note that $\eta < p$ for all $n \ge 1$ and for all $p \in (1,\infty)$, and that $\eta > 1$ so long as $p \ge 2$. Taking $m = \eta^{*_\tau}$, we summarize this discussion in the following lemma:

Lemma 3.2 *Let $\varepsilon \in (0, s/p)$ with $\varepsilon < 1-s$ and $p \ge 2$. Define $\eta = \frac{np+\varepsilon p^2}{n+sp+\varepsilon p}$. Then*

$$\left(\fint_B |u(x)-(u)_B|^p\,dx\right)^{1/p} \le \frac{CR^{s+\varepsilon}}{\varepsilon^{1/\eta}}\left(\fint_{\mathcal{B}} U^\eta\,d\nu\right)^{1/\eta},$$

where $C = C(n,s,p)$. The same inequality holds when the ball B is replaced by a cube Q with sides of length R and with $\mathcal{B}$ replaced by $Q \times Q$.

Recall that $G(x,y,U) = U^p + A(x,y)U^q$. We have the following L^1_{loc} estimate for G that will lead us to a scale-invariant reverse Hölder's inequality. The statement is precisely [6, Proposition 4.3], and its proof can be found in the same paper.

Proposition 3.3 *Let $p \in [2,\infty)$, and let $\varepsilon < 1-s$ with $\varepsilon \in (0, \min\{s(\frac{tq}{sp}-\frac{1}{p'}), \frac{s}{p}\})$. (This choice is possible by Assumption A2). Let η be given by the formula in (3.5). Let $B = B_R(x_0)$ be a ball with $R \le 1$. Then there exists a constant C depending* *only on* `data` *such that for any solution* $u \in W^{s,p}(\mathbb{R}^n) \cap L^\infty(\mathbb{R}^n)$ *to* (1.1) *and for any* $\sigma \in (0, 1)$

$$\begin{aligned}\left(\fint_{\frac{1}{4}\mathcal{B}} G(x,y,U)\,\mathrm{d}\nu\right)^{1/p} &\le \frac{C}{\varepsilon^{1/\eta-1/p}}\Bigg[\frac{1}{\sigma}\left(\fint_{\mathcal{B}} U^\eta\,\mathrm{d}\nu\right)^{1/\eta} \\ &\quad + \sigma\sum_{k=0}^{\infty}\left(2^{-k(\frac{sp}{p-1}-s-\varepsilon)} + 2^{-k(\frac{tq}{p-1}-s-\varepsilon)}\right)\left(\fint_{2^k\mathcal{B}} U^\eta\,\mathrm{d}\nu\right)^{1/\eta}\Bigg] \\ &\quad + \frac{C[\varepsilon\nu(\mathcal{B})]^{\frac{\theta}{p-1}}}{\varepsilon^{(1/p_*-1/p')\frac{1}{p-1}}}\left[\left(\fint_{\mathcal{B}} F^{p_*}\,\mathrm{d}\nu\right)^{1/p_*}\right]^{1/(p-1)},\end{aligned} \tag{3.6}$$

where

$$\theta := \frac{s-\varepsilon(p-1)}{n+\varepsilon p} > 0\,.$$

Remark 3.4 We make some remarks. The upper bound in (3.6) can be simplified down to just one series. Since $sp \ge tq$

$$2^{-k(\frac{sp}{p-1}-s-\varepsilon)} \le 2^{-k(\frac{tq}{p-1}-s-\varepsilon)}, \qquad k \in \mathbb{Z}_+\,, \tag{3.7}$$

so we can replace the infinite series on the right-hand side of (3.6) with

$$\frac{C\sigma}{\varepsilon^{1/\eta-1/p}}\sum_{k=0}^{\infty}\alpha_k\left(\fint_{2^k\mathcal{B}} U^\eta\,\mathrm{d}\nu\right)^{1/\eta},$$

where

$$\alpha_k := 2^{-k(\frac{tq}{p-1}-s-\varepsilon)}\,. \tag{3.8}$$

Moreover, in the case $a \equiv 0$, one simply takes $\alpha_k = 2^{-k(\frac{s}{p-1}-\varepsilon)}$. In any case, since $\varepsilon \le \min\{s(\frac{tq}{sp}-\frac{1}{p'}), \frac{s}{p}\}$, the series $\sum_{k=0}^{\infty}\alpha_k < \infty$, and as a consequence,

$$\begin{aligned}\sum_{k=0}^{\infty}\alpha_k\left(\fint_{2^k\mathcal{B}} U^\eta\,\mathrm{d}\nu\right)^{1/\eta} &\le \sum_{k=0}^{\infty}\alpha_k\left(\fint_{2^k\mathcal{B}} U^p\,\mathrm{d}\nu\right)^{1/p} \\ &= C(\varepsilon,p,s)\sum_{k=0}^{\infty}\alpha_k\left(\fint_{2^k B}\int_{2^k B}\frac{|u(y)-u(x)|^p}{|x-y|^{n+sp}}\,\mathrm{d}x\mathrm{d}y\right)^{1/p} \\ &\le R^{-n/p-\epsilon}C(\varepsilon,p,s)\left(\int_{\mathbb{R}^n}\int_{\mathbb{R}^n}\frac{|u(y)-u(x)|^p}{|x-y|^{n+sp}}\mathrm{d}x\mathrm{d}y\right)^{1/p} < \infty.\end{aligned}$$

The following corollary establishes a genuine scale-invariant reverse Hölder inequality for an appropriately scaled version of the integrand G. This quantity will satisfy a self-improving result.

Corollary 3.4.1 *Let* $\varepsilon \in \left(0, \min\{s(\frac{tq}{sp} - \frac{1}{p'}), \frac{s}{p}\}\right)$. *(This choice is possible by Assumption A2). Let* $B = B_R(x_0)$ *be a ball with* $R \le 1$. *Define* $H(x, y, U) := G(x, y, U)^{(p-1)/p}$. *Then there exists a constant C depending only on* `data` *such that for any solution* $u \in W^{s,p}(\mathbb{R}^n) \cap L^\infty(\mathbb{R}^n)$ *to* (1.1) *and for any* $\sigma \in (0, 1)$

$$\begin{aligned}\left(\fint_{\frac{1}{4}\mathcal{B}} H(x, y, U)^{p'}\, \mathrm{d}\nu\right)^{1/p'} &\le \frac{C}{\sigma \varepsilon^{1/\gamma - 1/p'}} \left(\fint_{\mathcal{B}} H(x, y, U)^{\gamma}\, \mathrm{d}\nu\right)^{1/\gamma} \\ &\quad + \frac{C\sigma}{\varepsilon^{1/\gamma-1/p'}} \sum_{k=0}^{\infty} \alpha_k \left(\fint_{2^k\mathcal{B}} H(x, y, U)^{\gamma}\, \mathrm{d}\nu\right)^{1/\gamma}, \\ &\quad + \frac{C[\nu(\mathcal{B})]^{\theta}}{\varepsilon^{1/p_* - 1/p'}} \left(\fint_{\mathcal{B}} F^{p_*}\, \mathrm{d}\nu\right)^{1/p_*},\end{aligned} \tag{3.9}$$

where $\gamma := \frac{\eta}{p-1} = p' \cdot \frac{n+\varepsilon p}{n+sp+\varepsilon p} < p'$ *and* $\theta := \frac{s-\varepsilon(p-1)}{n+\varepsilon p}$.

Remark 3.5 If $a \equiv 0$, one can see from careful inspection of the proofs they need not assume $u \in L^\infty(\mathbb{R}^n)$ in Proposition 3.3 and Corollary 3.4.1.

4 Proof of the Level Set Estimate

We are now ready to state and prove the level set estimate. It is stated precisely in Proposition 4.1 below, and this section is devoted to its proof. We first define the following: using the notation of Corollary 3.4.1, for any $x_0 \in \mathbb{R}^n$ and $R > 0$, set

$$\Theta(x_0, R) := \Upsilon_0(x_0, R) + Tail(x_0, R) + \Psi_1(x_0, R), \tag{4.1}$$

where

$$\begin{aligned}\Upsilon_0(x_0, R) &:= \left(\fint_{\mathcal{B}(x_0,R)} F^{p_*+\delta_f}\, \mathrm{d}\nu\right)^{1/(p_*+\delta_f)}, \text{ with } \delta_f \in (0, \delta_0) \text{ to be determined,} \\ Tail(x_0, R) &:= \sum_{k=0}^{\infty} 2^{-k(\frac{tq}{p-1} - s - \varepsilon)} \left(\fint_{\mathcal{B}(x_0,2^kR)} H^{\gamma}\, \mathrm{d}\nu\right)^{1/\gamma},\end{aligned} \tag{4.2}$$

and, for any constant $M \ge 1$,

$$\Psi_M(x_0, R) := \left(\fint_{\mathcal{B}(x_0,R)} H^{p'} \, d\nu \right)^{1/p'} + M \frac{[\nu(\mathcal{B}(x_0, R))]^\theta}{\varepsilon^{1/p_* - 1/p'}} \left(\fint_{\mathcal{B}(x_0,R)} F^{p_*} \, d\nu \right)^{1/p_*} ; \tag{4.3}$$

we write Ψ_M with $M = 1$ as Ψ_1.

Proposition 4.1 *Assume* (A1), (A2), *and* (A3). *Assume that* $\varepsilon > 0$ *satisfies*

$$\varepsilon \in (0, s/p), \qquad \varepsilon < s\left(\frac{tq}{sp} - \frac{1}{p'}\right), \qquad \varepsilon < 1 - s. \tag{A4}$$

Let $u \in W^{s,p}(\mathbb{R}^n) \cap L^\infty(\mathbb{R}^n)$ *be a bounded weak solution to* (1.1), *and let* $f \in L^{p_*+\delta_0}(\mathbb{R}^n)$ *for given* $\delta_0 > 0$. *Let* U *and* F *be as in* (3.4). *Let* $\mathcal{B}(x_0, \varrho_0) \subset \mathbb{R}^{2n}$ *with* $0 < \varrho_0 \le 1$, *and let* α *and* β *be such that* $\varrho_0 < \beta < \alpha < \frac{3}{2}\varrho_0$ *so that we have*

$$\mathcal{B}(x_0, \varrho_0) \subset \mathcal{B}(x_0, \beta) \subset \mathcal{B}(x_0, \alpha) \subset \mathcal{B}\big(x_0, \frac{3}{2}\varrho_0\big).$$

Then there exist constants $C_\alpha = C_\alpha(\texttt{data}) > 0$, $C_f = C_f(\texttt{data}, \varepsilon) \ge 1$, *and* $\kappa_f = \kappa_f(\texttt{data}, \varepsilon) \in (0, 1)$, *with positive constants*

$$\vartheta := \frac{3(p' - \gamma)}{\gamma}, \qquad \vartheta_f := (p_* + \delta_f)\left(\frac{p_*\theta}{1 - p_*\theta}\right), \qquad \widetilde{\vartheta}_f := \frac{p_*(1 + \theta\delta_f)}{1 - p_*\theta}, \tag{4.4}$$

such that

$$\begin{aligned} \frac{1}{\lambda^{p'}} \int_{\mathcal{B}(x_0,\beta)\cap\{H>\lambda\}} H^{p'} \, d\nu &\le \frac{C_\alpha}{\varepsilon^\vartheta \lambda^\gamma} \int_{\mathcal{B}(x_0,\alpha)\cap\{H>\lambda\}} H^\gamma \, d\nu \\ &+ \frac{C_f \lambda_0^{\vartheta_f}}{\lambda^{\widetilde{\vartheta}_f}} \int_{\mathcal{B}(x_0,\alpha)\cap\{F>\kappa_f\lambda\}} F^{p_*} \, d\nu \end{aligned} \tag{4.5}$$

for every $\lambda \ge \lambda_0$, *where* λ_0 *is defined as*

$$\lambda_0 := \frac{C_a}{\varepsilon}\left(\frac{\varrho_0}{\alpha - \beta}\right)^{2n+p} \Theta(x_0, 2\varrho_0), \tag{4.6}$$

and where $C_a = C_a(\texttt{data})$ *and* Θ *has been defined in* (4.1). *(See also* (4.2) *and* (4.3)*).*

For $p = 2$, this proposition is proved in Section 5 of [4]. Our proof is essentially the same as the proof found in [4]. However, we are writing the proof to make sure that the choice of the other parameters in (4.4) is correctly made and to emphasize the robustness of the arguments in [4] and how they can be used for more general nonlinear operators. As it has been explained in [4], the main difficulty in proving (4.5) is that the reverse Hölder inequality (3.9) only holds on diagonal balls of the type $\mathcal{B}(x_0, x_0, R)$. Thus maximal function arguments cannot be used, and we must resort to more direct arguments. We use a Calderón–Zygmund decomposition to decompose the level set $\{H > \lambda\}$ into dyadic cubes. These cubes are then sorted into cubes situated on or near the diagonal (called "diagonal" cubes) and cubes far from the diagonal (called "off-diagonal" cubes). What is meant by "far from" will be quantified below. The level set estimate for the diagonal cubes is handled using the reverse Hölder inequality (3.9). It turns out that Sobolev functions automatically satisfy a type of reverse Hölder inequality on off-diagonal cubes, and we use this to obtain the level set estimate for said cubes.

4.1 Vitali Covering

Just as in [4], we begin with an exit-time argument. The goal is to cover the portion of the set $\{H > \lambda\}$ that lies on or near the diagonal $\{(x, x) : x \in \mathbb{R}^n\}$. Let $\kappa \in (0, 1]$ be a constant that will be chosen later, in (4.64); all arguments in the paper up until then are independent of the choice of κ. Define

$$\lambda_1 := \frac{1}{\kappa} \sup_{\frac{\alpha-\beta}{40^n} \le R \le \frac{\varrho_0}{2}} \sup_{x \in B(x_0, \beta)} \{\Psi_M(x, R) + \Upsilon_0(x, R) + Tail(x, R)\}. \tag{4.7}$$

For the same κ and for $\lambda \ge \lambda_1$, define the "diagonal" level set of the functional Ψ_M by

$$D_{\kappa\lambda} := \left\{ (x, x) \in \mathcal{B}(x_0, \beta) : \sup_{0 < R < \frac{\alpha-\beta}{40^n}} \Psi_M(x, R) > \kappa\lambda \right\}. \tag{4.8}$$

Then by definition of λ_1, we have

$$\Psi_M(x, R) \le \kappa\lambda_1 \le \kappa\lambda, \qquad \text{for each } (x, x) \in \mathcal{B}(x_0, \beta), \quad R \in \left[\frac{\alpha-\beta}{40^n}, \frac{\varrho_0}{2}\right], \tag{4.9}$$

and so it follows that for every (x, x) in the diagonal level set $D_{\kappa\lambda}$ there exists an exit time $R(x) \in (0, \frac{\alpha-\beta}{40^n})$ such that

$$\Psi_M(x, R(x)) \ge \kappa\lambda \qquad \text{while at the same time} \qquad \sup_{R(x) < R < \frac{\alpha-\beta}{40^n}} \Psi_M(x, R) \le \kappa\lambda. \tag{4.10}$$

Thus the collection $\{\mathcal{B}(x, 2R(x))\}$ forms a cover of $D_{\kappa\lambda}$, so by the Vitali covering theorem we can find a countable subcollection $\{\mathcal{B}(x_j, 2R(x_j))\}$ such that

$$\bigcup_{(x,x) \in D_{\kappa\lambda}} \mathcal{B}(x, 2R(x)) \subset \bigcup_j \mathcal{B}(x_j, 10R(x_j)), \quad \mathcal{B}(x_j, 2R(x_j)) \text{ are mutually disjoint}. \tag{4.11}$$

We hereafter use the abbreviations

$$\mathcal{B}_j := \mathcal{B}(x_j, R(x_j)), \qquad \sigma\mathcal{B}_j := \mathcal{B}(x_j, \sigma R(x_j)), \quad \sigma > 0. \tag{4.12}$$

The quantity $\sum \nu(\mathcal{B}_j)$ is treated by the diagonal estimates in the next section. Note that since $x_j \in \mathcal{B}(x_0, \beta)$ and $R(x_j) \le \frac{\alpha-\beta}{40^n}$, we have $10\mathcal{B}_j \subset \mathcal{B}(x_0, \alpha)$ for every j. By (4.10) and by the doubling property of the measure ν in (3.2), we also have

$$\sum_j \int_{10\mathcal{B}_j} H^{p'}\,\mathrm{d}\nu \le \sum_j \nu(10\mathcal{B}_j)[\Psi_M(x_j, 10R(x_j))]^{p'} \le 10^{n+\varepsilon p}\kappa^{p'}\lambda^{p'} \sum_j \nu(\mathcal{B}_j). \tag{4.13}$$

4.2 Analysis on the Diagonal

By (4.10), it follows that at least one of two inequalities hold: either

$$\left(\fint_{\mathcal{B}_j} H^{p'}\,\mathrm{d}\nu\right)^{1/p} \ge \frac{\kappa\lambda}{2} \tag{4.14}$$

or

$$\frac{M[\nu(\mathcal{B}_j)]^{\theta}}{\varepsilon^{1/p_*-1/p'}}\left(\fint_{\mathcal{B}_j} F^{p_*}\,\mathrm{d}\nu\right)^{1/p_*} \ge \frac{\kappa\lambda}{2}. \tag{4.15}$$

Case 1 If (4.14) occurs, then by the reverse Hölder inequality (3.9)

$$\begin{aligned}\kappa\lambda \le{}& \tfrac{C}{\sigma\varepsilon^{1/\gamma-1/p'}}\left(\fint_{4\mathcal{B}_j} H^{\gamma}\,\mathrm{d}\nu\right)^{1/\gamma} + \tfrac{\sigma}{\varepsilon^{1/\gamma-1/p'}}\textstyle\sum_{k=0}^{\infty}\alpha_k\left(\fint_{2^{k+2}\mathcal{B}_j} H^{\gamma}\,\mathrm{d}\nu\right)^{1/\gamma}\\ &+\tfrac{C[\nu(\mathcal{B}_j)]^{\theta}}{\varepsilon^{1/p_*-1/p'}}\left(\fint_{4\mathcal{B}_j} F^{p_*}\,\mathrm{d}\nu\right)^{1/p_*},\end{aligned} \tag{4.16}$$

where $\sigma \in (0,1]$ has yet to be chosen and $C = C(\mathtt{data})$. Choose the unique $m \in \mathbb{Z}_+$ such that $2^{-m}\varrho_0 \le R(x_j) < 2^{-m+1}\varrho_0$. Since $R(x_j) < \frac{\alpha-\beta}{40^n}$ and $0 < \alpha-\beta < \varrho_0/2$, we have $m \ge 3$. Further, $\frac{\alpha-\beta}{40^n} \le \frac{\varrho_0}{2\cdot 40^n} \le 2^{m-1}R(x_j)$, so by (4.7)

$$\Upsilon_0(x_j, 2^{m-1}R(x_j)) + Tail(x_j, 2^{m-1}R(x_j)) \le \kappa\lambda_1. \tag{4.17}$$

This allows us to estimate $Tail$. The first $m-2$ terms can be handled by the exit-time condition (4.10); that is,

$$\left(\fint_{2^k\mathcal{B}_j} H^\gamma \,\mathrm{d}\nu\right)^{1/\gamma} \leq \kappa\lambda \quad \text{if } 1 \leq k \leq m-2\,. \tag{4.18}$$

Then by (4.17) and (4.18), and recalling that $\alpha_k = 2^{-k(\frac{tq}{p-1}-s-\varepsilon)}$,

$$\begin{aligned}
&\sum_{k=0}^{\infty} \alpha_k \left(\fint_{2^{k+2}\mathcal{B}_j} H^\gamma \,\mathrm{d}\nu\right)^{1/\gamma} \\
&= \frac{1}{\alpha_2}\sum_{k=2}^{\infty} \alpha_k \left(\fint_{2^{k}\mathcal{B}_j} H^\gamma \,\mathrm{d}\nu\right)^{1/\gamma} \\
&= \frac{1}{\alpha_2}\sum_{k=2}^{m-2} \alpha_k \left(\fint_{2^{k}\mathcal{B}_j} H^\gamma \,\mathrm{d}\nu\right)^{1/\gamma} + \frac{1}{\alpha_2}\sum_{k=0}^{\infty} \alpha_{k+m-1} \left(\fint_{2^{k+m-1}\mathcal{B}_j} H^\gamma \,\mathrm{d}\nu\right)^{1/\gamma} \\
&\leq \frac{1}{\alpha_2}\left[\kappa\lambda\sum_{k=2}^{m-2}\alpha_k + \alpha_{m-1} Tail(x_j, 2^{m-1}R(x_j))\right] \\
&\leq \frac{1}{\alpha_2}\left[\kappa\lambda\sum_{k=2}^{m-2}\alpha_k + \alpha_{m-1}\kappa\lambda_1\right] \\
&\leq \kappa\lambda\sum_{k=0}^{\infty}\alpha_k \leq \frac{4^{\frac{tq}{p-1}-s-\varepsilon}\kappa\lambda}{(\frac{tq}{p-1}-s-\varepsilon)\ln(2)} \leq \frac{4^q\kappa\lambda}{s(p'-1)\left(\frac{tq}{sp}-\frac{1}{p'}\right)\ln(2)} := C_1\kappa\lambda\,,
\end{aligned} \tag{4.19}$$

where in the last line we used (2.1) and the bound $\varepsilon < s(\frac{tq}{sp}-\frac{1}{p'})$ in (A4). The constant C_1 depends only on data. Now using the fact that $m \geq 3$, we gain that $2R(x_j) \leq \frac{1}{2}\varrho_0$, so by the exit-time condition (4.10)

$$\frac{C[\nu(\mathcal{B}_j)]^\theta}{\varepsilon^{1/p_*-1/p'}}\left(\fint_{4\mathcal{B}_j} F^{p_*}\,\mathrm{d}\nu\right)^{1/p_*} \leq C\frac{\Psi_M(x_j, 2R(x_j))}{M} \leq \frac{C_2\kappa\lambda}{M}\,, C_2{=}C_2(\mathtt{data})\,. \tag{4.20}$$

Combining (4.19) and (4.20) in (4.16) gives

$$\kappa\lambda \leq \frac{C}{\sigma\varepsilon^{1/\gamma-1/p'}}\left(\fint_{4\mathcal{B}_j} H^\gamma\,\mathrm{d}\nu\right)^{1/\gamma} + \frac{C_1\sigma\kappa\lambda}{\varepsilon^{1/\gamma-1/p'}} + \frac{C_2\kappa\lambda}{M}\,. \tag{4.21}$$

Now we choose $\sigma \in (0, 1)$ and $M \geq 1$ and absorb the last two terms. We set

$$\sigma := \frac{\varepsilon^{1/\gamma - 1/p'}}{4C_1}, \qquad M := 4C_2, \tag{4.22}$$

and so we have for $C = C(\mathtt{data})$

$$\kappa\lambda \le \frac{C}{\varepsilon^{2/\gamma - 2/p'}} \left(\fint_{4\mathcal{B}_j} H^\gamma \, \mathrm{d}\nu \right)^{1/\gamma} \quad \Rightarrow \quad \nu(\mathcal{B}_j) \le \frac{C}{\varepsilon^{2-(2\gamma)/p'}(\kappa\lambda)^\gamma} \int_{4\mathcal{B}_j} H^\gamma \, \mathrm{d}\nu \,. \tag{4.23}$$

Now, let $\widetilde{\kappa} > 0$ be a constant, to be fixed in a moment. Using the doubling property (3.2),

$$\begin{aligned} &\frac{C}{\varepsilon^{2-(2\gamma)/p'}(\kappa\lambda)^\gamma} \int_{4\mathcal{B}_j} H^\gamma \, \mathrm{d}\nu \\ &\le \frac{C}{\varepsilon^{2-(2\gamma)/p'}(\kappa\lambda)^\gamma} \int_{4\mathcal{B}_j \cap \{H \le \widetilde{\kappa}\kappa\lambda\}} H^\gamma \, \mathrm{d}\nu + \frac{C}{\varepsilon^{2-(2\gamma)/p'}(\kappa\lambda)^\gamma} \int_{4\mathcal{B}_j \cap \{H > \widetilde{\kappa}\kappa\lambda\}} H^\gamma \, \mathrm{d}\nu \\ &\le \frac{C_3 \nu(\mathcal{B}_j)\widetilde{\kappa}^\gamma}{\varepsilon^{2-(2\gamma)/p'}} + \frac{C_3}{\varepsilon^{2-(2\gamma)/p'}(\kappa\lambda)^\gamma} \int_{4\mathcal{B}_j \cap \{H > \widetilde{\kappa}\kappa\lambda\}} H^\gamma \, \mathrm{d}\nu \,, \end{aligned} \tag{4.24}$$

where $C_3 = C_3(\mathtt{data})$. Choose

$$\widetilde{\kappa} = \frac{\varepsilon^{2/\gamma - 2/p'}}{(2C_3)^{1/\gamma}}, \tag{4.25}$$

and then substituting (4.24) into $(4.23)_2$ and absorbing the term give

$$\nu(\mathcal{B}_j) \le \frac{C_4}{\varepsilon^{2-2\gamma/p'}\kappa^\gamma\lambda^\gamma} \int_{4\mathcal{B}_j \cap \{\widetilde{\kappa}\kappa\lambda\}} H^\gamma \, \mathrm{d}\nu \,, \qquad C_4 = C_4(\mathtt{data}). \tag{4.26}$$

Case 2 If (4.15) occurs, then

$$\left(\frac{\kappa\lambda}{2}\right)^{p_*} \le \frac{M^{p_*}[\nu(\mathcal{B}_j)]^{p_*\theta - 1}}{\varepsilon^{1-p_*/p'}} \int_{\mathcal{B}_j} F^{p_*} \, \mathrm{d}\nu \,,$$

which implies

$$\nu(\mathcal{B}_j) \le \left(\frac{2M}{\varepsilon^{1/p_* - 1/p'}\kappa\lambda} \right)^{p_*/(1-p_*\theta)} \left(\int_{\mathcal{B}_j} F^{p_*} \, \mathrm{d}\nu \right)^{1/(1-p_*\theta)}, \tag{4.27}$$

where we used the inequality $p_*\theta < 1$ that follows from the definition of θ that $p_*\theta \le \frac{nsp}{(n+sp)(n+\varepsilon p)} < \frac{p}{p+1}$. Moreover, since $p \ge 2$, we have that $3 \le \frac{1}{1-p_*\theta} = 1 + \frac{p_*\theta}{1-p_*\theta}$. This decomposition of the exponent will allow us to incorporate the level set of F in the integrand and then remove the exponent on the integral. To that end, with a constant $\widehat{\kappa} \in (0, 1)$ to be determined, we split the integral as

$$\left(\int_{\mathcal{B}_j} F^{p_*}\,d\nu\right)^{\frac{1}{1-p_*\theta}} \leq \left(\int_{\mathcal{B}_j\cap\{F>\widehat{\kappa}\kappa\lambda\}} F^{p_*}\,d\nu + (\widehat{\kappa}\kappa\lambda)^{p_*}\nu(\mathcal{B}_j)\right)^{\frac{1}{1-p_*\theta}} . \tag{4.28}$$

Now, since $\varrho_0 \leq 1$,

$$\nu(\mathcal{B}(x_0, 2\varrho_0) \leq \frac{C(n)2^{n+\varepsilon p}}{\varepsilon} := L = L(n, p, \varepsilon)\,.$$

Then by noting that $\mathcal{B}_j \subset \mathcal{B}(x_0, 2\varrho_0)$, we can estimate

$$[\nu(\mathcal{B}_j)]^{\frac{1}{1-p_*\theta}} \leq [\nu(\mathcal{B}(x_0, 2\varrho_0))]^{\frac{p_*\theta}{1-p_*\theta}}\,\nu(\mathcal{B}_j) \leq L^{\frac{p_*\theta}{1-p_*\theta}}\,\nu(\mathcal{B}_j)\,.$$

Then using the elementary inequality $(a+b)^r \leq 2^{r-1}(a^r + b^r)$ for any $r \geq 1$ in (4.28) and using the estimate above for $[\nu(\mathcal{B}_j)]^{1/(1-p_*\theta)}$, we have

$$\begin{aligned}\left(\int_{\mathcal{B}_j} F^{p_*}\,d\nu\right)^{\frac{1}{1-p_*\theta}} &\leq 2^{p_*\theta/(1-p_*\theta)}\left(\int_{\mathcal{B}_j\cap\{F>\widehat{\kappa}\kappa\lambda\}} F^{p_*}\,d\nu\right)^{\frac{1}{1-p_*\theta}} \\ &\quad + (2(L+1))^{\frac{p_*\theta}{1-p_*\theta}}\,(\widehat{\kappa}\kappa\lambda)^{\frac{p_*\theta}{1-p_*\theta}}\,\nu(\mathcal{B}_j)\,.\end{aligned} \tag{4.29}$$

Combining (4.27) and (4.29) and using that $\theta \leq 1$,

$$\begin{aligned}\nu(\mathcal{B}_j) \leq &\left(\tfrac{2^2 M}{\varepsilon^{1/p_*-1/p'}\kappa\lambda}\right)^{p_*/(1-p_*\theta)}\left(\textstyle\int_{\mathcal{B}_j\cap\{F>\widehat{\kappa}\kappa\lambda\}} F^{p_*}\,d\nu\right)^{1/(1-p_*\theta)} \\ &+\left(\tfrac{2^2 M(L+1)\widehat{\kappa}}{\varepsilon^{1/p_*-1/p'}}\right)^{p_*/(1-\theta p_*)}\nu(\mathcal{B}_j)\,.\end{aligned}$$

Now set $\widehat{\kappa} \in (0, 1)$ to satisfy

$$\left(\frac{4M(L+1)\widehat{\kappa}}{\varepsilon^{1/p_*-1/p'}}\right)^{p_*/(1-p_*\theta)} \leq \frac{1}{2} \quad\Rightarrow\quad \widehat{\kappa} \leq \left(\frac{1}{2}\right)^{(1-p_*\theta)/p_*}\frac{\varepsilon^{1/p_*-1/p'}}{4M(L+1)}. \tag{4.30}$$

Then

$$\nu(\mathcal{B}_j) \leq 2\left(\frac{4M}{\varepsilon^{1/p_*-1/p'}\kappa\lambda}\right)^{\frac{p_*}{1-p_*\theta}}\left(\int_{\mathcal{B}_j\cap\{F>\widehat{\kappa}\kappa\lambda\}} F^{p_*}\,d\nu\right)^{\frac{1}{1-p_*\theta}} .$$

Writing $\left(\int_{\mathcal{B}_j\cap\{F>\widehat{\kappa}\kappa\lambda\}} F^{p_*}\,\mathrm{d}\nu\right)^{\frac{1}{1-p_*\theta}} = \left(\int_{\mathcal{B}_j\cap\{F>\widehat{\kappa}\kappa\lambda\}} F^{p_*}\,\mathrm{d}\nu\right)\left(\int_{\mathcal{B}_j\cap\{F>\widehat{\kappa}\kappa\lambda\}} F^{p_*}\,\mathrm{d}\nu\right)^{\frac{p_*\theta}{1-p_*\theta}}$, let us transfer some of the "decay" from the integrand to the cutoff λ_0, and in so doing, remove the exponent from the integral.

$$\begin{aligned}\int_{\mathcal{B}_j\cap\{F>\widehat{\kappa}\kappa\lambda\}} F^{p_*}\,\mathrm{d}\nu &\le (\widehat{\kappa}\kappa\lambda)^{p_*}\int_{\mathcal{B}_j\cap\{F>\widehat{\kappa}\kappa\lambda\}}\left(\frac{F}{\widehat{\kappa}\kappa\lambda}\right)^{p_*+\delta_f}\mathrm{d}\nu\\ &\le \frac{\nu(2^{m-1}\mathcal{B}_j)}{(\widehat{\kappa}\kappa\lambda)^{\delta_f}}\fint_{2^{m-1}\mathcal{B}_j} F^{p_*+\delta_f}\,\mathrm{d}\nu\\ &\le \frac{\nu(\mathcal{B}(x_0,2\varrho_0))}{(\widehat{\kappa}\kappa\lambda)^{\delta_f}}[\Upsilon_0(x_j,2^{m-1}R(x_j))]^{p_*+\delta_f} \le \frac{L\lambda_1^{p_*+\delta_f}}{(\widehat{\kappa}\kappa\lambda)^{\delta_f}}\,.\end{aligned}$$

Therefore, powering the above inequality by $\frac{p_*\theta}{1-p_*\theta}$ on both sides, we obtain that

$$\nu(\mathcal{B}_j) \le \frac{C_5\lambda_1^{(p_*+\delta_f)\theta p_*/(1-p_*\theta)}}{(\widehat{\kappa}\kappa\lambda)^{(1+\theta\delta_f)p_*/(1-p_*\theta)}}\int_{\mathcal{B}_j\cap\{F>\widehat{\kappa}\kappa\lambda\}} F^{p_*}\,\mathrm{d}\nu\,, \tag{4.31}$$

where

$$C_5 := 2\left(\frac{4M(L+1)}{\varepsilon^{1/p_*-1/p'}}\right)^{p_*/(1-p_*\theta)}\,. \tag{4.32}$$

Combining (4.26) and (4.31),

$$\nu(\mathcal{B}_j) \le \frac{C_4}{\varepsilon^{2-2\gamma/p'}\kappa^\gamma\lambda^\gamma}\int_{4\mathcal{B}_j\cap\{H>\widetilde{\kappa}\kappa\lambda\}} H^\gamma\,\mathrm{d}\nu + \frac{C_5\lambda_1^{\vartheta_f}}{(\widehat{\kappa}\kappa\lambda)^{\widetilde{\vartheta}_f}}\int_{\mathcal{B}_j\cap\{F>\widehat{\kappa}\kappa\lambda\}} F^{p_*}\,\mathrm{d}\nu\,,$$

and since $\{\mathcal{B}_j\}$ is a disjoint collection whose members are all contained in $\mathcal{B}(x_0,\alpha)$, we have

$$\begin{aligned}\textstyle\sum_j \nu(\mathcal{B}_j) \le &\tfrac{C_4}{\varepsilon^{2-2\gamma/p'}\kappa^\gamma\lambda^\gamma}\textstyle\int_{\mathcal{B}(x_0,\alpha)\cap\{H>\widetilde{\kappa}\kappa\lambda\}} H^\gamma\,\mathrm{d}\nu\\ &+\tfrac{C_5\lambda_1^{\vartheta_f}}{(\widehat{\kappa}\kappa\lambda)^{\vartheta_f}}\textstyle\int_{\mathcal{B}(x_0,\alpha)\cap\{F>\widehat{\kappa}\kappa\lambda\}} F^{p_*}\,\mathrm{d}\nu\,,\end{aligned} \tag{4.33}$$

where the constants $C_4 = C_4(\mathtt{data})$, $C_5(\mathtt{data},\varepsilon)$, $\widetilde{\kappa}$, and $\widehat{\kappa}$ have been chosen to satisfy (4.26), (4.32), (4.25), and (4.30), respectively. The constant $\kappa \in (0,1]$ defined in (4.64) will be fixed in the course of the off-diagonal estimates below.

4.3 Analysis Off the Diagonal

The analysis far from the diagonal is much more technical. We begin with defining the collections of dyadic cubes and summarizing their properties that will be used repeatedly. We then define two-dimensional constants that will be used throughout the analysis of cubes far from the diagonal. In Sect. 4.3.3, we recall the classical Calderón–Zygmund decomposition and adapt it to decompose the level set $\{H > \lambda\}$ into the aforementioned dyadic cubes. In Sect. 4.3.4, we analyze the cubes near the diagonal and show they can be covered by the collection $\{\mathcal{B}_j\}$. In the remaining sections, we treat the cubes far from the diagonal using the "almost-reverse" Hölder inequality in Lemma 4.6, a careful choice of the constant κ, and combinatorial information about the cubes coming from their size and distance from the diagonal.

4.3.1 Dyadic Cubes

The following contains information regarding cubes arising from a Calderón–Zygmund decomposition of Euclidean space. The dyadic cubes considered here are centered at x_0, and their size has been changed so as to be compatible with the starting ball $\mathcal{B}(x_0, \beta)$. We will consider cubes with side length 2^{-k} for integers $k \geq k_0$, where

$$k_0 := \left\lfloor -\log_2\left(\frac{\alpha-\beta}{n40^{n+1}}\right)\right\rfloor + 1\,. \tag{4.34}$$

Here, $\lfloor\cdot\rfloor$ denotes the floor function. For each $k \geq k_0$, let $\mathcal{C}_k$ be the disjoint collection of half-open cubes in $\mathbb{R}^n$ centered at x_0 with side length 2^{-k} whose closures intersect $\overline{B}(x_0, \frac{1}{2}(\alpha-\beta))$. That is,

$$\mathcal{C}_k{:=}\{x_0{+}2^{-k}z{+}[0,2^{-k})^n : z\in\mathbb{Z}^n,\,(x_0{+}2^{-k}z{+}[0,2^{-k})^n)\cap\overline{B}(x_0,\frac{1}{2}(\alpha{-}\beta))\neq\emptyset\}.$$

Then by (4.34), we have

$$B(x_0,\beta)\subset\bigcup_{K\in\mathcal{C}_k}K\subset B(x_0,\alpha)\,. \tag{4.35}$$

Note that every cube $K \in \mathcal{C}_{k+1}$ has a unique "predecessor" $\widetilde{K} \in \mathcal{C}_k$ such that $K \subset \widetilde{K}$. With these cubes, we can define dyadic cubes on $\mathbb{R}^{2n}$, denoted by

$$\Delta_k := \{\mathcal{K} := K_1\times K_2 : K_1, K_2\in\mathcal{C}_k\}\,,\qquad \Delta := \bigcup_{k\geq k_0}\Delta_k\,. \tag{4.36}$$

Denote the diagonal cubes $\Delta_k^d := \{K \times K \,:\, K \in \Delta_k\}$. Then

$$\mathcal{B}(x_0, \beta) \subset \bigcup_{\mathcal{K} \in \Delta_k} \mathcal{K} \subset \mathcal{B}(x_0, \alpha)\,.$$

Note that these product cubes in $\mathbb{R}^{2n}$ satisfy all of the same properties as the cubes. Numerous times a cube $\mathcal{K} \in \Delta$ will be given; there exists $K_1, K_2 \in \mathcal{C}_k$ such that $\mathcal{K} = K_1 \times K_2$, and we denote

$$k(\mathcal{K}) = k\,.$$

We denote the cube projections for a cube $\mathcal{K} = K_1 \times K_2$ by

$$\pi_1 \mathcal{K} := K_1 \times K_1\,, \qquad \text{and} \qquad \pi_2 \mathcal{K} := K_2 \times K_2\,.$$

Proposition 4.2 *Let $\mathcal{K} = K_1 \times K_2 \in \Delta$. The following hold:*

- $\pi_1 \mathcal{K}, \pi_2 \mathcal{K} \in \Delta$.
- $\nu(\pi_1 \mathcal{K}) = \nu(\pi_2 \mathcal{K})$ *and* $k(\mathcal{K}) = k(\pi_1 \mathcal{K}) = k(\pi_2 \mathcal{K})$.
- *If* $\mathcal{H} \in \Delta$ *and* $\mathcal{H} \subset \mathcal{K}$, *then* $k(\mathcal{K}) \leq k(\mathcal{H})$.
- *If* $\widetilde{\mathcal{K}} = \widetilde{K}_1 \times \widetilde{K}_2$ *is the predecessor of* $\mathcal{K}$, *then*

$$dist(\widetilde{K}_1, \widetilde{K}_2) \leq \operatorname{dist}(K_1, K_2)\,. \tag{4.37}$$

- *The following hold:*

$$\operatorname{dist}(K_1, K_2) = \frac{1}{\sqrt{2}} \operatorname{dist}(\pi_1 \mathcal{K}, \pi_2 \mathcal{K})\,. \tag{4.38}$$

$$\operatorname{dist}(\mathcal{K}, \operatorname{diag}) = \frac{1}{\sqrt{2}} \operatorname{dist}(K_1, K_2) = \frac{1}{2} \operatorname{dist}(\pi_1 \mathcal{K}, \pi_2 \mathcal{K})\,. \tag{4.39}$$

Note that for two sets $\mathcal{A}_1$ and $\mathcal{A}_2 \subset \mathbb{R}^{2n}$ the definition of $\operatorname{dist}(\mathcal{A}_1, \mathcal{A}_2)$ *uses the traditional Euclidean norm.*

4.3.2 Two Constants That Depend On The Dimension

In the analysis that follows, we will often compare the sizes of the cubes with their distance to the diagonal. This leads us to define two constants that depend on the dimension and independent of ε that will be used repeatedly. The existence of these constants follows from geometric arguments.

Lemma 4.3 *There exists a constant C_{dd} depending only on n such that for $h \in \{1, 2\}$*

$$\begin{aligned} C_{dd} \geq \sup_{\mathcal{K}\in\Delta} &\left\{ \tfrac{1}{\varepsilon} \left(\tfrac{\operatorname{dist}(K_1,K_2)}{2^{-k}} \right)^{n-\varepsilon p} \tfrac{\nu(\mathcal{K})}{\nu(\pi_h\mathcal{K})} \right\} \\ &+ \sup_{\substack{\mathcal{K}\in\Delta \\ \operatorname{dist}(K_1,K_2)\geq 2^{-k}}} \left\{ \varepsilon \left(\tfrac{\operatorname{dist}(K_1,K_2)}{2^{-k}} \right)^{\varepsilon p-n} \tfrac{\nu(\pi_h\mathcal{K})}{\nu(\mathcal{K})} \right\}, \end{aligned} \tag{4.40}$$

where $k = k(\mathcal{K})$.

Proof Without loss of generality, assume $\operatorname{dist}(K_1, K_2) > 0$. By definition of ν,

$$\nu(\pi_h\mathcal{K}) \geq \frac{C(n)(2^{-k})^{n+\varepsilon p}}{\varepsilon}, \qquad \text{while} \qquad \nu(\mathcal{K}) \leq \frac{2^{(-k)2n}}{\operatorname{dist}(K_1, K_2)^{n-\varepsilon p}}.$$

Thus the first quantity is bounded;

$$\frac{1}{\varepsilon} \left(\frac{\operatorname{dist}(K_1, K_2)}{2^{-k}} \right)^{n-\varepsilon p} \frac{\nu(\mathcal{K})}{\nu(\pi_h\mathcal{K})} \leq C(n).$$

On the other hand, if $x \in K_1$ and $y \in K_2$, then

$$\begin{aligned} \operatorname{dist}(K_1, K_2) \leq |x-y| &\leq \operatorname{dist}(K_1, K_2) + 2\sqrt{n}(2^{-k}) \\ &\leq 2\sqrt{n}(\operatorname{dist}(K_1, K_2) + 2^{-k}). \end{aligned}$$

Thus by definition of ν,

$$\nu(\mathcal{K}) \geq \frac{(2^{-k})^{2n}}{(2\sqrt{n})^{n-\varepsilon p}(\operatorname{dist}(K_1, K_2) + 2^{-k})^{n-\varepsilon p}}.$$

Since

$$\nu(\pi_h\mathcal{K}) = \frac{C(n)}{\varepsilon}(2^{-k})^{n+\varepsilon p},$$

we have

$$\begin{aligned} &\varepsilon \left(\tfrac{\operatorname{dist}(K_1,K_2)}{2^{-k}} \right)^{\varepsilon p-n} \tfrac{\nu(\pi_h\mathcal{K})}{\nu(\mathcal{K})} \\ &\leq c(n) \left(\tfrac{\operatorname{dist}(K_1,K_2)}{2^{-k}} \right)^{\varepsilon p-n} \tfrac{(\operatorname{dist}(K_1,K_2)+2^{-k})^{n-\varepsilon p}}{(2^{-k})^{n-\varepsilon p}} \\ &\leq c(n) \left(\tfrac{\operatorname{dist}(K_1,K_2)}{2^{-k}} \right)^{\varepsilon p'-n} \tfrac{(\operatorname{dist}(K_1,K_2))^{n-\varepsilon p}}{(2^{-k})^{n-\varepsilon p}} \leq C(n), \end{aligned}$$

where in the second inequality we used that $\mathrm{dist}(K_1, K_2) \geq 2^{-k}$. Thus the second term is bounded by a dimensional constant as well, and so (4.40) holds for some constant C_{dd} depending only on n. □

Lemma 4.4 *There exists a constant C_{ddd} depending only on n such that*

$$\sup\left\{\frac{\nu(\widetilde{\mathcal{K}})}{\nu(\mathcal{K})} : \widetilde{\mathcal{K}} \text{ is the predecessor of } \mathcal{K},\ \mathrm{dist}(\widetilde{K_1}, \widetilde{K_2}) \geq 2^{-k(\mathcal{K})}\right\} \leq C_{ddd}\,. \tag{4.41}$$

Proof The triangle inequality gives

$$|x-y| \leq 2\sqrt{n}2^{-k(\mathcal{K})+1} + \mathrm{dist}(\widetilde{K_1}, \widetilde{K_2}) \leq 8\sqrt{n}\,\mathrm{dist}(\widetilde{K_1}, \widetilde{K_2})\,, \tag{4.42}$$

whenever $x \in K_1$, $y \in K_2$, and $\mathrm{dist}(\widetilde{K_1}, \widetilde{K_2}) \geq 2^{-k(\mathcal{K})}$. Therefore,

$$\begin{aligned}
\nu(\widetilde{\mathcal{K}}) &\leq \mathrm{dist}(\widetilde{K_1}, \widetilde{K_2})^{-n-\varepsilon p}|\widetilde{K_1}||\widetilde{K_2}| \\
&= 4^n\,\mathrm{dist}(\widetilde{K_1}, \widetilde{K_2})^{-n-\varepsilon p}\int_{K_1}\int_{K_2} \mathrm{d}x\,\mathrm{d}y \\
&\overset{(4.42)}{\leq} 4^n 8\sqrt{n}\int_{K_1}\int_{K_2}\frac{1}{|x-y|^{n-\varepsilon p}}\,\mathrm{d}x\,\mathrm{d}y = C(n)\nu(\mathcal{K})\,.
\end{aligned}$$

□

4.3.3 Calderón–Zygmund Covering and Sorting of Dyadic Cubes

Here we write down a version of the classical Calderón–Zygmund decomposition adapted for use in our context. The proof is virtually identical to the classical proof, taking into account that the measure is doubling and absolutely continuous with respect to Lebesgue measure; see [4, 12].

Theorem 4.5 *Let $Q_0 \subset \mathbb{R}^{2n}$ be a cube and let $V \geq 0$ be a function in $L^1(Q_0)$. Let $\widetilde{\lambda}$ be a number such that*

$$⨍_{Q_0} V\,\mathrm{d}\nu \leq \widetilde{\lambda}\,.$$

Then there exists a collection of at most countable cubes $\{Q_i\}$ that are pairwise disjoint with sides parallel to those of Q_0 such that

$$\widetilde{\lambda} < ⨍_{Q_i} V\,\mathrm{d}\nu \qquad \text{and} \qquad ⨍_{\widetilde{Q}_i} V\,\mathrm{d}\nu \leq \widetilde{\lambda} \quad \text{for every } Q_i\,,$$

where $\widetilde{Q}_i$ is the predecessor of Q_i, and

$$V \le \widetilde{\lambda} \quad \textit{almost everywhere in } Q_0 \setminus \bigcup_i Q_i \,.$$

We will use this theorem to cover the level set of H. Define

$$\lambda_2 := \max\left\{\lambda_1,\ \sup_{\mathcal{K}\in\Delta_{k_0}} \left(\fint_{\mathcal{K}} H^{p'}\,\mathrm{d}\nu\right)^{1/p'}\right\}. \tag{4.43}$$

Recall that λ_1 and k_0 have been defined in (4.7) and (4.34), respectively. Note that by (4.35) the cubes $\{\mathcal{K}\}_{\mathcal{K}\in\Delta_{k_0}}$ cover $\mathcal{B}(x_0,\beta)$. For $\lambda \ge \lambda_2$, apply Theorem 4.5 with $Q_0 = \mathcal{K}_0$ for each and every cube $\mathcal{K}_0 \in \Delta_{k_0}$. We obtain a pairwise disjoint family of cubes $Q_i(\mathcal{K}_0)$ such that

$$\lambda^{p'} < \fint_{Q_i(\mathcal{K}_0)} H^{p'}\,\mathrm{d}\nu \qquad \text{and} \qquad \fint_{\widetilde{Q}_i(\mathcal{K}_0)} H^{p'}\,\mathrm{d}\nu \le \widetilde{\lambda}^{p'} \quad \text{for every } Q_i(\mathcal{K}_0)\,, \tag{4.44}$$

where $\widetilde{Q}_i(\mathcal{K}_0)$ denotes the predecessor of $Q_i(\mathcal{K}_0)$, and

$$H \le \lambda \quad \text{a.e. in } \mathcal{K}_0 \setminus \bigcup_i Q_i(\mathcal{K}_0)\,.$$

Thus, we get a countable collection of disjoint dyadic cubes

$$\mathcal{H}_\lambda := \bigcup_{\mathcal{K}_0\in\Delta_{k_0}} \{Q_i(\mathcal{K}_0)\} = \{\mathcal{K}\}$$

that satisfy

$$\lambda^{p'} < \fint_{\mathcal{K}} H^{p'}\,\mathrm{d}\nu \quad \text{and} \quad \fint_{\widetilde{\mathcal{K}}} H^{p'}\,\mathrm{d}\nu \le \lambda^{p'} \qquad \text{for every } \mathcal{K}\in\mathcal{H}_\lambda, \tag{4.45}$$

where $\widetilde{\mathcal{K}}$ denotes the predecessor of $\mathcal{K}$ and such that

$$H \le \lambda \quad \text{a.e. in } \mathcal{B}(x_0,\alpha) \setminus \bigcup_{\mathcal{K}\in\mathcal{H}_\lambda} \mathcal{K}\,. \tag{4.46}$$

4.3.4 Nearly Diagonal Cubes

It turns out that the cubes whose distance to the diagonal is smaller than their size can be covered by diagonal balls $\mathcal{B}_j$ chosen in the exit-time argument above, see (4.10)–(4.11). This leaves us to deal with the cubes that are "far" from the diagonal in the next section.

To begin, we define the family of "nearly diagonal" cubes

$$\mathcal{H}_\lambda^d := \{\mathcal{K} \in \mathcal{H}_\lambda \,:\, \operatorname{dist}(\widetilde{K}_1, \widetilde{K}_2) < 2^{-k(\mathcal{K})}\,, \quad \widetilde{\mathcal{K}} = \widetilde{K}_1 \times \widetilde{K}_2 \text{ is the predecessor of } \mathcal{K}\}\,.$$

With $\mathcal{K} \in \mathcal{H}_\lambda^d$, let $(\widetilde{x}, \widetilde{x}) \in \text{Diag}$ such that $\operatorname{dist}((\widetilde{x}, \widetilde{x}), \widetilde{\mathcal{K}}) = \operatorname{dist}(\text{Diag}, \widetilde{\mathcal{K}})$ and a diagonal ball $\mathcal{B}(\widetilde{x}, R)$ such that

$$R \geq \frac{5\sqrt{n}}{2} \operatorname{dist}(\widetilde{K}_1, \widetilde{K}_2) + 5\sqrt{n} 2^{-k(\mathcal{K})+1}\,.$$

Using (4.39) for $\widetilde{\mathcal{K}}$, it follows that $\widetilde{\mathcal{K}} \subset \mathcal{B}(\widetilde{x}, R)$. Thus, the diagonal ball $\mathcal{B} \equiv \mathcal{B}(\widetilde{x}, 24\sqrt{n}2^{-k(\mathcal{K})})$ satisfies $\mathcal{B} \subset \mathcal{K}$. Now note that by (3.3) in Theorem 3.1 there exists a constant $C_d = C_d(n, p)$ such that

$$1 \leq \frac{\nu(\mathcal{B})}{\nu(\mathcal{K})} \leq \frac{C_d}{\varepsilon}\,.$$

Thus if $\mathcal{K} \in \mathcal{H}_\lambda^d$, then the lower bound in (4.45) gives

$$\lambda^{p'} < \fint_{\mathcal{K}} H^{p'}\, d\nu \leq \frac{\nu(\mathcal{B})}{\nu(\mathcal{K})} \fint_{\mathcal{B}} H^{p'}\, d\nu \leq \frac{C_d}{\varepsilon} \fint_{\mathcal{B}} H^{p'}\, d\nu\,.$$

By choosing the number $\kappa \in (0, 1]$ introduced in (4.7) to satisfy

$$\kappa \in (0, \kappa_0]\,, \qquad \kappa_0 := \frac{\varepsilon^{1/p'}}{(2C_d)^{1/p'}}\,, \tag{4.47}$$

we have therefore proved that

$$\begin{gathered}\text{For every } \mathcal{K} \in \mathcal{H}_\lambda^d \text{ there exists } \mathcal{B}^{\mathcal{K}} = B^{\mathcal{K}} \times B^{\mathcal{K}} \\ \text{such that } \kappa^{p'}\lambda^{p'} < \fint_{\mathcal{B}^{\mathcal{K}}} H^{p'}\, d\nu \text{ with } \mathcal{K} \subset \mathcal{B}^{\mathcal{K}}\,.\end{gathered}$$

Denote the center of $\mathcal{B}^{\mathcal{K}}$ by $\widetilde{x}$. Then by the choice (4.34) for the lower bound on $k(\mathcal{K})$, it follows that the radius of the ball $\mathcal{B} \equiv \mathcal{B}(\widetilde{x}, 24\sqrt{n}2^{-k(\mathcal{K})})$ is smaller than $\frac{\alpha-\beta}{40^n}$. Therefore, we can apply the exit-time condition (4.10) to obtain that $(\widetilde{x}, \widetilde{x}) \in C_{\kappa\lambda}$, and then $\mathcal{B}^{\mathcal{K}} \subset \mathcal{B}(\widetilde{x}, R(\widetilde{x}))$. By (4.11), it follows that

$$\bigcup_{\mathcal{K} \in \mathcal{H}_\lambda^d} \mathcal{K} \subset \bigcup_j 10\mathcal{B}_j\,.. \tag{4.48}$$

4.3.5 Off-diagonal Reverse Hölder Inequalities

Since the nearly diagonal cubes $\mathcal{H}_\lambda^d$ have been covered by the diagonal cover, we need to only consider the off-diagonal cubes

$$\mathcal{H}_\lambda^{nd} := \{\mathcal{K} \in \mathcal{H}_\lambda \, : \, \mathrm{dist}(\widetilde{K}_1, \widetilde{K}_2) \geq 2^{-k(\mathcal{K})}, \\ \widetilde{\mathcal{K}} = \widetilde{K}_1 \times \widetilde{K}_2 \text{ is the predecessor of } \mathcal{K}\} \, . \tag{4.49}$$

By (4.37), we also have

$$\mathrm{dist}(K_1, K_2) \leq 2^{-k(\mathcal{K})} \qquad \text{for every } \mathcal{K} \in \mathcal{H}_\lambda^{nd} \, .$$

Our objective now is to categorize and estimate sums of the measures of cubes in $\mathcal{H}_\lambda^{nd}$. We will use the following lemma to do so. This lemma states that for off-diagonal cubes an "almost reverse Hölder inequality" holds automatically regardless of whether the function u solves an equation. However, diagonal cubes appear in the estimate, which must be treated by a combinatorial argument in subsequent sections.

Lemma 4.6 *Let $k \geq k_0$ and let $\mathcal{K} = K_1 \times K_2 \in \Delta_k$. Then there exists a constant $C_{nd} = C_{nd}(\mathtt{data})$ independent of ε such that if* $\mathrm{dist}(K_1, K_2) \geq 2^{-k}$, *then*

$$\left(\fint_{\mathcal{K}} H^{p'} \,\mathrm{d}\nu\right)^{1/p'} \leq C_{nd} \left(\fint_{\mathcal{K}} H^{\gamma} \,\mathrm{d}\nu\right)^{1/\gamma} + \frac{C_{nd}}{\varepsilon^{1/\gamma}} \left(\frac{2^{-k}}{\mathrm{dist}(K_1, K_2)}\right)^{(p-1)(s+\varepsilon)} \\ \left[\left(\fint_{\pi_1 \mathcal{K}} H^{\gamma} \,\mathrm{d}\nu\right)^{1/\gamma} + \left(\fint_{\pi_2 \mathcal{K}} H^{\gamma} \,\mathrm{d}\nu\right)^{1/\gamma}\right], \tag{4.50}$$

with γ defined as in Corollary 3.4.1. In particular, (4.50) *holds whenever $\mathcal{K} \in \mathcal{H}_\lambda^{nd}$.*

Proof First, there exist $x_1 \in \overline{K_1}$ and $y_1 \in \overline{K_2}$ such that $\mathrm{dist}(K_1, K_2) = |x_1 - y_1|$. Then for any $(x, y) \in \mathcal{K}$,

$$\begin{aligned} |x - y| &\leq \mathrm{dist}(K_1, K_2) + |x_1 - x| + |y_1 - y| \\ &\leq \mathrm{dist}(K_1, K_2) + 2\sqrt{n} 2^{-k} \\ &\leq 3\sqrt{n}\, \mathrm{dist}(K_1, K_2) \, , \end{aligned}$$

since $\mathrm{dist}(K_1, K_2) \geq 2^{-k}$. Therefore,

$$1 \leq \frac{|x - y|}{\mathrm{dist}(K_1, K_2)} \leq 3\sqrt{n} \quad \text{for all } (x, y) \in \mathcal{K} \, ; \tag{4.51}$$

the first inequality is a consequence of the definition of $\mathrm{dist}(K_1, K_2)$. Next, by definition of ν, we have

$$\frac{1}{C(n, p)} \frac{4^{-nk}}{\mathrm{dist}(K_1, K_2)^{n - \varepsilon p}} \leq \nu(\mathcal{K}) \leq C(n, p) \frac{4^{-nk}}{\mathrm{dist}(K_1, K_2)^{n - \varepsilon p}} \, . \tag{4.52}$$

We therefore have, using (4.51) and (4.52),

$$
\begin{aligned}
\left(\fint_{\mathcal{K}} H^{p'}\,\mathrm{d}\nu\right)^{1/p'} &= \left(\frac{1}{\nu(\mathcal{K})}\int_{K_1}\int_{K_2} \frac{|u(x)-u(y)|^p}{|x-y|^{n+sp}} + a(x,y)\frac{|u(x)-u(y)|^q}{|x-y|^{n+tq}}\,\mathrm{d}y\,\mathrm{d}x\right)^{1/p'} \\
&\leq C\Bigg[\left(\frac{\operatorname{dist}(K_1,K_2)^{n-\varepsilon p-(n+sp)}}{4^{-nk}}\int_{K_1}\int_{K_2}|u(x)-u(y)|^p\,\mathrm{d}y\,\mathrm{d}x\right) \\
&\quad + \left(\frac{\operatorname{dist}(K_1,K_2)^{n-\varepsilon p-(n+tq)}}{4^{-nk}}\int_{K_1}\int_{K_2}|u(x)-u(y)|^q\,\mathrm{d}y\,\mathrm{d}x\right)\Bigg]^{1/p'} \\
&\leq C\operatorname{dist}(K_1,K_2)^{-(p-1)(s+\varepsilon)}\Bigg[\fint_{K_1}\fint_{K_2}|u(x)-u(y)|^p\,\mathrm{d}y\,\mathrm{d}x \\
&\quad + \operatorname{dist}(K_1,K_2)^{sp-tq}\fint_{K_1}\fint_{K_2}|u(x)-u(y)|^q\,\mathrm{d}y\,\mathrm{d}x\Bigg]^{1/p'},
\end{aligned} \tag{4.53}
$$

where $C = C(\mathtt{data})$. Using (4.35), we can estimate

$$
\operatorname{dist}(K_1,K_2)^{sp-tq} \leq (2\alpha)^{sp-tq} \leq (3\varrho_0)^{sp-tq} \leq 3^{sp-tq} \equiv C(\mathtt{data}),
$$

and we also have

$$
\begin{aligned}
\fint_{K_1}\fint_{K_2}|u(x)-u(y)|^q\,\mathrm{d}y\,\mathrm{d}x &\leq (2\,\|u\|_{L^\infty(\mathbb{R}^n)})^{q-p}\fint_{K_1}\fint_{K_2}|u(x)-u(y)|^p\,\mathrm{d}y\,\mathrm{d}x \\
&\leq C(\mathtt{data})\fint_{K_1}\fint_{K_2}|u(x)-u(y)|^p\,\mathrm{d}y\,\mathrm{d}x\,.
\end{aligned}
$$

Combining these two estimates with (4.53) yields

$$
\left(\fint_{\mathcal{K}} H^{p'}\,\mathrm{d}\nu\right)^{1/p'} \leq C\operatorname{dist}(K_1,K_2)^{-(p-1)(s+\varepsilon)}\left(\fint_{K_1}\fint_{K_2}|u(x)-u(y)|^p\,\mathrm{d}y\,\mathrm{d}x\right)^{1/p'}, \tag{4.54}
$$

where $C = C(\mathtt{data})$. Now, using Minkowski's inequality,

$$
\begin{aligned}
\left(\fint_{K_1}\fint_{K_2}|u(x)-u(y)|^p\,\mathrm{d}y\,\mathrm{d}x\right)^{1/p} &\leq \left(\fint_{K_1}|u(x)-(u)_{K_1}|^p\,\mathrm{d}x\right)^{1/p} \\
&\quad + \left(\fint_{K_2}|u(x)-(u)_{K_2}|^p\,\mathrm{d}x\right)^{1/p} + |(u)_{K_1}-(u)_{K_2}|\,.
\end{aligned}
$$

Then using the Sobolev embedding theorem in Lemma 3.2 adapted for U and ν and applied to cubes, we have

$$\left(⨍_{K_h} |u(x) - (u)_{K_1}|^p \, dx\right)^{1/p} \leq \frac{C2^{-k(s+\varepsilon)}}{\varepsilon^{1/\eta}} \left(⨍_{\pi_h \mathcal{K}} U^\eta \, d\nu\right)^{1/\eta}, \quad h \in \{1,2\}, C = C(n,s,p).$$

Next, using Hölder's inequality and repeatedly using (4.51) and (4.52),

$$\begin{aligned}
|(u)_{K_1} - (u)_{K_2}| &\leq ⨍_{K_1} ⨍_{K_2} |u(x) - u(y)| \, dy \, dx \\
&\leq \left(⨍_{K_1} ⨍_{K_2} |u(x) - u(y)|^\eta \, dy \, dx\right)^{1/\eta} \\
&\leq C \left(\frac{1}{\operatorname{dist}(K_1, K_2)^{n-\varepsilon p} \nu(\mathcal{K})} \int_{K_1} \int_{K_2} |u(x) - u(y)|^\eta \, dy \, dx\right)^{1/\eta} \\
&\leq C \left(⨍_{\mathcal{K}} |u(x) - u(y)|^\eta \, d\nu\right)^{1/\eta} \leq C \operatorname{dist}(K_1, K_2)^{s+\varepsilon} \left(⨍_{\mathcal{K}} U^\eta \, d\nu\right)^{1/\eta},
\end{aligned}$$

where $C = C(\mathtt{data})$. Combining the three above inequalities with (4.54) and using the inequality $(a+b+c)^r \leq 3^{r-1}(a^r + b^r + c^r)$ valid for $r \geq 1$ and $a, b, c \geq 0$ gives

$$\begin{aligned}
\left(⨍_{\mathcal{K}} H^{p'} \, d\nu\right)^{1/p'} &\leq C_{nd} \left(⨍_{\mathcal{K}} U^\eta \, d\nu\right)^{(p-1)/\eta} \\
+ \frac{C_{nd}}{\varepsilon^{(p-1)/\eta}} \left(\frac{2^{-k}}{\operatorname{dist}(K_1, K_2)}\right)^{(p-1)(s+\varepsilon)} \\
&\left[\left(⨍_{\pi_1 \mathcal{K}} U^\eta \, d\nu\right)^{(p-1)/\eta} + \left(⨍_{\pi_2 \mathcal{K}} U^\eta \, d\nu\right)^{(p-1)/\eta}\right].
\end{aligned}$$

To see that (4.50) follows, recall that $\gamma = \frac{\eta}{p-1}$ and that by definition $U^\eta \leq H^\gamma$ pointwise. □

Note that this lemma holds for all functions $u \in W^{s,p}(\mathbb{R}^n) \cap L^\infty(\mathbb{R}^n)$ and for every integer k, and in the case $a \equiv 0$, it holds for every $u \in W^{s,p}(\mathbb{R}^n)$. We now apply it in order to begin the level set estimate.

Corollary 4.6.1 *Let $k \geq k_0$ be an integer, and suppose that $\mathcal{K} \in \Delta_k$ satisfies* $\operatorname{dist}(K_1, K_2) \geq 2^{-k}$. *Assume that*

$$\left(⨍_{\mathcal{K}} H^{p'} \, d\nu\right)^{1/p'} \geq \lambda$$

and that the constant κ introduced in (4.7) *satisfies*

$$\kappa \in (0, \kappa_1], \qquad \kappa_1 := \frac{\varepsilon^{1/\gamma}}{2^{1/\gamma} 3 C_{nd}}, \tag{4.55}$$

where $C_{nd} \equiv C_{nd}(\texttt{data})$ has been defined in Lemma 4.6. Then

$$\begin{aligned}\nu(\mathcal{K}) \leq \tfrac{3^\gamma C_{nd}^\gamma}{\lambda^\gamma} \int_{\mathcal{K}\cap\{H>\kappa\lambda\}} H^\gamma \,\mathrm{d}\nu + \tfrac{3^\gamma C_{nd}^\gamma}{\varepsilon\lambda^\gamma} \left(\tfrac{2^{-k}}{\operatorname{dist}(K_1,K_2)}\right)^{\eta(s+\varepsilon)} \\ \left[\tfrac{\nu(\mathcal{K})}{\nu(\pi_1\mathcal{K})} \int_{\pi_1\mathcal{K}\cap\{H>\kappa\lambda\}} H^\gamma \,\mathrm{d}\nu + \tfrac{\nu(\mathcal{K})}{\nu(\pi_2\mathcal{K})} \int_{\pi_2\mathcal{K}\cap\{H>\kappa\lambda\}} H^\gamma \,\mathrm{d}\nu\right].\end{aligned} \tag{4.56}$$

In particular, (4.56) *holds whenever $\mathcal{K} \in \mathcal{H}_\lambda^{nd}$.*

Proof Applying the inequality $(a+b+c)^r \leq 3^{r-1}(a^r+b^r+c^r)$ to the conclusion (4.50) in Lemma 4.6 and recalling that $(p-1)\gamma = \eta$, we get

$$\frac{\lambda^\gamma}{3^{\gamma-1}C_{nd}^\gamma} \leq \fint_{\mathcal{K}} H^\gamma \,\mathrm{d}\nu + \frac{1}{\varepsilon}\left(\frac{2^{-k}}{\operatorname{dist}(K_1, K_2)}\right)^{\eta(s+\varepsilon)} \left(\fint_{\pi_1\mathcal{K}} H^\gamma \,\mathrm{d}\nu + \fint_{\pi_2\mathcal{K}} H^\gamma \,\mathrm{d}\nu\right). \tag{4.57}$$

To estimate the integrals on the right-hand side, we use (4.55) to get

$$\fint_E H^\gamma \,\mathrm{d}\nu \leq \kappa_1^\gamma \lambda^\gamma + \frac{1}{\nu(E)} \int_{E\cap\{H>\kappa\lambda\}} H^\gamma \,\mathrm{d}\nu$$

for $E \in \{\mathcal{K}, \pi_1\mathcal{K}, \pi_2\mathcal{K}\}$, and therefore,

$$\begin{aligned}\frac{\lambda^\gamma}{3^{\gamma-1}C_{nd}^\gamma} &\leq \tfrac{3\kappa_1^\gamma\lambda^\gamma}{\varepsilon} + \tfrac{1}{\nu(\mathcal{K})} \int_{\mathcal{K}\cap\{H>\kappa\lambda\}} H^\gamma \,\mathrm{d}\nu \\ +\tfrac{1}{\varepsilon}\left(\tfrac{2^{-k}}{\operatorname{dist}(K_1,K_2)}\right)^{\eta(s+\varepsilon)} \\ &\left[\tfrac{1}{\nu(\pi_1\mathcal{K})} \int_{\pi_1\mathcal{K}\cap\{H>\kappa\lambda\}} H^\gamma \,\mathrm{d}\nu + \tfrac{1}{\nu(\pi_2\mathcal{K})} \int_{\pi_2\mathcal{K}\cap\{H>\kappa\lambda\}} H^\gamma \,\mathrm{d}\nu\right].\end{aligned}$$

Now the estimate (4.56) follows by using (4.55) in the above estimate and absorbing terms. □

4.3.6 Collections of Off-diagonal Cubes

We now split the collection $\mathcal{H}_\lambda^{nd}$ into collections where the average of H^γ is large and where the average of H^γ is small. The splitting is chosen based on the first exit-time argument.

Consider

$$\mathcal{G}_\lambda^h := \left\{\mathcal{K} \in \mathcal{H}_\lambda^{nd} \, : \, \fint_{\pi_h\mathcal{K}} H^\gamma \,\mathrm{d}\nu \leq (10n)^{n+p}\kappa^\gamma\lambda^\gamma\right\} \tag{4.58}$$

and

$$\mathscr{B}^h_\lambda := \left\{ \mathcal{K} \in \mathcal{H}^{nd}_\lambda \, : \, \fint_{\pi_h \mathcal{K}} H^\gamma \, \mathrm{d}\nu > (10n)^{n+p} \kappa^\gamma \lambda^\gamma \right\} \tag{4.59}$$

for $h \in \{1, 2\}$ and κ and γ introduced in (4.7) and Corollary 3.4.1, respectively. We further define

$$\mathscr{G}_\lambda := \mathscr{G}^1_\lambda \cap \mathscr{G}^2_\lambda\,, \qquad \text{and} \qquad \mathscr{B}_\lambda := \mathscr{B}^1_\lambda \cup \mathscr{B}^2_\lambda\,.$$

We further split the set $\mathscr{B}_\lambda$ in order to remove cubes that are already covered by the diagonal balls in (4.10)–(4.11):

$$\mathscr{B}_{\lambda,d} := \left\{ \mathcal{K} \in \mathscr{B}_\lambda \, : \, \mathcal{K} \subset \bigcup_j 10\mathcal{B}_j \right\}, \qquad \mathscr{B}_{\lambda,nd} := \mathscr{B}_\lambda \setminus \mathscr{B}_{\lambda,d}\,. \tag{4.60}$$

Thus we have the decomposition into disjoint families

$$\mathcal{H}^{nd}_\lambda = \mathscr{G}_\lambda \cup \mathscr{B}_{\lambda,d} \cap \mathscr{B}_{\lambda,nd}\,.$$

It turns out that the family $\mathscr{G}_\lambda$ is “good” in the sense that the measures of cubes in this collection are estimated by the μ-measure of the level set $\{H > \kappa\lambda\}$, where $\mathrm{d}\mu := H^{p'} \mathrm{d}\nu$. The cubes belonging to $\mathscr{B}_{\lambda,nd}$ are “bad” because there is no available control of the size of $H^{p'}$ on diagonal cubes via the exit-time argument. They will instead be dealt with using combinatorial arguments, and the cutoff $(10n)^{n+p}\kappa^\gamma\lambda^\gamma$ is chosen precisely for this purpose.

Lemma 4.7 (First (Easier) Off-diagonal Estimate) *We have*

$$\sum_{\mathcal{K} \in \mathscr{G}_\lambda} \nu(\mathcal{K}) \leq \frac{6^\gamma C^\gamma_{nd}}{\lambda^\gamma} \int_{\mathcal{B}(x_0,\alpha) \cap \{H > \kappa\lambda\}} H^\gamma \, \mathrm{d}\nu$$

whenever the number $\kappa \in (0, 1]$ *satisfies*

$$\kappa \in (0, \kappa_2]\,, \qquad \kappa_2 := \frac{\varepsilon^{1/\gamma}}{8^{1/\gamma} 3 C_{nd} (10n)^{(n+p)/\gamma}}\,. \tag{4.61}$$

The constant C_{nd} *has been defined in Lemma 4.6.*

Proof Since the cubes $\mathcal{K} \in \mathscr{G}_\lambda$ are disjoint and since (4.35) holds, it suffices to show that

$$\nu(\mathcal{K}) \leq \frac{6^\gamma C^\gamma_{nd}}{\lambda^\gamma} \int_{\mathcal{K} \cap \{H > \kappa\lambda\}} H^\gamma \, \mathrm{d}\nu \tag{4.62}$$

for every $\mathcal{K} \in \mathscr{G}_\lambda$. Since $\mathcal{K} \in \mathscr{G}_\lambda$, using (4.61), we have

$$\begin{aligned} &\frac{3^\gamma C_{nd}^\gamma}{\varepsilon\lambda^\gamma}\left(\frac{2^{-k}}{\operatorname{dist}(K_1,K_2)}\right)^{\eta(s+\varepsilon)}\frac{\nu(\mathcal{K})}{\nu(\pi_h\mathcal{K})}\int_{\pi_h\mathcal{K}\cap\{H>\kappa\lambda\}} H^\gamma\,\mathrm{d}\nu \\ &\quad\le \frac{3^\gamma C_{nd}^\gamma}{\varepsilon\lambda^\gamma}\nu(\mathcal{K})\fint_{\pi_h\mathcal{K}} H^\gamma\,\mathrm{d}\nu \le \nu(\mathcal{K})\frac{3^\gamma C_{nd}^\gamma}{\varepsilon\lambda^\gamma}(10n)^{n+p}\kappa^\gamma\lambda^\gamma \le \frac{\nu(\mathcal{K})}{8}\,. \end{aligned} \tag{4.63}$$

Using this estimate for $h \in \{1, 2\}$ in (4.56) and then absorbing terms give (4.62). □

4.3.7 Determining κ

At this point, the constant κ introduced in (4.7) can be completely determined. We choose

$$\kappa := \min\{\kappa_0,\kappa_1,\kappa_2\} \equiv \min\left\{\frac{\varepsilon^{1/p'}}{(2C_d)^{1/p}},\frac{\varepsilon^{1/\gamma}}{2^{1/\gamma}3C_{nd}},\frac{\varepsilon^{1/\gamma}}{8^{1/\gamma}3C_{nd}(10n)^{(n+p)/\gamma}}\right\}, \tag{4.64}$$

so that conditions (4.47), (4.55), and (4.61) are satisfied. Since the constant C_d defined in Theorem 3.1 depends only on n and p, and since C_{nd} defined in Lemma 4.6 depends only on $\mathtt{data}$, we conclude that there exists $C_\kappa = C_\kappa(\mathtt{data})$ such that

$$\kappa \ge \frac{\varepsilon^{1/\gamma}}{C_\kappa}. \tag{4.65}$$

4.3.8 Summation in $\mathscr{B}_{\lambda,nd}$

Dealing with the cubes from $\mathscr{B}_{\lambda,nd}$ requires delicate estimates and combinatorial arguments. We will first set up notation designed to describe the cubes more precisely. We will then demonstrate that the choice of cutoff $(10n)^{n+p}\kappa^\gamma\lambda^\gamma$ gives us an upper bound on the distance of a cube from the diagonal.

Define the "problematic" projections of the "bad" off-diagonal set

$$\pi_h\mathscr{B}_\lambda := \{\pi_h\mathcal{K} : \mathcal{K} \in \mathscr{B}_\lambda^h\}, \quad h \in \{1,2\}.$$

Since all cubes belonging to the collection $\pi_1\mathscr{B}_\lambda \cup \pi_2\mathscr{B}_\lambda$ are dyadic cubes, it follows that a disjoint subfamily of cubes always exists. We denote this disjoint subfamily of $\pi_1\mathscr{B}_\lambda \cup \pi_2\mathscr{B}_\lambda$ by $\pi\mathscr{B}_\lambda$. By definition, all cubes of $\pi\mathscr{B}_\lambda$ belong to $\pi_1\mathscr{B}_\lambda \cup \pi_2\mathscr{B}_\lambda$ and are therefore diagonal cubes.

Lemma 4.8 *Let $\mathcal{K} \in \mathscr{B}_{\lambda,nd}$ be a cube such that $\pi_h\mathcal{K} \subset \mathcal{M}$ for some $\mathcal{M} \in \pi\mathscr{B}_\lambda$ and some $h \in \{1, 2\}$. Then* $\operatorname{dist}(K_1, K_2) \ge 2^{-k(\mathcal{M})}$.

Proof First, consider $\mathcal{M} \in \pi\mathscr{B}_\lambda$. Take the diagonal ball $\mathcal{B}(\mathcal{M}) \equiv \mathcal{B}(x_\mathcal{M}, 2^{-(k(\mathcal{M})+1)})$ with $x_\mathcal{M}$ being the center of $\mathcal{M}$. Thus,

$$\mathcal{B}(\mathcal{M}) \subset \mathcal{M} \subset \sqrt{n}\mathcal{B}(\mathcal{M}).$$

Thus, by (3.2), Hölder's inequality, and by the definition of $\pi\mathscr{B}_\lambda$,

$$\begin{aligned}(10n)^{(n+p)/\gamma}\kappa\lambda &< \left(\fint_\mathcal{M} H^\gamma \,\mathrm{d}\nu\right)^{1/\gamma} \\ &\le \left(\frac{\nu(10n\mathcal{B}(\mathcal{M}))}{\nu(\mathcal{B}(\mathcal{M}))}\fint_\mathcal{M} H^\gamma \,\mathrm{d}\nu\right)^{1/\gamma} \\ &\le (10n)^{(n+p)\gamma}\left(\fint_\mathcal{M} H^{p'} \,\mathrm{d}\nu\right)^{1/p'}.\end{aligned}$$

By the definition of $D_{\kappa\lambda}$ in (4.8), we have $(x_\mathcal{M}, x_\mathcal{M}) \in D_{\kappa\lambda}$, and then the exit-time condition (4.10) gives $\mathcal{B}(\mathcal{M}) \subset \mathcal{B}(x_\mathcal{M}, R(\mathcal{M}))$. We can use the exit-time condition since the radius of the ball $10n\mathcal{B}(\mathcal{M})$ is smaller than $\frac{\alpha-\beta}{40^n}$, which is in turn a consequence of the fact that $k(\mathcal{M}) + 1 \ge k_0$ and k_0 has been chosen as in (4.34). Therefore, by (4.11),

$$10n\mathcal{B}(\mathcal{M}) \subset \bigcup_j 10\mathcal{B}_j. \tag{4.66}$$

Now, assume by contradiction that $\operatorname{dist}(K_1, K_2) < 2^{-k(\mathcal{M})}$. We will show that

$$\mathcal{K} \subset 10n\mathcal{B}(\mathcal{M}), \tag{4.67}$$

which then contradicts the assumption $\mathcal{M} \in \mathscr{B}_{\lambda,nd}$ by (4.66). To show (4.67), we use Proposition 4.2 and that $\pi_h\mathcal{K} \subset \mathcal{M}$ to get

$$\operatorname{dist}(\mathcal{K}, \mathcal{M}) \le \operatorname{dist}(\mathcal{K}, \pi_h\mathcal{K}) = \operatorname{dist}(K_1, K_2) \le 2^{-k(\mathcal{M})}.$$

Again using Proposition 4.2, we have $k(\mathcal{K}) = k(\pi_h\mathcal{K})$ and $k(\mathcal{K}) \ge k(\mathcal{M})$. Therefore, since $\mathcal{M} \subset \sqrt{n}\mathcal{B}(\mathcal{M})$ and the radius of $\mathcal{B}(\mathcal{M})$ is $2^{-(k(\mathcal{M})+1)}$, then (4.67) must hold. □

Lemma 4.9 (Second (Harder) Off-diagonal Estimate) *There exists a constant $C = C(\texttt{data})$ such that the estimate*

$$\sum_{\mathcal{K}\in\mathscr{B}_{\lambda,nd}} \nu(\mathcal{K}) \le \frac{C}{\lambda^\gamma}\int_{\mathcal{B}(x_0,\alpha)\cap\{H>\kappa\lambda\}} H^\gamma \,\mathrm{d}\nu \tag{4.68}$$

holds, where κ is as in (4.64).

Proof

Step 1: Classification We classify cubes from $\mathscr{B}_{\lambda,nd}$ according to the location of their projections, their size, and their distance from the diagonal. Lemma 4.8, in summary, correlates the distance of a cube from the diagonal with the location of its projection. This allows us to consider only cubes in $\pi\mathscr{B}_\lambda$.

We will partition $\mathscr{B}_{\lambda,nd}$ into suitable disjoint subfamilies. Define the collections

$$\mathscr{B}^h_{\lambda,nd} := \mathscr{B}_{\lambda,nd} \cap \mathscr{B}^h_\lambda\,, \qquad h \in \{1,2\}.$$

For every $\mathcal{M} \in \pi\mathscr{B}_\lambda$, set

$$\mathscr{B}^h_{\lambda,nd}(\mathcal{M}) := \{\mathcal{K} \in \mathscr{B}_{\lambda,nd} \ : \ \pi_h\mathcal{K} \subset \mathcal{M}\}\,, \quad h \in \{1,2\}\,.$$

Thus, we have the decomposition into disjoint subcollections

$$\mathscr{B}^h_{\lambda,nd} = \bigcup_{\mathcal{M}\in\pi\mathscr{B}_\lambda} \mathscr{B}^h_{\lambda,nd}(\mathcal{M})\,. \tag{4.69}$$

These subcollections are disjoint in the sense that for $\mathcal{M}_1,\ \mathcal{M}_2 \in \pi\mathscr{B}_\lambda$, $\mathscr{B}^h_{\lambda,nd}(\mathcal{M}_1) \cap \mathscr{B}^h_{\lambda,nd}(\mathcal{M}_2) \neq \emptyset$ implies $\mathcal{M}_1 = \mathcal{M}_2$. This follows from the definition of $\mathscr{B}^h_{\lambda,nd}(\mathcal{M})$ and from the fact that all elements in $\pi\mathscr{B}_\lambda$ are pairwise disjoint.

Next we classify cubes according to their size. For each $\mathcal{K} \in \mathscr{B}^h_{\lambda,nd}(\mathcal{M})$, we have $k(\mathcal{K}) = k(\pi_h\mathcal{K}) \geq k(\mathcal{M})$, so we can define the collections

$$[\mathscr{B}^h_{\lambda,nd}(\mathcal{M})]_i := \{\mathcal{K} \in \mathscr{B}^h_{\lambda,nd}(\mathcal{M}) \ : \ k(\mathcal{K}) = i + k(\mathcal{M})\}\,, \qquad h \in \{1,2\}\,, \quad i \in \mathbb{Z}_+\,.$$

We again have a decomposition in mutually disjoint subcollections

$$\mathscr{B}^h_{\lambda,nd}(\mathcal{M}) = \bigcup_{i\in\mathbb{Z}_+} [\mathscr{B}^h_{\lambda,nd}(\mathcal{M})]_i\,,$$

in the sense that $[\mathscr{B}^h_{\lambda,nd}(\mathcal{M})]_{i_1} \cap [\mathscr{B}^h_{\lambda,nd}(\mathcal{M})]_{i_2} \neq \emptyset$ implies that $i_1 = i_2$.

We further classify the cubes according to their distance from the diagonal. We now use the combinatorial information obtained from Lemma 4.8. Take $\mathcal{M} \in \pi\mathscr{B}_\lambda$. If $\mathcal{K} \in \mathscr{B}^h_{\lambda,nd}(\mathcal{M})$, that is, if $\pi_h\mathcal{K} \subset \mathcal{M}$, then by Lemma 4.8 it follows that $\mathrm{dist}(K_1, K_2) \geq 2^{-k(\mathcal{M})}$. This leads us to the definition of the subcollections

$$[\mathscr{B}^h_{\lambda,nd}(\mathcal{M})]_{i,j} := \{\mathcal{K} \in [\mathscr{B}^h_{\lambda,nd}(\mathcal{M})]_i \ : \ 2^{j-k(\mathcal{M})} \leq \mathrm{dist}(K_1,K_2) < 2^{j+1-k(\mathcal{M})}\}\,,$$
$$h \in \{1,2\}\,, \quad i,j \in \mathbb{Z}_+\,.$$

We yet again have the decomposition in mutually disjoint subcollections

$$\mathscr{B}^h_{\lambda,nd}(\mathcal{M}) = \bigcup_{i,j\in\mathbb{Z}_+} [\mathscr{B}^h_{\lambda,nd}(\mathcal{M})]_{i,j},$$

and these are mutually disjoint in the sense that $[\mathscr{B}^h_{\lambda,nd}(\mathcal{M})]_{i_1,j_1} \cap [\mathscr{B}^h_{\lambda,nd}(\mathcal{M})]_{i_2,j_2} \neq \emptyset$ implies that $(i_1, j_1) = (i_2, j_2)$. In summary, we have the decomposition

$$\mathscr{B}^h_{\lambda,nd} = \bigcup_{\mathcal{M}\in\pi\mathscr{B}_\lambda} \bigcup_{i,j\in\mathbb{Z}_+} [\mathscr{B}^h_{\lambda,nd}(\mathcal{M})]_{i,j}\,.$$

Step 2: Summation and Further Partitioning Fix $\mathcal{M} \in \pi\mathscr{B}_\lambda$. Our goal in this step is to show that for $h \in \{1, 2\}$

$$\begin{aligned}&\tfrac{1}{\varepsilon}\textstyle\sum_{\mathcal{K}\in\mathscr{B}^h_{\lambda,nd}(\mathcal{M})} \frac{\nu(\mathcal{K})}{\nu(\pi_h\mathcal{K})}\left(\frac{2^{-k}}{\mathrm{dist}(K_1,K_2)}\right)^{\eta(s+\varepsilon)} \int_{\pi_h\mathcal{K}\cap\{H>\kappa\lambda\}} H^\gamma\,\mathrm{d}\nu\\ &\le \tfrac{C(\mathtt{data})}{s^2}\int_{\mathcal{M}\cap\{H>\kappa\lambda\}} H^\gamma\,\mathrm{d}\nu\,.\end{aligned} \tag{4.70}$$

To begin, note that for $\mathcal{K} \in \mathscr{B}^h_{\lambda,nd}$, we have $\mathrm{dist}(K_1, K_2) \ge 2^{-k(\mathcal{K})}$, and so we can apply the inequality from Lemma 4.3 to get that

$$\frac{1}{\varepsilon}\frac{\nu(\mathcal{K})}{\nu(\pi_h\mathcal{K})} \le C_{dd}\left(\frac{2^{-k}}{\mathrm{dist}(K_1, K_2)}\right)^{n-\varepsilon p}$$

for $h \in \{1, 2\}$, and further if $\mathcal{K} \in [\mathscr{B}^h_{\lambda,nd}]_{i,j}$, then

$$\frac{2^{-k(\mathcal{K})}}{\mathrm{dist}(K_1, K_2)} = \frac{1}{2^i}\frac{2^{-k(\mathcal{M})}}{\mathrm{dist}(K_1, K_2)} \le \frac{1}{2^{i+j}}\,.$$

With these, we can estimate

$$\begin{aligned}&\frac{1}{\varepsilon}\sum_{\mathcal{K}\in\mathscr{B}^h_{\lambda,nd}(\mathcal{M})} \frac{\nu(\mathcal{K})}{\nu(\pi_h\mathcal{K})}\left(\frac{2^{-k}}{\mathrm{dist}(K_1, K_2)}\right)^{\eta(s+\varepsilon)} \int_{\pi_h\mathcal{K}\cap\{H>\kappa\lambda\}} H^\gamma\,\mathrm{d}\nu\\ &\le C_{dd}\sum_{\mathcal{K}\in\mathscr{B}^h_{\lambda,nd}(\mathcal{M})}\left(\frac{2^{-k}}{\mathrm{dist}(K_1, K_2)}\right)^{n+\eta(s+\varepsilon)-\varepsilon p} \int_{\pi_h\mathcal{K}\cap\{H>\kappa\lambda\}} H^\gamma\,\mathrm{d}\nu\end{aligned}$$

$$
\begin{aligned}
&= C_{dd} \sum_{i,j=0}^{\infty} \sum_{\mathcal{K} \in [\mathscr{B}^h_{\lambda,nd}(\mathcal{M})]_{i,j}} \left(\frac{2^{-k}}{\operatorname{dist}(K_1, K_2)} \right)^{n+\eta(s+\varepsilon)-\varepsilon p} \int_{\pi_h \mathcal{K} \cap \{H > \kappa\lambda\}} H^\gamma \, d\nu \\
&\leq C_{dd} \sum_{i,j=0}^{\infty} \left(\frac{1}{2^{i+j}} \right)^{n+\eta(s+\varepsilon)-\varepsilon p} \sum_{\mathcal{K} \in [\mathscr{B}^h_{\lambda,nd}(\mathcal{M})]_{i,j}} \int_{\pi_h \mathcal{K} \cap \{H > \kappa\lambda\}} H^\gamma \, d\nu \, .
\end{aligned}
\tag{4.71}
$$

In order to evaluate the last sum, we must further partition $[\mathscr{B}^h_{\lambda,nd}(\mathcal{M})]_{i,j}$. In order to do so, note that for each $i \in \mathbb{Z}_+$ the cube $\mathcal{M}$ contains precisely 2^{2ni} disjoint dyadic cubes belonging to $\Delta_{i+k(\mathcal{M})}$ and precisely 2^{ni} disjoint dyadic diagonal cubes from $\Delta^d_{i+k(\mathcal{M})}$; see the definition (4.36) and the text immediately following. Thus, $\mathcal{M}$ contains at most 2^{ni} dyadic cubes from the class $\Delta^d_{i+k(\mathcal{M})} \cap (\pi_1 \mathscr{B}_\lambda \cap \pi_2 \mathscr{B}_\lambda)$. In any case, we consider all the diagonal cubes from $\mathcal{M}$ in $\Delta^d_{i+k(\mathcal{M})}$ and label them as

$$
\{\widetilde{\mathcal{M}} \in \Delta^d_{i+k(\mathcal{M})} \, : \, \widetilde{\mathcal{M}} \subset \mathcal{M}\} \equiv \{\mathcal{M}^m_i \, : \, 1 \leq m \leq 2^{ni}\} \, ,
$$

so that

$$
\sum_{m=1}^{2^{ni}} \int_{\mathcal{M}^m_i \cap \{H > \kappa\lambda\}} H^\gamma \, d\nu \leq \int_{\mathcal{M} \cap \{H > \kappa\lambda\}} H^\gamma \, d\nu \, . \tag{4.72}
$$

For any $\mathcal{K} \in [\mathscr{B}^h_{\lambda,nd}(\mathcal{M})]_{i,j}$ with $h \in \{1, 2\}$, there exists a unique diagonal cube from the collection Δ^d_k, which we denote by $\mathcal{M}^m_i(\mathcal{K})$, such that $\pi_h \mathcal{K} = \mathcal{M}^m_i(\mathcal{K})$. Then we can split $[\mathscr{B}^h_{\lambda,nd}(\mathcal{M})]_{i,j}$ into subsets

$$
[\mathscr{B}^h_{\lambda,nd}(\mathcal{M})]_{i,j,m} := \{\mathcal{K} \in [\mathscr{B}^h_{\lambda,nd}(\mathcal{M})]_{i,j} \, : \, \pi_h \mathcal{K} = \mathcal{M}^m_i\} \, , \qquad m \in \{1, \dots, 2^{ni}\} \, .
$$

Since $\mathscr{B}^1_{\lambda,nd}$ is a family of dyadic cubes, if $\mathcal{K}_1, \mathcal{K}_2 \in [\mathscr{B}^h_{\lambda,nd}(\mathcal{M})]_{i,j,m}$ and $\mathcal{K}_1 \neq \mathcal{K}_2$, then $\pi_2 \mathcal{K}_1 \cap \pi_2 \mathcal{K}_2 = \emptyset$, else the two cubes would coincide.

Thus for each $i, j \geq 0$, and for $m \in \{1, \dots, 2^{ni}\}$, the number of cubes $\mathcal{K}$ of side length $2^{i+k(\mathcal{M})}$ with $\operatorname{dist}(K_1, K_2) \in [2^{j-k(\mathcal{M})}, 2^{j+1-k(\mathcal{M})})$ that project into $\mathcal{M}^m_i$ is bounded from above by $C(n) \left(\frac{\text{length of half-open interval defining the distance from the diagonal}}{\text{cube side length}} \right)^n$; that is,

$$
\#[\mathscr{B}^h_{\lambda,nd}(\mathcal{M})]_{i,j,m} \leq C(n) 2^{n(i+j)} \, , \qquad h \in \{1, 2\} \, . \tag{4.73}
$$

Thus, using (4.72) and (4.73), we estimate

$$\sum_{\mathcal{K}\in[\mathscr{B}^h_{\lambda,nd}(\mathcal{M})]_{i,j}} \int_{\pi_h\mathcal{K}\cap\{H>\kappa\lambda\}} H^\gamma \, d\nu = \sum_{m=1}^{2^{ni}} \sum_{\mathcal{K}\in[\mathscr{B}^h_{\lambda,nd}(\mathcal{M})]_{i,j,m}} \int_{\mathcal{M}_i^m\cap\{H>\kappa\lambda\}} H^\gamma \, d\nu$$
$$\le C(n)2^{n(i+j)} \sum_{m=1}^{2^{ni}} \int_{\mathcal{M}_i^m\cap\{H>\kappa\lambda\}} H^\gamma \, d\nu$$
$$\le C(n)2^{n(i+j)} \int_{\mathcal{M}\cap\{H>\kappa\lambda\}} H^\gamma \, d\nu\,.$$

Then using the series estimate (2.1), it follows that

$$\sum_{i,j=0}^{\infty}\left(\frac{1}{2^{i+j}}\right)^{n+\eta(s+\varepsilon)-\varepsilon p} \sum_{\mathcal{K}\in[\mathscr{B}^h_{\lambda,nd}(\mathcal{M})]_{i,j}} \int_{\pi_h\mathcal{K}\cap\{H>\kappa\lambda\}} H^\gamma \, d\nu$$
$$\le C(n)\sum_{i,j=0}^{\infty}\left(\frac{1}{2^{i+j}}\right)^{\eta(s+\varepsilon)-\varepsilon p} \int_{\mathcal{M}\cap\{H>\kappa\lambda\}} H^\gamma \, d\nu$$
$$\le C(n)\left[\frac{2^{\eta(s+\varepsilon)-\varepsilon p}}{\ln(2)(\eta(s+\varepsilon)-\varepsilon p)}\right]^2 \int_{\mathcal{M}\cap\{H>\kappa\lambda\}} H^\gamma \, d\nu$$
$$\le \frac{C(n,p)}{[\eta(s+\varepsilon)-\varepsilon p]^2}\int_{\mathcal{M}\cap\{H>\kappa\lambda\}} H^\gamma \, d\nu \le \frac{C(n,p)}{s^2}\int_{\mathcal{M}\cap\{H>\kappa\lambda\}} H^\gamma \, d\nu\,.$$

In the last line, we used that $\eta > 1$ and that $\varepsilon < \frac{s}{p}$, which gives $\eta(s+\varepsilon) - \varepsilon p > \frac{s}{p}$. Combining this inequality with (4.3.8) results in (4.70).

Step 3: Further Summation Let $\mathcal{K} \in \mathscr{B}^1_{\lambda,nd}$. Then, either $\mathcal{K} \in \mathscr{G}^2_\lambda$ or $\mathcal{K} \in \mathscr{B}^2_\lambda$ (see (4.58) and (4.59)). If $\mathcal{K} \in \mathscr{G}^2_\lambda$, then using the almost-reverse Hölder inequality (4.56) along with the estimate (4.63) for cubes in $\mathscr{G}^2_\lambda$ and reabsorbing terms,

$$\nu(\mathcal{K}) \le \tfrac{6^\gamma C^\gamma_{nd}}{\lambda^\gamma}\int_{\mathcal{K}\cap\{H>\kappa\lambda\}} H^\gamma \, d\nu$$
$$+\tfrac{6^\gamma C^\gamma_{nd}}{\varepsilon\lambda^\gamma}\left(\tfrac{2^{-k}}{\operatorname{dist}(K_1,K_2)}\right)^{\eta(s+\varepsilon)} \tfrac{\nu(\mathcal{K})}{\nu(\pi_1\mathcal{K})}\int_{\pi_1\mathcal{K}\cap\{H>\kappa\lambda\}} H^\gamma \, d\nu\,.$$

On the other hand, if $\mathcal{K} \in \mathscr{B}^2_\lambda$, then we can use the almost-reverse Hölder inequality(4.56) to obtain

$$\nu(\mathcal{K}) \le \tfrac{3^\gamma C^\gamma_{nd}}{\lambda^\gamma}\int_{\mathcal{K}\cap\{H>\kappa\lambda\}} H^\gamma \, d\nu + \tfrac{3^\gamma C^\gamma_{nd}}{\varepsilon\lambda^\gamma}\left(\tfrac{2^{-k}}{\operatorname{dist}(K_1,K_2)}\right)^{\eta(s+\varepsilon)}$$
$$\left[\tfrac{\nu(\mathcal{K})}{\nu(\pi_1\mathcal{K})}\int_{\pi_1\mathcal{K}\cap\{H>\kappa\lambda\}} H^\gamma \, d\nu + \tfrac{\nu(\mathcal{K})}{\nu(\pi_2\mathcal{K})}\int_{\pi_2\mathcal{K}\cap\{H>\kappa\lambda\}} H^\gamma \, d\nu\right].$$

A similar reasoning holds for $\mathcal{K} \in \mathscr{B}_\lambda^2$. Summing over the cubes $\mathcal{K} \in \mathscr{B}_{\lambda,nd} = \mathscr{B}_{\lambda,nd}^1 \cup \mathscr{B}_{\lambda,nd}^2$ gives us

$$\begin{aligned}\sum_{\mathcal{K}\in\mathscr{B}_{\lambda,nd}} \nu(\mathcal{K}) \leq \frac{6^\gamma C_{nd}^\gamma}{\lambda^\gamma} &\sum_{\mathcal{K}\in\mathscr{B}_{\lambda,nd}} \int_{\mathcal{K}\cap\{H>\kappa\lambda\}} H^\gamma \, \mathrm{d}\nu \\ &+ \frac{6^\gamma C_{nd}^\gamma}{\varepsilon\lambda^\gamma} \sum_{\mathcal{K}\in\mathscr{B}_{\lambda,nd}^1} \left(\frac{2^{-k}}{\operatorname{dist}(K_1, K_2)}\right)^{\eta(s+\varepsilon)} \frac{\nu(\mathcal{K})}{\nu(\pi_1\mathcal{K})} \int_{\pi_1\mathcal{K}\cap\{H>\kappa\lambda\}} H^\gamma \, \mathrm{d}\nu \\ &+ \frac{6^\gamma C_{nd}^\gamma}{\varepsilon\lambda^\gamma} \sum_{\mathcal{K}\in\mathscr{B}_{\lambda,nd}^2} \left(\frac{2^{-k}}{\operatorname{dist}(K_1, K_2)}\right)^{\eta(s+\varepsilon)} \frac{\nu(\mathcal{K})}{\nu(\pi_2\mathcal{K})} \int_{\pi_2\mathcal{K}\cap\{H>\kappa\lambda\}} H^\gamma \, \mathrm{d}\nu \, .\end{aligned} \tag{4.74}$$

The point of this argument beginning in Step 3 is that terms involving the projections $\pi_h\mathcal{K}$ appear if and only if $\mathcal{K} \in \mathscr{B}_{\lambda,nd}^h$ for $h \in \{1, 2\}$. We will now argue that the last two terms in the above inequality coincide.

For a cube $\mathcal{K} = K_1 \times K_2 \in \Delta$, define

$$\operatorname{Symm}(\mathcal{K}) := K_2 \times K_1 \, .$$

Then by definition

$$\pi_1(\mathcal{K}) = \pi_1(K_1 \times K_2) = \pi_2(K_2 \times K_1) = \pi_2(\operatorname{Symm}(\mathcal{K})) \, . \tag{4.75}$$

Further, by symmetry of H, we have

$$\int_{\mathcal{K}} H^r \, \mathrm{d}\nu = \int_{\operatorname{Symm}(\mathcal{K})} H^r \, \mathrm{d}\nu \, , \qquad r \in [1, \infty)$$

and

$$\int_{\mathcal{K}\cap\{H>\kappa\lambda\}} H^r \, \mathrm{d}\nu = \int_{\operatorname{Symm}(\mathcal{K})\cap\{H>\kappa\lambda\}} H^r \, \mathrm{d}\nu \, , \qquad r \in [1, \infty) \, . \tag{4.76}$$

Therefore, by definition of $\mathscr{B}_\lambda^h$,

$$\mathcal{K} \in \mathscr{B}_\lambda^1 \quad \Leftrightarrow \quad \operatorname{Symm}(\mathcal{K}) \in \mathscr{B}_\lambda^2 \, ,$$

and vice versa, and so

$$\mathcal{K} \in \mathscr{B}_{\lambda,nd}^1 \quad \Leftrightarrow \quad \operatorname{Symm}(\mathcal{K}) \in \mathscr{B}_{\lambda,nd}^2 \, , \tag{4.77}$$

and vice versa. It follows from (4.75)–(4.77) that the last two terms in (4.74) coincide. Recalling the decomposition (4.69), (4.74) can be written as

$$\sum_{\mathcal{K}\in\mathscr{B}_{\lambda,nd}} \nu(\mathcal{K}) \le \frac{C}{\lambda^\gamma} \sum_{\mathcal{K}\in\mathscr{B}_{\lambda,nd}} \int_{\mathcal{K}\cap\{H>\kappa\lambda\}} H^\gamma \,\mathrm{d}\nu$$

$$+ \frac{C}{\varepsilon\lambda^\gamma} \sum_{\mathcal{M}\in\pi\mathscr{B}_\lambda} \sum_{\mathcal{K}\in\mathscr{B}^1_{\lambda,nd}(\mathcal{M})} \left(\frac{2^{-k}}{\mathrm{dist}(K_1,K_2)}\right)^{\eta(s+\varepsilon)} \frac{\nu(\mathcal{K})}{\nu(\pi_1\mathcal{K})} \int_{\pi_1\mathcal{K}\cap\{H>\kappa\lambda\}} H^\gamma \,\mathrm{d}\nu\,,$$

for a constant $C \equiv C(\texttt{data})$. Using the estimate (4.70) proved in Step 2 yields

$$\sum_{\mathcal{K}\in\mathscr{B}_{\lambda,nd}} \nu(\mathcal{K}) \le \frac{C}{\lambda^\gamma} \sum_{\mathcal{K}\in\mathscr{B}_{\lambda,nd}} \int_{\mathcal{K}\cap\{H>\kappa\lambda\}} H^\gamma \,\mathrm{d}\nu + \frac{C}{\lambda^\gamma} \sum_{\mathcal{M}\in\pi\mathscr{B}_\lambda} \int_{\mathcal{M}\cap\{H>\kappa\lambda\}} H^\gamma \,\mathrm{d}\nu\,.$$

Then since the collections $\mathscr{B}_{\lambda,nd}$ and $\pi\mathscr{B}_\lambda$ are comprised of mutually disjoint dyadic cubes all of which are contained in $\mathcal{B}(x_0,\alpha)$ (see (4.35)), we can estimate

$$\sum_{\mathcal{K}\in\mathscr{B}_{\lambda,nd}} \int_{\mathcal{K}\cap\{H>\kappa\lambda\}} H^\gamma \,\mathrm{d}\nu + \sum_{\mathcal{M}\in\pi\mathscr{B}_\lambda} \int_{\mathcal{M}\cap\{H>\kappa\lambda\}} H^\gamma \,\mathrm{d}\nu \le 2 \int_{\mathcal{B}(x_0,\alpha)\cap\{H>\kappa\lambda\}} H^\gamma \,\mathrm{d}\nu.$$

Thus (4.68) is proved, and the proof of Lemma 4.9 is complete. □

4.3.9 Off-diagonal Conclusion

The next lemma summarizes the off-diagonal estimate obtained for H.

Lemma 4.10 *The inequality*

$$\int_{\mathcal{B}(x_0,\beta)\cap\{H>\lambda\}} H^{p'} \,\mathrm{d}\nu \le 10^{n+p}\kappa^{p'}\lambda^{p'} \sum_j \nu(\mathcal{B}_j) + C\lambda^{p'-\gamma} \int_{\mathcal{B}(x_0,\alpha)\cap\{H>\kappa\lambda\}} H^\gamma \,\mathrm{d}\nu \tag{4.78}$$

holds for a constant $C \equiv C(\texttt{data})$, while κ has been chosen in (4.64) *and satisfies* (4.65).

Proof We have the decomposition $\mathcal{H}_\lambda = \mathcal{H}^d_\lambda \cup \mathcal{H}^{nd}_\lambda$ with $\mathcal{H}^{nd}_\lambda = \mathscr{G}_\lambda \cup \mathscr{B}_{\lambda,d} \cup \mathscr{B}_{\lambda,nd}$. Recall from the relation (4.48) and from the definition (4.60) that

$$\bigcup_{\mathcal{K}\in\mathcal{H}^d_\lambda} \mathcal{K} \cup \bigcup_{\mathcal{K}\in\mathscr{B}_{\lambda,d}} \mathcal{K} \subset \bigcup_j 10\mathcal{B}_j\,.$$

Thus

$$\bigcup_{\mathcal{K}\in\mathcal{H}_\lambda} \subset \bigcup_j 10\mathcal{B}_j \cup \bigcup_{\mathcal{K}\in\mathcal{G}_\lambda} \mathcal{K} \cup \bigcup_{\mathcal{K}\in\mathcal{B}_{\lambda,nd}} \mathcal{K}\,.$$

Now, with this and with (4.46), we estimate

$$\int_{\mathcal{B}(x_0,\beta)\cap\{H>\lambda\}} H^{p'}\,\mathrm{d}\nu \le \sum_j \int_{10\mathcal{B}_j\cap\{H>\lambda\}} H^{p'}\,\mathrm{d}\nu + \sum_{\mathcal{K}\in\mathcal{G}_\lambda\cup\mathcal{B}_{\lambda,nd}} \int_{\mathcal{K}\cap\{H>\lambda\}} H^{p'}\,\mathrm{d}\nu\,.$$

Now, by the choice of Calderón–Zygmund splitting in (4.44), for $\mathcal{K}\in\mathcal{G}_\lambda\cup\mathcal{B}_{\lambda,nd}\subset\mathcal{H}_\lambda^{nd}$

$$\fint_{\mathcal{K}} H^{p'}\,\mathrm{d}\nu \le \frac{\nu(\widetilde{\mathcal{K}})}{\nu(\mathcal{K})}\fint_{\widetilde{\mathcal{K}}} H^{p'}\,\mathrm{d}\nu \le \frac{\nu(\widetilde{\mathcal{K}})}{\nu(\mathcal{K})}\lambda^{p'}\,.$$

Since $\mathcal{K}\in\mathcal{H}_\lambda^{nd}$, it follows from the definition (4.49) that $\operatorname{dist}(\widetilde{\mathcal{K}}_1,\widetilde{\mathcal{K}}_2)\ge 2^{-k(\mathcal{K})}$, so we can use the geometric estimate (4.41) from Lemma 4.4 to obtain the bound $\frac{\nu(\widetilde{\mathcal{K}})}{\nu(\mathcal{K})}\le C_{ddd}$. In summary,

$$\mathcal{K}\in\mathcal{G}_\lambda\cup\mathcal{B}_{\lambda,nd} \quad\Rightarrow\quad \int_{\mathcal{K}\cap\{H>\lambda\}} H^{p'}\,\mathrm{d}\nu \le C_{ddd}\lambda^{p'}\nu(\mathcal{K})\,.$$

Using this inequality in conjunction with (4.13) results in

$$\int_{\mathcal{B}(x_0,\beta)\cap\{H>\lambda\}} H^{p'}\,\mathrm{d}\nu \le 10^{n+\varepsilon p}\kappa^{p'}\lambda^{p'}\sum_j \nu(\mathcal{B}_j) + C_{ddd}\lambda^{p'}\sum_{\mathcal{K}\in\mathcal{G}_\lambda\cup\mathcal{B}_{\lambda,nd}}\nu(\mathcal{K})\,.$$

Then (4.78) follows from the conclusions of Lemmas 4.7 and 4.9. □

4.4 *Conclusion of the Proof*

We now come to the conclusion of the proof of Proposition 4.1. We start by combining the diagonal estimate (4.33) with the off-diagonal estimate (4.78). We use the elementary estimate

$$\int_{\mathcal{B}(x_0,\beta)\cap\{H>\widetilde{\kappa}\kappa\lambda\}} H^{p'}\,\mathrm{d}\nu \le \lambda^{p'-\gamma}\int_{\mathcal{B}(x_0,\beta)\cap\{H>\widetilde{\kappa}\kappa\lambda\}} H^{\gamma}\,\mathrm{d}\nu + \int_{\mathcal{B}(x_0,\beta)\cap\{H>\lambda\}} H^{p'}\,\mathrm{d}\nu\,;$$

recall that $\widetilde{\kappa},\kappa\in(0,1]$. Then (4.33) and (4.78) give, after some elementary algebraic manipulations, the estimate

$$\int_{\mathcal{B}(x_0,\beta)\cap\{H>\widetilde{\kappa}\kappa\lambda\}} H^{p'}\,\mathrm{d}\nu \le \frac{C(\mathtt{data})}{\varepsilon^{2-2\gamma/p'}(\widetilde{\kappa}\kappa)^{p'-\gamma}}(\widetilde{\kappa}\kappa\lambda)^{p'-\gamma}\int_{\mathcal{B}(x_0,\alpha)\cap\{H>\widetilde{\kappa}\kappa\lambda\}} H^{\gamma}\,\mathrm{d}\nu$$
$$+\frac{C_5\lambda_1^{\vartheta_f}}{\widehat{\kappa}^{p'}(\widehat{\kappa}\kappa\lambda)^{\widetilde{\vartheta}_f-p'}}\int_{\mathcal{B}(x_0,\alpha)\cap\{F>\widehat{\kappa}\kappa\lambda\}} F^{p_*}\,\mathrm{d}\nu\,. \tag{4.79}$$

Recall that C_5 is defined in (4.32). We can reformulate this estimate as

$$\int_{\mathcal{B}(x_0,\beta)\cap\{H>\lambda\}} H^{p'}\,\mathrm{d}\nu \le \frac{C}{\varepsilon^{2-2\gamma/p'}(\widetilde{\kappa}\kappa)^{p'-\gamma}}\lambda^{p'-\gamma}\int_{\mathcal{B}(x_0,\alpha)\cap\{H>\lambda\}} H^{\gamma}\,\mathrm{d}\nu$$
$$+\frac{C_6\lambda_1^{\vartheta_f}}{\lambda^{\widetilde{\vartheta}_f-p'}}\int_{\mathcal{B}(x_0,\alpha)\cap\{F>\widehat{\kappa}\lambda/\widetilde{\kappa}\}} F^{p_*}\,\mathrm{d}\nu\,. \tag{4.80}$$

The constants C and C_6 satisfy the following dependencies:

$$C \equiv C(\mathtt{data})\,, \qquad C_6 \equiv C_6(\mathtt{data},\varepsilon)\,.$$

Since (4.79) holds for all $\lambda \ge \lambda_2$, where λ_2 has been defined in (4.43), we have that (4.80) holds for all $\lambda \ge \widetilde{\kappa}\kappa\lambda_2$. Recall again that $\widetilde{\kappa}, \widehat{\kappa}, \kappa \in (0,1]$ have been defined in (4.25), (4.30), and (4.64), respectively.

In order to conclude with the level set estimate (4.5), we need to estimate several constants. We need to obtain the specific dependence on ε of the constant appearing in front of the first integrand on the right-hand side. By using (4.25) and (4.64), we see that we can choose $\widetilde{\kappa}, \kappa$ to satisfy

$$\widetilde{\kappa}\kappa = \frac{\varepsilon^{3/\gamma-2/p'}}{C}\,,$$

where $C \equiv C(\mathtt{data})$. With this choice, we can estimate the constant appearing in front of the second integral in (4.80) and therefore arrive at a choice of ϑ exactly as in (4.4). We next set

$$\kappa_f := \widehat{\kappa}/\widetilde{\kappa}\,,$$

and making note of (4.30) we can additionally choose $\widehat{\kappa}$ small so that $\kappa_f \in (0,1)$.

Last, we need to find an upper bound on the numbers λ_1 and λ_2 defined in (4.7) and (4.43), respectively, so that the level set estimate (4.5) can be verified for the range prescribed by (4.6). If $x \in B(x_0,\beta)$ and $\frac{\alpha-\beta}{40^n} \le R \le \frac{\varrho_0}{2}$, then $\mathcal{B}(x,R) \subset \mathcal{B}(x_0,2\varrho_0)$. Recalling the doubling property (3.2), whenever $\widetilde{H}$ is a ν-integrable function, we can estimate

$$\fint_{\mathcal{B}(x,R)} \widetilde{H}\,\mathrm{d}\nu \le \frac{\nu(\mathcal{B}(x_0,2\varrho_0))}{\nu(\mathcal{B}(x,R))}\fint_{\mathcal{B}(x_0,2\varrho_0)} \widetilde{H}\,\mathrm{d}\nu \le C\left(\frac{\varrho_0}{\alpha-\beta}\right)^{n+\varepsilon p}\fint_{\mathcal{B}(x_0,2\varrho_0)} \widetilde{H}\,\mathrm{d}\nu,$$

where $C = C(n)$. Applying this inequality to $H^{p'}$, F^{p_*}, H^{γ}, and $F^{p_*+\delta_f}$, as well as on different balls $2^k\mathcal{B}(x, R) \subset 2^k\mathcal{B}(x_0, 2\varrho_0)$, we get

$$\begin{aligned} &\kappa^{-1}\big\{\Psi_M(x, R) + \Upsilon_0(x, R) + Tail(x, R)\big\} \\ &\quad \le \frac{C}{\varepsilon^{1/\gamma}}\left(\frac{\varrho_0}{\alpha-\beta}\right)^{n+\varepsilon p}\big\{\Psi_M(x_0, 2\varrho_0) + \Upsilon_0(x_0, 2\varrho_0) + Tail(x_0, 2\varrho_0)\big\} \\ &\quad \le \frac{C}{\varepsilon^{1/\gamma}}\left(\frac{\varrho_0}{\alpha-\beta}\right)^{n+\varepsilon p}\Theta(x_0, 2\varrho_0) \le \frac{C}{\varepsilon}\left(\frac{\varrho_0}{\alpha-\beta}\right)^{2n+p}\Theta(x_0, 2\varrho_0)\,, \end{aligned} \tag{4.81}$$

where $C \equiv C(\mathtt{data})$. We also used (4.64) to remove dependence on κ, (4.22) to remove dependence on M, that $\varepsilon < 1$, and that $\frac{\varrho_0}{\alpha-\beta} \ge 2$. Recall also that Θ has been defined in (4.1). Thus we have obtained the desired upper bound on λ_1. To estimate λ_2, note that for $\mathcal{K} = K_1 \times K_2 \in \Delta_{k_0}$ with k_0 as in (4.34) we have $\mathcal{K} \subset \mathcal{B}(x_0, \alpha) \subset \mathcal{B}(x_0, 2\varrho_0)$, and therefore,

$$\nu(\mathcal{K}) \ge \frac{C}{\varrho_0^{n+\varepsilon p}}\int_{K_1}\int_{K_2} \mathrm{d}y\,\mathrm{d}x = \frac{C(\alpha-\beta)^{2n}}{\varrho_0^{n+\varepsilon p}}\,, \qquad C \equiv C(n, p)\,.$$

Thus, for any cube $\mathcal{K} \in \Delta_{k_0}$, we can estimate

$$\begin{aligned} \left(⨍_{\mathcal{K}} H^{p'}\,\mathrm{d}\nu\right)^{1/p'} &\le \left(\tfrac{\nu(\mathcal{B}(x_0,2\varrho_0))}{\nu(\mathcal{K})} ⨍_{\mathcal{B}(x_0,2\varrho_0)} H^{p'}\,\mathrm{d}\nu\right)^{1/p'} \\ &\le \tfrac{C}{\varepsilon^{1/p'}}\left(\tfrac{\varrho_0}{\alpha-\beta}\right)^{2n/p'}\left(⨍_{\mathcal{B}(x_0,2\varrho_0)} H^{p'}\,\mathrm{d}\nu\right)^{1/p'}. \end{aligned} \tag{4.82}$$

Then using (4.81) and (4.82) (also using that $\varepsilon < 1$ and $\frac{\varrho_0}{\alpha-\beta} \ge 2$), we get that

$$\lambda_2 \le \frac{C}{\varepsilon}\left(\frac{\varrho_0}{\alpha-\beta}\right)^{2n+p}\Theta(x_0, 2\varrho_0)\,,$$

where $C \equiv C(\mathtt{data})$. We therefore take λ_0 as in (4.6) so that $\lambda_0 \ge \max\{\lambda_1, \lambda_2\}$. We finally arrive at (4.5) with λ prescribed as in (4.6), and the proof of Proposition 4.1 is complete.

Remark 4.11 Note that in the proof of Proposition 4.1 the off-diagonal analysis in Section 4.3 does not make use of assumption (3.9). Thus, the conclusions of Lemma 4.10 hold for general fractional Sobolev functions u, and not just solutions to (1.1).

References

1. Richard F Bass and Moritz Kassmann. Hölder continuity of harmonic functions with respect to operators of variable order. *Communications in Partial Differential Equations*, 30(8):1249–1259, 2005.
2. Cristiana De Filippis and Giampiero Palatucci. Hölder regularity for nonlocal double phase equations. *Journal of Differential Equations*, 267(1):547–586, 2019.
3. Tuomo Kuusi, Giuseppe Mingione, and Yannick Sire. A fractional Gehring lemma, with applications to nonlocal equations. *Rendiconti Lincei-Matematica e Applicazioni*, 25(4):345–358, 2014.
4. Tuomo Kuusi, Giuseppe Mingione, and Yannick Sire. Nonlocal self-improving properties. *Anal. PDE*, 8(1):57–114, 2015.
5. Tadele Mengesha and James M. Scott. A fractional Korn-type inequality for smooth domains and a regularity estimate for nonlinear nonlocal systems of equations, *Commun. Math. Sci.* 20(2):405–423, 2022.
6. Tadele Mengesha and James M. Scott. Self-improving inequalities for bounded weak solutions to nonlocal double-phase equations. *Communications on Pure and Applied Analysis*, 21(1):183–212, 2020.
7. Giuseppe Mingione. The singular set of solutions to non-differentiable elliptic systems. *Archive for Rational Mechanics and Analysis*, 166(4):287–301, 2003.
8. Eleonora Di Nezza, Giampiero Palatucci, and Enrico Valdinoci. Hitchhiker's guide to the fractional Sobolev spaces. *Bull. Sci. Math.*, 136(5):521–573, 2012.
9. Simon Nowak, Regularity theory for nonlocal equations with VMO coefficients. *To appear in Ann. Inst. H. Poincaré Anal. Non Linéaire, arXiv:2101.11690.*
10. Simon Nowak, Improved S obolev regularity for linear nonlocal equations with VMO coefficients, *Math. Ann.* 385, 1323–1378, (2023). https://doi.org/10.1007/s00208-022-02369-w
11. Armin Schikorra. Nonlinear commutators for the fractional p-Laplacian and applications. *Mathematische Annalen*, 366(1-2):695–720, 2016.
12. Elias M. Stein. *Singular integrals and differentiability properties of functions*. Princeton Mathematical Series, No. 30. Princeton University Press, Princeton, N.J., 1970.

An Overview of Almost Minimizers of Bernoulli-Type Functionals

Mariana Smit Vega Garcia

Abstract In David and Toro (Calc Var 54(1):455–524, 2015) and David et al. (Adv Math 350:1109–1192, 2019), the authors studied almost minimizers for functionals of the type first studied by Alt and Caffarelli (J Reine Angew Math325:105–144, 1981) and Alt, Caffarelli and Friedman (Trans Am Math Soc 282:431–461, 1984). In this chapter, we study the regularity of almost minimizers to energy functionals with variable coefficients (as opposed to Alt and Caffarelli (J Reine Angew Math325:105–144, 1981), Alt, Caffarelli and Friedman (Trans Am Math Soc 282:431–461, 1984), David et al. (Adv Math 350:1109–1192, 2019), and David and Toro (Calc Var 54(1):455–524, 2015) that deal only with the "Laplacian" setting). We prove Lipschitz regularity up to, and across, the free boundary, fully generalizing the results of David and Toro (Calc Var 54(1):455–524, 2015) to the variable coefficient setting.

Keywords Almost minimizer · Free boundary problem · Alt–Caffarelli functional · Alt–Caffarelli–Friedman

2010 Mathematics Subject Classification Primary 35R35

1 Introduction

In this chapter, we provide an overview of almost minimizers of Bernoulli-type functionals, with a focus on the results addressing the variable coefficient setting in [10]. We start Sect. 2 with a history of the minimizing problems that inspired the

M.S.V.G. has been partially supported by the NSF grant DMS 2054282.

M. Smit Vega Garcia (✉)
Department of Mathematics, Western Washington University, Bellingham, WA, USA
e-mail: Mariana.SmitVegaGarcia@wwu.edu

T. Mengesha, A. J. Salgado (eds.), *A³N²M: Approximation, Applications, and Analysis of Nonlocal, Nonlinear Models*, The IMA Volumes in Mathematics and its Applications 165, https://doi.org/10.1007/978-3-031-34089-5_8

study of almost minimizers, following the ground-breaking papers of Alt–Caffarelli and Alt–Caffarelli–Friedman, [3, 4]: given a domain $\Omega \subset \mathbb{R}^n$ and functions $q_\pm \in L^\infty(\Omega)$, one minimizes

$$J(u) = \int_\Omega \Big(|\nabla u(x)|^2 + q_+^2(x)\chi_{\{u>0\}}(x) + q_-^2(x)\chi_{\{u\leq 0\}}(x) \Big)\, dx$$

among all $u \in \{u \in L^1_{\mathrm{loc}}(\Omega) \ : \ \nabla u \in L^2(\Omega)\}$ with $u = u_0$ on $\partial\Omega$, for a given u_0. An overview of such minimizers and important results concerning them is described in Sect. 2.1.

In Sect. 2.2, we introduce almost minimizers, first in the context of the Laplacian, as considered by Anzellotti in [1], and then in the context of Bernoulli-type functionals. We also mention related literature concerning almost minimizers to other energy functionals and quasi-minimizers. We discuss in more detail the papers of De Silva–Savin [14, 15], along with the work of David–Toro and David–Engelstein–Toro [9, 17], which more directly relate to the work [10].

In Sect. 2.3, we introduce and motivate almost minimizers to a variable coefficient version of the Bernoulli-type functional previously described: one considers almost minimizers to

$$J_\Omega(u) = \int_\Omega \Big(\langle A(x)\nabla u(x), \nabla u(x)\rangle + q_+^2(x)\chi_{\{u>0\}}(x) + q_-^2(x)\chi_{\{u<0\}}(x) \Big)\, dx,$$

where $A(x) = (a_{ij}(x))$ with $a_{ij} \in C^{0,\alpha}(\Omega)$, $A(x)$ symmetric, uniformly elliptic, and $q_\pm \in L^\infty(\Omega)$. In Sect. 2.3 we discuss related literature, in particular the papers [25], which considers almost minimizers to a related variable coefficient energy, [30], which assumes a priori Lipschitz regularity of almost minimizers to prove free boundary regularity, [24], where the notion of ω-almost minimizers is adapted to the framework of multiple-valued functions in the sense of Almgren, and [31], where the author considers vectorial "quasi-minimizers."

In Sect. 3, we introduce our precise notions of almost minimizers in the variable coefficient setting. In Sect. 3.1, we address basic facts regarding the change of coordinates that are used throughout [10]; in Sect. 3.2, we address the connection between the "multiplicative" almost minimizers used in [9, 17, 25] and the "additive" almost minimizers introduced in [10]. In Sect. 4, we briefly discuss how the continuity of almost minimizers is obtained in [10], and in Sect. 5, we discuss the $C^{1,\beta}$ regularity of almost minimizers in $\{u > 0\}$ and $\{u < 0\}$. In Sect. 6, we discuss the bulk of the technical results needed to obtain local Lipschitz regularity for both the one-phase and two-phase problems. In Sect. 7, we discuss the first main result of [10]: the local Lipschitz continuity of almost minimizers of the one-phase problem. A sketch of the proof is provided. In Sect. 8, we discuss an analogue of the Alt–Caffarelli–Friedman monotonicity formula for variable coefficient almost minimizers. Finally, in Sect. 9, we provide a sketch of the proof of the second main result of [10], the local Lipschitz continuity for two-phase almost minimizers.

2 History

2.1 Minimizers

Before we discuss almost minimizers of Bernoulli-type functionals, let us first introduce the minimizing problems from which they arose. Let $\Omega \subset \mathbb{R}^n$ be a bounded, connected, Lipschitz domain, $q_\pm \in L^\infty(\Omega)$, and define the following set of admissible functions:

$$K(\Omega) = \left\{ u \in L^1_{\text{loc}}(\Omega) \; : \; \nabla u \in L^2(\Omega) \right\}.$$

In [3] and [4], the authors considered the problem of minimizing

$$J(u) = \int_\Omega \left(|\nabla u(x)|^2 + q_+^2(x)\chi_{\{u>0\}}(x) + q_-^2(x)\chi_{\{u\le 0\}}(x) \right) dx$$

among all $u \in K(\Omega)$ with $u = u_0$ on $\partial\Omega$, where $u_0 \in K(\Omega)$. When $q_- \equiv 0$ and $u_0 \ge 0$, one obtains a so-called *one-phase* problem; otherwise, one obtains a *two-phase* problem.

One-Phase Problem The paper [3] proved very important results, opening the way to the study of more general Bernoulli-type functionals. First, minimizers of the one-phase problem were proved to exist. Moreover, if u is such a minimizer, then $u \ge 0$, u is subharmonic in Ω, and $\Delta u = 0$ in $\{u > 0\}$. In terms of optimal regularity, it was proved in [3] that $u \in C^{0,1}_{\text{loc}}(\Omega)$. It was also shown that if $\exists c_+ > 0$ such that $q_+ \ge c_+$, then given $x \in \{u > 0\}$,

$$\frac{u(x)}{\text{dist}(x, \partial\{u > 0\})} \approx 1.$$

Furthermore, $\{u > 0\} \cap \Omega$ is a set of locally finite perimeter, and $\partial\{u > 0\} \cap \Omega$ is $(n-1)$-rectifiable.

Two-Phase Problem The ground-breaking paper [4] introduced several important ideas. First, the authors proved that minimizers of the two-phase problem exist. Moreover, it was shown that if u is a minimizer of the two-phase problem, then $u^\pm$ are subharmonic in Ω and $\Delta u = 0$ in $\{u > 0\} \cup \{u < 0\}$. As in the case of the one-phase problem, minimizers of the two phase have the optimal regularity $C^{0,1}_{\text{loc}}(\Omega)$, and if there exist $c_\pm > 0$ such that $q_\pm \ge c_\pm$, then for any $x \in \{u^\pm > 0\}$,

$$\frac{u^\pm(x)}{\text{dist}(x, \partial\{u^\pm > 0\})} \approx 1.$$

Furthermore, the positive and negative phases $\{u^\pm > 0\} \cap \Omega$ are of locally finite perimeter.

Free Boundary In the case of the one-phase problem, the free boundary is defined as $\Gamma = \partial\{u > 0\}$. In terms of its regularity, one can show (see [3, 4, 7, 22, 32]) that if $q_+ \in C^{0,\gamma}(\Omega)$ and $q_+ \geq c_+ > 0$, then:

- If $n = 2, 3, 4$: Γ is a $C^{1,\beta}$ $(n-1)$-dimensional submanifold.
- If $n \geq 5$, then $\Gamma = \mathcal{R} \cup \mathcal{S}$, where $\mathcal{R}$ is a $C^{1,\beta}$ $(n-1)$-dimensional submanifold, and $\mathcal{S}$ is a closed set of Hausdorff dimension less than or equal to $n-5$.

In the case of the two-phase problem, the free boundary is defined as $\Gamma = \partial\{u > 0\} \cup \partial\{u < 0\}$. In this case, one can show (see [3, 4, 7, 22, 32]) that:

- If $n = 2, 3, 4$: Γ is a $C^{1,\beta}_{\text{loc}}$ $(n-1)$-dimensional submanifold.
- If $n \geq 5$, then $\Gamma = \mathcal{R} \cup \mathcal{S}$, where $\mathcal{R}$ is a $C^{1,\beta}_{\text{loc}}$ $(n-1)$-dimensional submanifold, and $\mathcal{S}$ is a closed set of Hausdorff dimension less than or equal to $n-5$.

Moreover, a very interesting result De Silva–Jerison [12] shows the existence of a non-smooth minimizer for J in $\mathbb{R}^7$ such that Γ is a cone. Consequently, $\mathcal{S} \neq \emptyset$ when $n \geq 7$.

Intimately connected to the energy functionals studied by Alt and Caffarelli [3], Alt et al. [4] is [13]. Here, the authors consider local minimizers of

$$\int_\Omega \left(|\nabla u|^2 + q_+^2 \chi_{\{u>0\}} + q_-^2 \chi_{\{u<0\}} \right) dx.$$

We say that $x_0 \in \Gamma$ is a *two-phase point* if $x_0 \in \partial\{u > 0\} \cap \partial\{u < 0\} \cap \Omega$. The collection of all two-phase free boundary points is called the two-phase free boundary.

Assuming $x_0 \in \Gamma$ is a two-phase point, the authors of [13] proved that $\exists r_0 > 0$ such that

$$\partial\{u > 0\} \cap B(x_0, r_0), \qquad \partial\{u < 0\} \cap B(x_0, r_0)$$

are $C^{1,\eta}$ graphs for some $\eta > 0$.

Moreover, $\partial\{u > 0\} \cap \Omega = \text{Reg}(\partial\{u > 0\}) \cup \text{Sing}(\partial\{u > 0\})$, where:

- $\text{Reg}(\partial\{u > 0\})$ is a relatively open subset of $\partial\{u > 0\} \cap \Omega$ and is locally $C^{1,\eta}$ for some $\eta > 0$. Also, the two-phase free boundary is regular.
- $\text{Sing}(\partial\{u > 0\})$ is a closed subset of $\partial\{u > 0\} \cap \Omega$ of Hausdorff dimension at most $n-5$. Also, $\exists n^* \in [5, 7]$ such that:
  - If $n < n^*$, then $\text{Sing}(\partial\{u > 0\}) = \emptyset$.
  - If $n = n^*$, then $\text{Sing}(\partial\{u > 0\})$ is locally finite in Ω.
  - If $n > n^*$, then $\text{Sing}(\partial\{u > 0\})$ is a closed $(n-n^*)$-rectifiable subset of $\partial\{u > 0\} \cap \Omega$, with locally finite $\mathcal{H}^{n-n^*}$ measure.

2.2 Almost Minimizers

Having discussed motivating minimizing problems, we start our discussion of almost minimizers with the simplest type of almost minimizers, considered in the context of the Laplacian. For this, we need to first introduce local minimizers.

Let $\Omega \subset \mathbb{R}^n$ be a bounded domain. Given $u \in W^{1,2}(\Omega)$, consider its Dirichlet energy

$$J_\Omega(u) = \int_\Omega |\nabla u(x)|^2 dx.$$

Definition 2.1 We say $u \in W^{1,2}_{\text{loc}}(\Omega)$ is a *local minimizer* of J_Ω if for any ball $B(x_0, r) \Subset \Omega$, we have

$$J_{B(x_0,r)}(u) \le J_{B(x_0,r)}(v) \tag{2.1}$$

for any $v \in u + W^{1,2}_0(B(x_0, r))$.

Notice that local minimality implies the Laplacian is equal to zero.

In [2], Anzellotti introduced the concept of *almost minimizers* for energy functionals. Intuitively, a function is an almost minimizer for J_Ω if certain types of error that depend on r are allowed to be present in (2.1). To formalize this, let us specify what we require of the error:

Definition 2.2 Given $r_0 > 0$, a function $\omega : (0, r_0) \to [0, \infty)$ is a modulus of continuity, or gauge function, if $\omega(r)$ is monotone non-decreasing and $\omega(0+) = \lim_{r \to 0+} \omega(r) = 0$.

Definition 2.3 Let $\Omega \subset \mathbb{R}^n$ be an open set. Given $r_0 > 0$ and a gauge function $\omega(r)$ defined on $(0, r_0)$, we say $u \in W^{1,2}_{\text{loc}}(\Omega)$ is a multiplicative almost minimizer, or ω-multiplicative almost minimizer for the functional J_Ω, if for any ball $B(x_0, r) \Subset \Omega$ with $0 < r < r_0$, we have

$$J_{B(x_0,r)}(u) \le (1 + \omega(r)) J_{B(x_0,r)}(v) \quad \text{for any } v \in u + W^{1,2}_0(B(x_0, r)). \tag{2.2}$$

Heuristically, the energy of u on $B(x_0, r)$ might not be minimal among all competitors $v \in u + W^{1,2}_0(B(x_0, r))$, but *almost minimal.*

Almost minimizers have been receiving increasing attention, due to three important facts. First, they can be viewed as perturbations of minimizers: they are natural to consider once the presence of noise or lower order terms in a problem is taken into account. Second, minimizers with certain constraints, for example, with fixed volume, or solutions of the obstacle problem, can be realized as almost minimizers of unconstrained problems [2]. Third, the study of almost minimizers requires an analysis from a different point of view, leading to the development of different techniques that also lead to results about minimizers.

Almost minimizers of energy functionals of the type

$$\int_{\Omega} f(x, u, Du)dx$$

were studied in [16, 18, 23, 29]. Related notions were also studied in the context of geometric measure theory in [1, 5, 6, 24]. It is also worth mentioning that the concept of almost minimizers is related to that of quasi-minimizers, introduced in [26, 27], see also the quasi-minimizers from [31].

For energy functionals exhibiting free boundaries, almost minimizers have been addressed only more recently in [9, 10, 14, 15, 17, 25, 30], and in the case of obstacle-type problems, in [19] (for the thin obstacle problem), [20] (for the fractional obstacle problem), and [21] (for the variable coefficient thin obstacle problem).

We briefly describe the results of [9, 14, 15, 17], as this will enlighten their differences to [10], which we will describe in more detail in the rest of this chapter.

In [14], the authors considered thin one-phase almost minimizers, that is, almost minimizers of the energy

$$E_{\Omega}(u) = \int_{\Omega} |\nabla u(x)|^2 dx + \mathcal{H}^n(\{(x, 0) \in \Omega : u(x, 0) > 0\}).$$

The authors prove optimal $C^{0,\frac{1}{2}}_{\text{loc}}(\Omega)$ regularity of almost minimizers, along with partial regularity of the free boundary. More precisely, they show that the free boundary $\Gamma(u) = \partial\{u > 0\}$ is $C^{1,\alpha}$ regular outside a closed singular set of Hausdorff dimension $n - 3$. This allows them to prove the $C^{1,\alpha}$ regularity of Lipschitz free boundaries.

In [15], the authors addressed almost minimizers of the one-phase free boundary problem given by the Alt–Caffarelli functional. That is, they studied almost minimizers of the energy

$$J_{\Omega}(u) = \int_{\Omega} \left(|\nabla u(x)|^2 + \chi_{\{u>0\}}(x) \right) dx \qquad \text{for } u \geq 0.$$

Their first main theorem proves the optimal $C^{0,1}_{\text{loc}}(\Omega)$ Lipschitz regularity of almost minimizers. Regarding the free boundary $\Gamma(u) = \partial\{u > 0\}$, the authors prove an improvement of flatness result that leads them to showing the free boundary is locally $C^{1,\alpha}$ regular outside a closed singular set of Hausdorff dimension $n - 5$, also allowing them to prove the $C^{1,\alpha}$ regularity of Lipschitz free boundaries.

In the case of [17] and [9], the authors considered the following energy:

$$J_{\Omega}(u) = \int_{\Omega} \left(|\nabla u(x)|^2 + q_+^2(x)\chi_{\{u>0\}}(x) + q_-^2(x)\chi_{\{u<0\}} \right) dx,$$

where $q_{\pm} \in L^{\infty}(\Omega)$. In [17], the authors proved optimal regularity $C^{0,1}_{\text{loc}}(\Omega)$ of almost minimizers. To understand the challenges involved in the proof of optimal regularity for almost minimizers, it is important to notice that in [3, 4], the following facts were key ingredients in the proof of the optimal $C^{0,1}_{\mathrm{loc}}(\Omega)$ regularity for minimizers:

(a) If u is a minimizer, then $u^{\pm}$ are subharmonic in Ω.
(b) If u is a minimizer, then u is harmonic in $\{u^{\pm} > 0\}$.
(c) In the case of the two-phase problem, one uses the Alt–Caffarelli monotonicity formula. This formula says that the functional

$$\Phi(r) = \frac{1}{r^4}\left(\int_{B_r(x)} \frac{|\nabla u^+(y)|^2}{|x-y|^{n-2}}dy\right)\left(\int_{B_r(x)} \frac{|\nabla u^-(y)|^2}{|x-y|^{n-2}}dy\right)$$

is monotone non-decreasing as a function of r.

In contrast, when working with almost minimizers, one does not have (a), (b), or (c), as almost minimizers do not solve partial differential equations as minimizers do. To bypass this challenge, the use of good comparison functions is required. As the goal is to prove $C^{0,1}_{\mathrm{loc}}(\Omega)$ regularity of almost minimizers, it suffices to control, in a careful way, $\omega(x,s) = \left(⨍_{B(x,s)} |\nabla u|^2\right)^{\frac{1}{2}}$ for $s \in (0,r)$ if $B(x,r) \subset \Omega$.

Through comparisons with harmonic replacements and an iteration scheme, the authors proved in [17] the optimal regularity for almost minimizers of the one-phase problem. In the case of the two-phase problem, more was needed. First, the authors of [17] analyzed the interplay between

$$m(x,r) = \frac{1}{r}⨍_{\partial B(x,r)} u, \quad n(x,r) = \frac{1}{r}⨍_{\partial B(x,r)} |u|, \quad \omega(x,r) = \left(⨍_{B(x,r)} |\nabla u|^2\right)^{\frac{1}{2}}. \tag{2.3}$$

This analysis, together with an almost Alt–Caffarelli–Friedman monotonicity formula, allowed them to prove the optimal regularity of almost minimizers of the two-phase problem. More precisely, the almost Alt–Caffarelli–Friedman monotonicity formula proved in [17] states that if u is an almost minimizer for J in Ω, then $\exists \delta > 0$ such that for $K \Subset \Omega$, $\exists r_K > 0$, $C_K > 0$ such that for all $x \in \Gamma(u) \cap K$, for all $0 < s < r < r_K$,

$$\Phi(s) \leq \Phi(r) + C_K r^{\delta},$$

where $\Phi(r)$ is defined as in [4].

In terms of the free boundary of almost minimizers, in [17], the authors first proved a non-degeneracy result. Assuming u is an almost minimizer of the one-phase problem, $q_+ \in L^{\infty}(\Omega) \cap C(\Omega)$ and $q_+ \geq c_+ > 0$, [17] proved that there exists $\eta > 0$ so that if $x_0 \in \Gamma$ and $B(x_0, 2r_0) \subset \Omega$, then for $0 < r < r_0$,

$$\frac{1}{r}⨍_{\partial B(x_0,r)} u^+ \geq \eta$$

and $u(x) \geq \frac{\eta}{4}\text{dist}(x, \partial\{u > 0\})$, for $x \in B(x_0, r_0) \cap \{u > 0\}$. Under the same assumptions, [17] also show $\{u > 0\} \subset \Omega$ is "locally" NTA. Moreover, if $x_0 \in \Gamma$ with $B(x_0, 2r_0) \subset \Omega$, then there exists an Ahlfors regular measure μ_0 supported in $B(x_0, r_0) \cap \Gamma$. Furthermore, Γ is $(n-1)$-uniformly rectifiable, and $\{u > 0\} \cap \Omega$ is a set of locally finite perimeter.

With respect to the two-phase problem, assume $q_\pm \in L^\infty(\Omega) \cap C(\overline{\Omega})$ and $q_\pm \geq c_0 > 0$. In [9], the authors proved that Γ is locally Ahlfors regular and uniformly rectifiable. In the case of the one-phase problem, [9] assumed $q_+ \in L^\infty \cap C^\gamma(\Omega)$ and $q_+ \geq c_+ > 0$. They proved that $\Gamma = \mathcal{R} \cup \mathcal{S}$, where $\mathcal{R}$ is a $C^{1,\beta}$ $(n-1)$-dimensional submanifold and $\mathcal{S}$ is a closed set with $\mathcal{H}^{n-1}(\mathcal{S}) = 0$, following the result for minimizers. Furthermore, $\mathcal{S} = \emptyset$ when $n = 2, 3, 4$. In terms of the dimension of the singular set, let k^* be the smallest natural number such that there exists a stable one-homogeneous globally defined minimizer $u : \mathbb{R}^{k^*} \to \mathbb{R}$ that is not the half-plane solution. From the work of Caffarelli–Jerison–Kenig, Jerison–Savin, and De Silva–Jerison, one knows that $4 < k^* \leq 7$. Let u be an almost minimizer of J^+ in Ω, with $q_+ \in L^\infty \cap C^\gamma(\Omega)$ and $q_+ \geq c_+ > 0$. David et al. [9] proved that if $s > n - k^*$, then $\mathcal{H}^s(\Gamma^+ \setminus \mathcal{R}) = 0$.

We also mention the recent preprint [11], where the authors construct a family of minimizers to an Alt–Caffarelli–Friedman-type functional whose free boundaries contain branch points in the strict interior of the domain. They also give an example showing that branch points in the free boundary of almost minimizers of the same functional can have very little structure.

2.3 Almost Minimizers with Variable Coefficients

In contrast to the energy considered in [9, 15, 17], and [10] dealt with a variable coefficient version of this energy:

$$J_\Omega(u) = \int_\Omega \Big(\langle A(x)\nabla u(x), \nabla u(x)\rangle + q_+^2(x)\chi_{\{u>0\}}(x) + q_-^2(x)\chi_{\{u<0\}}(x) \Big)\, dx,$$

where $A(x) = (a_{ij}(x))$ with $a_{ij} \in C^{0,\alpha}(\Omega)$, $A(x)$ symmetric, uniformly elliptic, and $q_\pm \in L^\infty(\Omega)$. The point of this generalization is to allow anisotropic energies that depend mildly on the point of the domain, so that in particular the classes of minimizers considered in [10] are essentially invariant by $C^{1+\alpha}$ diffeomorphisms.

Almost minimizers to functionals of Alt–Caffarelli or Alt–Caffarelli–Friedman type with variable coefficients arise naturally in measure-penalized minimization problems for Dirichlet eigenvalues of elliptic operators (e.g., the Laplace–Beltrami operator on a manifold; see [28] for a treatment of the analogous measure-constrained problem).

In the following sections, we describe the developments in [10]. These address the regularity of almost minimizers to energy functionals with variable coefficients (as opposed to [3, 4, 9, 15, 17], which deal only with the "Laplacian" setting). First, we discuss the connection of [10] to other literature.

Variable coefficient problems have been studied before: Caffarelli, in [8], proved regularity for solutions to a more general free boundary problem. De Queiroz and Tavares, in [25], provided the first results for almost minimizers with variable coefficients: the authors proved regularity away from the free boundary for almost minimizers to the same functionals considered in [10] (they considered a slightly broader class of functionals, of which the functionals from [10] are a limiting case). More precisely, the authors prove in [17] regularity away from the free boundary $\Gamma(u) = \partial\{u > 0\} \cup \{u < 0\}$ for almost minimizers of

$$J_\Omega(u) = \int_\Omega \Big(\langle A(x)\nabla u(x), \nabla u(x)\rangle + q_+(x)(u^+(x))^\gamma + q_-(x)(u^-(x))^\gamma \Big)\, dx,$$

where $A(x) = (a_{ij}(x))$ with $a_{ij} \in C^{0,\alpha}_{\text{loc}}(\Omega)$, $A(x)$ is symmetric, uniformly elliptic, and $q_\pm \in L^\infty(\Omega)$ for $0 \le \gamma \le 1$ and $q_\pm \ge q_0 > 0$. The work of [10] differs from that of [25] in two ways: first, the definition of almost minimizing in [10] is, a priori, broader than that considered in [9, 14, 15, 17, 19, 25] (for more discussion, see Sect. 3.2 below). Second, and more significantly, [10] prove Lipschitz regularity up to, and across, the free boundary, in contrast to [25], thus fully generalizing the results of [17] to the variable coefficient setting.

Another connected paper is [30]. There, the authors assume a priori Lipschitz regularity of almost minimizers to the functionals studied in [10]. With this assumption, they prove $C^{1,\alpha}$ regularity of the free boundary, in dimension two, for almost minimizers of the constrained one-phase Alt–Caffarelli and the two-phase Alt–Caffarelli–Friedman functionals for an energy with variable coefficients. The paper [10] shows (as alluded to in their paper) that the a priori Lipschitz assumption is redundant. Note that the class of almost minimizers considered in [30] is equivalent to the one considered in [10].

Besides including the notion of almost minimizers from [9, 14, 17, 25], or [15], the definition of almost minimizers from [10] also connects to the work of [24]. There, the authors extend the notion of ω-minimizers introduced by Anzellotti in [1], to the framework of multiple-valued functions in the sense of Almgren, and prove Hölder regularity of Dirichlet multiple-valued (c, α)-almost minimizers.

Finally, in [31], the author studies the regularity of shape optimizers for variable coefficient divergence form elliptic operators (e.g., the domain of area one that minimizes the sum of the first k Dirichlet eigenvalues of a given operator). In particular, Trey adapts the approach of [17] to prove results of similar flavor to the ones from [10], but for vectorial "quasi-minimizers" of (3.2), with the additional property that they are solutions of a divergence form elliptic PDE with right-hand side. The results from [10] neither imply nor are implied by those of [31, Theorem 1.2], due to the different notions of "minimization" used and to the presence of an underlying PDE in [31].

3 Preliminaries

As in [10], consider a bounded domain $\Omega \subset \mathbb{R}^n$, $n \geq 2$, and study the regularity of almost minimizers of the functional

$$J(u) = \int_\Omega \big\langle A(x)\nabla u(x), \nabla u(x)\big\rangle + q_+^2(x)\chi_{\{u>0\}}(x) + q_-^2(x)\chi_{\{u<0\}}(x), \tag{3.1}$$

where $q_+, q_- \in L^\infty(\Omega)$ are bounded real-valued functions and $A \in C^{0,\alpha}(\Omega; \mathbb{R}^{n\times n})$ is a Hölder continuous function with values in symmetric, uniformly positive definite matrices. Let $0 < \lambda \leq \Lambda < \infty$ be such that $\lambda|\xi|^2 \leq A(x)\xi \cdot \xi \leq \Lambda|\xi|^2$ for all $x \in \Omega$.

We will also consider the situation where $u \geq 0$ and $q_- \equiv 0$, and

$$J^+(u) = \int_\Omega \langle A\nabla u, \nabla u\rangle + q_+^2(x)\chi_{\{u>0\}}, \tag{3.2}$$

where q_+ and A are as above.

Definition 3.1 (Definition 1 of Almost Minimizers, with Balls) Set

$$K_{\text{loc}}(\Omega) = \Big\{u \in L^1_{\text{loc}}(\Omega) : \forall\, B(x,r) \Subset \Omega,\ \nabla u \in L^2(B(x,r))\Big\}, \tag{3.3}$$

$$K^+_{\text{loc}}(\Omega) = \big\{u \in K_{\text{loc}}(\Omega) : u(x) \geq 0 \text{ almost everywhere on } \Omega\big\}, \tag{3.4}$$

and let constants $\kappa \in (0, +\infty)$ and $\alpha \in (0, 1]$ be given.

We say that u is a (κ, α)-almost minimizer for J_B^+ in Ω if $u \in K^+_{\text{loc}}(\Omega)$ and $\forall B(x,r) \Subset \Omega$ and $\forall v \in L^1(B(x,r))$ such that $\nabla v \in L^2(B(x,r))$ and $v = u$ on $\partial B(x,r)$,

$$J^+_{B,x,r}(u) \leq J^+_{B,x,r}(v) + \kappa r^{n+\alpha}, \tag{3.5}$$

where

$$J^+_{B,x,r}(v) = \int_{B(x,r)} \Big(\big\langle A(x)\nabla v(x), \nabla v(x)\big\rangle + q_+^2(x)\,\chi_{\{v>0\}}(x)\Big)\,dx. \tag{3.6}$$

Similarly, we say that u is a (κ, α)-almost minimizer for J_B in Ω if $u \in K_{\text{loc}}(\Omega)$ and $\forall B(x,r) \Subset \Omega$ and $\forall v \in L^1(B(x,r))$ such that $\nabla v \in L^2(B(x,r))$ and $v = u$ on $\partial B(x,r)$,

$$J_{B,x,r}(u) \leq J_{B,x,r}(v) + \kappa r^{n+\alpha}, \tag{3.7}$$

where

$$J_{B,x,r}(v)=\int_{B(x,r)}\Big(\big\langle A(x)\nabla v(x),\nabla v(x)\big\rangle+q_+^2(x)\chi_{\{v>0\}}(x)+q_-^2(x)\chi_{\{v<0\}}(x)\Big)\,dx. \tag{3.8}$$

This definition differs from the one found in [17] (or [25]), even when $A = I$ (see Sect. 3.2). Let us already comment that the definition given by (3.7) is more general than that of [17].

When working with variable coefficients, it is very convenient to work with a definition of almost minimizers that considers ellipsoids instead of balls. For this effect, we define

$$\begin{aligned} T_x(y) &= A^{-1/2}(x)(y-x)+x, \qquad T_x^{-1}(y) = A^{1/2}(x)(y-x)+x,\\ E_x(x,r) &= T_x^{-1}(B(x,r)). \end{aligned} \tag{3.9}$$

Note that

$$E_x(x,r)\subset B(x,\Lambda^{1/2}r) \qquad \text{and} \qquad B(x,r)\subset E_x(x,\lambda^{-1/2}r). \tag{3.10}$$

For completeness, we include here what it means to be an almost minimizer with respect to ellipsoids.

Definition 3.2 (Definition 2 of Almost Minimizers, with Ellipsoids) Let

$$K_{\mathrm{loc}}(\Omega,E)=\Big\{u\in L^1_{\mathrm{loc}}(\Omega)\,:\,\nabla u\in L^2(E_x(x,r))\text{ for }\overline{E}_x(x,r)\subset\Omega\Big\} \tag{3.11}$$

and

$$K^+_{\mathrm{loc}}(\Omega,E)=\big\{u\in K_{\mathrm{loc}}(\Omega,E)\,:\,u(x)\ge 0\text{ almost everywhere on }\Omega\big\}. \tag{3.12}$$

We say that u is a (κ,α)-almost minimizer for J_E^+ in Ω if $u\in K^+_{\mathrm{loc}}(\Omega,E)$ and

$$J^+_{E,x,r}(u)\le J^+_{E,x,r}(v)+\kappa r^{n+\alpha} \tag{3.13}$$

for every ellipsoid such that $E_x(x,r)\Subset\Omega$ and every $v\in L^1(E_x(x,r))$ such that $\nabla v\in L^2(E_x(x,r))$ and $v=u$ on $\partial E_x(x,r)$, where

$$J^+_{E,x,r}(v)=\int_{E_x(x,r)}\Big(\big\langle A(x)\nabla v(x),\nabla v(x)\big\rangle+q_+^2(x)\chi_{\{v>0\}}(x)\Big)\,dx. \tag{3.14}$$

Similarly, we say that u is a (κ,α)-almost minimizer for J_E in Ω if $u\in K_{\mathrm{loc}}(\Omega,E)$ and

$$J_{E,x,r}(u)\le J_{E,x,r}(v)+\kappa r^{n+\alpha} \tag{3.15}$$

for every ellipsoid with $E_x(x,r) \Subset \Omega$ and every $v \in L^1(E_x(x,r))$ such that $\nabla v \in L^2(E_x(x,r))$ and $v = u$ on $\partial E_x(x,r)$, where

$$J_{E,x,r}(v) = \int_{E_x(x,r)} \Big(\langle A(x)\nabla v(x), \nabla v(x)\rangle + q_+^2(x)\chi_{\{v>0\}}(x) + q_-^2(x)\chi_{\{v<0\}}(x) \Big). \tag{3.16}$$

Remark 3.1 The following summarizes the relationship between Definitions 3.1 and 3.2:

- When $A = I$, both definitions coincide.
- For a general matrix A, if u is a (κ, α)-almost minimizer for J_B in Ω according to Definition 3.1, then it satisfies (3.13) in Definition 3.2 (with constant $\Lambda^{(n+\alpha)/2}\kappa$ and exponent α) whenever x and r are such that $\overline{B}(x, \Lambda^{1/2}r) \subset \Omega$.
- If u is a (κ, α)-almost minimizer for J_E in Ω according to Definition 3.2, then it satisfies (3.5) in Definition 3.1 (with constant $\lambda^{-(n+\alpha)1/2}\kappa$ and exponent α) whenever x and r are such that $\overline{B}(x, \Lambda^{1/2}\lambda^{-1/2}r) \subset \Omega$.

Given that we are mostly interested in the regularity of almost minimizers away from $\partial\Omega$, these definitions are essentially equivalent. Bearing this in mind, we work with almost minimizers according to Definition 3.2, recalling that such functions satisfy (3.5) when $\overline{B}(x, \Lambda^{1/2}\lambda^{-1/2}r) \subset \Omega$. We will most often not write "(κ, α)-almost minimizer," but only "almost minimizer," and we will drop the subscripts B and E from the energy functional.

Notation We write $B(x,r) = \{y \in \mathbb{R}^n : |y - x| < r\}$ and $\partial B(x,r) = \{y \in \mathbb{R}^n : |y - x| = r\}$. We will consider $A \in C^{0,\alpha}(\Omega; \mathbb{R}^{n\times n})$ a Hölder continuous function with values in symmetric, uniformly positive definite matrices, and $0 < \lambda \le \Lambda < \infty$ such that $\lambda|\xi|^2 \le A(x)\xi \cdot \xi \le \Lambda|\xi|^2$ for all $x \in \Omega$. Additionally, $q_\pm \in L^\infty(\Omega)$ will be bounded real-valued functions. We also refer to

$$\begin{aligned} &T_x(y) = A^{-1/2}(x)(y - x) + x, \quad T_x^{-1}(y) = A^{1/2}(x)(y - x) + x, \\ &E_x(x,r) = T_x^{-1}(B(x,r)). \end{aligned} \tag{3.17}$$

Moreover, we will write

$$\begin{aligned} &u_x(y) = u(T_x^{-1}(y)), \quad (q_x)_\pm(y) = q_\pm(T_x^{-1}(y)), \\ &A_x(y) = A^{-1/2}(x)A(T_x^{-1}(y))A^{-1/2}(x). \end{aligned} \tag{3.18}$$

Notice that $T_x(x) = x$ and $A_x(x) = I$.

3.1 Coordinate Changes

Compared to [17] and [9], the proofs of [10] use two new ingredients: the good invariance properties of our notion with respect to bijective affine transformations, and the fact that the slow variations of A allow us to make approximations by freezing the coefficients. We take care of the first part in this subsection.

Many of our proofs use the affine mapping T_x to transform our almost minimizer u into another one u_x, which corresponds to a new matrix function $A_x(y)$ that coincides with the identity at x. In this subsection, we check that our notion of almost minimizer behaves well under bijective affine transformations. Our second definition, with ellipsoids, is more adapted to this.

Lemma 3.1 *Let u be a (κ, α)-almost minimizer for J_E (or J_E^+) in $\Omega \subset \mathbb{R}^n$. Let $T : \mathbb{R}^n \to \mathbb{R}^n$ be an injective affine mapping, and denote by S the linear map corresponding to T (i.e., $Tx = Sx + z_0$ for some $z_0 \in \mathbb{R}^n$). Also let $0 < a \le b < +\infty$ be such that $a|\xi| \le |S\xi| \le b|\xi|$ for $\xi \in \mathbb{R}^n$. Then define functions u_T, $q_{T,+}$, $q_{T,-}$ on $\Omega_T = T(\Omega)$ by*

$$u_T(y) = u(T^{-1}(y)) \text{ and } q_{T,\pm}(y) = q_\pm(T^{-1}(y)) \text{ for } y \in \Omega_T, \tag{3.19}$$

and a matrix-valued function A_T by

$$A_T(y) = SA(T^{-1}y)S^t \text{ for } y \in \Omega_T, \tag{3.20}$$

where S^t is the transposed matrix of S.

Then u_T is a $(\tilde{\kappa}, \alpha)$-almost minimizer of $J_{E,T}$ (or $J_{E,T}^+$) in Ω_T, according to Definition 3.2, where $J_{E,S}$ (or $J_{E,S}^+$) is defined in terms of A_T and the $q_{T,\pm}$, i.e.,

$$J_{E,S}(v) = \int \langle A_T(y)\nabla v(y), \nabla v(y)\rangle + q_{T,+}^2(y)\chi_{\{v>0\}}(y) + q_{T,-}^2(y)\chi_{\{v<0\}}(y)\, dy, \tag{3.21}$$

and $\tilde{\kappa} = \kappa|\det T|$.

This lemma says that under an affine change of variables, almost minimizers are transformed to almost minimizers for a modified functional. Its proof (see [10]) shows why Definition 3.2 is natural. Lemma 3.1 is applied almost exclusively in the following circumstances: let $x \in \Omega$, and take $S = A^{-1/2}(x)$. In this case, $T(y) = x + S(y - x)$, we recognize the affine mapping T_x from (3.17), and then $u_T = u_x$ and $A_T(y) = A_x(y)$ (from (3.18)). The advantage is that $A_T(x_0) = I$, and simpler competitors can be used in the definition of almost minimizer.

3.2 "Additive" Almost Minimizers

Let us now address the differences between the definition of almost minimizers used in [10] under (3.7) or (3.15) and the definition of an almost minimizer in [17] (similarly used in [9, 14, 25]). Recall that when $A = I$, being an almost minimizer for J_E is equivalent to being an almost minimizer for J_B, and that in [17] (with $A = I$), u was an almost minimizer for J_E if, instead of satisfying (3.15) for all admissible v, it satisfied

$$J_{E,x,r}(u) \leq (1 + \kappa r^{\alpha}) J_{E,x,r}(v) \tag{3.22}$$

(and similarly for J_E^+). Here A is variable, and we stick to J_E (J_B would work the same way). Let almost minimizers in the sense of (3.15) be **additive almost minimizers**, whereas almost minimizers in the sense of (3.22) are **multiplicative** almost minimizers.

In [10], we prove results for additive minimizers, first showing that multiplicative almost minimizers are also additive almost minimizers. To conclude this, we first need to show that multiplicative almost minimizers, in the variable coefficient setting, obey a certain decay property. This is done through the next Lemma (see the proof in [10]). With this result in hand, we showed in [10] that every multiplicative almost minimizer is actually an additive almost minimizer, reducing our study to the case of additive minimizers, see Lemma 3.3 below.

Lemma 3.2 *Let u be a multiplicative almost minimizer for J_E in Ω. Then $\exists C > 0$ such that if $x \in \Omega$ and $r > 0$ are such that $\overline{E}_x(x,r) \subset \Omega$, then for $0 < s \leq r$,*

$$\left(⨍_{E_x(x,s)} |\nabla u|^2 \right)^{1/2} \leq C \left(⨍_{E_x(x,r)} |\nabla u|^2 \right)^{1/2} + C \log(r/s). \tag{3.23}$$

Lemma 3.3 *Let u be a multiplicative almost minimizer of J_E in Ω with constant κ and exponent α, and let $\widetilde{\Omega} \subset\subset \Omega$ be an open subset of Ω whose closure is a compact subset of Ω. Then u is an additive almost minimizer of J_E in $\widetilde{\Omega}$, with exponent $\alpha/2$ and a constant $\widetilde{\kappa}$ that depends on the constants for J, u, and $\widetilde{\Omega}$.*

Given this result, we restrict ourselves to working with additive almost minimizers and refer to them simply as almost minimizers.

4 Continuity of Almost Minimizers

Given the equivalence between almost minimizers of J_B and J_E, we omit the subscript. Inspired by the ideas of [17], we first prove in [10] the continuity of almost minimizers for J and J^+:

Theorem 4.1 *Almost minimizers of J are continuous in Ω. Moreover, if u is an almost minimizer for J and $\overline{B}(x_0, 2r_0) \subset \Omega$, then there exists a constant $C > 0$ such that for $x, y \in B(x_0, r_0)$*

$$|u(x) - u(y)| \le C|x - y| \left(1 + \log\left(\frac{2r_0}{|x - y|}\right)\right). \tag{4.1}$$

A simple consequence of Theorem 4.1 is:

Corollary 4.1 *If u is an almost minimizer for J, then for each compact $K \subset \Omega$, there exists a constant $C_K > 0$ such that for $x, y \in K$,*

$$|u(x) - u(y)| \le C_K|x - y| \left(1 + \left|\log\frac{1}{|x - y|}\right|\right). \tag{4.2}$$

5 Almost Minimizers are $C^{1,\beta}$ in $\{u > 0\}$ and in $\{u < 0\}$

While our final goal is to prove Lipschitz regularity also across the free boundary, we first prove in [10] Lipschitz bounds away from the free boundary. Note that since u is continuous, $\{u > 0\}$ and $\{u < 0\}$ are open sets.

Theorem 5.1 *Let u be an almost minimizer for J (or J^+) in Ω. Then u is locally Lipschitz in $\{u > 0\}$ and in $\{u < 0\}$.*

This is done by estimating the energy of the almost minimizer u in comparison to that of the harmonic replacements of u on $B(x, r)$, first under the assumption that $A(x) = I$, together with careful iterations. Finally, a careful analysis of how to address the case $A(x) \neq I$, which involves applying the previous case to u_x (defined in 3.18), completes the proof.

We then improve Theorem 5.1 and prove that u is $C^{1,\beta}$ away from the free boundary. Initially, we wanted bounds on averages of $|\nabla u|^2$, and now we want to be more precise and control the variations of ∇u. Our main tool is a (more careful) comparison with the harmonic approximation $(u_x)^*_r$, where u_x was defined in 3.18 and $(u_x)^*_r$ denotes the harmonic function on $B(x, r)$ with boundary data u_x.

Theorem 5.2 *Let u be an almost minimizer for J in Ω and set $\beta = \frac{\alpha}{n+2+\alpha}$. Then u is of class $C^{1,\beta}$ locally in $\{u > 0\}$ and in $\{u < 0\}$.*

6 Estimates Toward Lipschitz Continuity

In this section, we discuss some technical results needed to obtain local Lipschitz regularity for both the one-phase and two-phase problems.

Define the quantities

$$b(x,r) = \fint_{\partial B(x,r)} u_x \quad \text{and} \quad b^+(x,r) = \fint_{\partial B(x,r)} |u_x|, \tag{6.1}$$

where we recall that $u_x = u \circ T_x^{-1}$ and T_x is the affine mapping from (3.17). We sometimes write $b(u_x, x, r)$ and $b^+(u_x, x, r)$ to stress the dependence on u_x.

One has to distinguish two types of pairs (x, r), for which we have to use different estimates. For constants $\tau \in (0, 10^{-2})$, $C_0 \geq 1$, $C_1 \geq 3$, and $r_0 > 0$, we study the class $\mathcal{G}(\tau, C_0, C_1, r_0)$ of pairs $(x, r) \in \Omega \times (0, r_0]$ such that

$$E_x(x, 2r) \subset \Omega, \tag{6.2}$$

$$C_0\tau^{-n}(1 + r^\alpha \omega(u_x, x, r)^2)^{1/2} \leq r^{-1}|b(x,r)| \tag{6.3}$$

(recall the definition of ω from (2.3)) and

$$b^+(x,r) \leq C_1 |b(x,r)|. \tag{6.4}$$

We force $r \leq r_0$ to have uniform estimates. We end up choosing τ very small, so (6.3) says that the quantity $r^{-1}|b(x,r)|$ is as large as we want. This quantity has the same dimensionality of the expected variation of u on $B(x,r)$. In addition, (6.4) says that b accounts for a significant part of b^+, which measures the average size of $|u|$. We mostly expect this to happen only far from the free boundary, and the next lemmas go in that direction.

One of the challenges of the variable coefficient setting stems from the fact that one has to consider different centers x and at the same time also consider various ellipsoids $E_z(z, \rho)$, with z near x, with different orientations. Set

$$k = \frac{1}{6}\lambda^{1/2}\Lambda^{-1/2}, \tag{6.5}$$

which we choose like this so that

$$E_z(z, kr) \subset B(z, \Lambda^{1/2}kr) \subset E_x(x, r/2) \text{ whenever } x \in \Omega \text{ and } z \in E_x(x, r/3). \tag{6.6}$$

We start with a self-improvement lemma.

Lemma 6.1 *Assume u is an almost minimizer for J in Ω. For each choice of constants $C_1 \geq 3$ and r_0, there is a constant $\tau_1 \in (0, 10^{-2})$ (which depends only on $n, \kappa, \alpha, r_0, C_1, \lambda$ and Λ), such that if $(x, r) \in \mathcal{G}(\tau, C_0, C_1, r_0)$ for some* *choice of* $\tau \in (0, \tau_1)$ *and* $C_0 \geq 1$, *then for each* $z \in E_x(x, \tau r/3)$, *we can find* $\rho_z \in (\tau k r/2, \tau k r)$ *such that* $(z, \rho_z) \in \mathcal{G}(\tau, 10C_0, 3, r_0)$. *Here k is defined as in* (6.5) *and satisfies* (6.6).

Lemma 6.2 *Let* u, x, r *satisfy the hypothesis of Lemma 6.1; in particular* $(x, r) \in \mathcal{G}(\tau, C_0, C_1, r_0)$ *for some* $C_0 \geq 1$, $C_1 \geq 3$, *and* $\tau \leq \tau_1$. *Recall that* $b(x, r) \neq 0$ *by* (6.3). *If* $b(x, r) > 0$, *then*

$$u \geq 0 \text{ on } E_x(x, \tau r/3) \text{ and } u > 0 \text{ almost everywhere on } E_x(x, \tau r/3). \tag{6.7}$$

Similarly, if $b(x, r) < 0$, *then*

$$u \leq 0 \text{ on } E_x(x, \tau r/3) \text{ and } u < 0 \text{ almost everywhere on } E_x(x, \tau r/3). \tag{6.8}$$

For the next lemma, we use Lemma 6.2 to get some regularity for u near a point x such that $(x, r) \in \mathcal{G}(\tau, C_0, C_1, r_0)$, with the same method as for the local regularity of u away from the free boundary.

Lemma 6.3 *There exist constants* $k_1 \in (0, k/2)$, *depending only on* λ *and* Λ, *and* $\tau_2 \in (0, \tau_1)$, *with* τ_1 *as in Lemmas 6.1 and 6.2 with the following properties. Let u be an almost minimizer for J in* Ω. *Let* $(x, r) \in \mathcal{G}(\tau, C_0, C_1, r_0)$ *for some* $\tau \in (0, \tau_2)$ *and* $C_0 \geq 1$. *Then for* $z \in B(x, \tau r/10)$ *and* $s \in (0, k_1 \tau r)$,

$$\omega(u, z, s) \leq C\left(\tau^{-\frac{n}{2}}\omega(u_x, x, r) + r^{\frac{\alpha}{2}}\right), \tag{6.9}$$

and for $y, z \in B(x, \tau r/10)$,

$$|u(y) - u(z)| \leq C\left(\tau^{-\frac{n}{2}}\omega(u_x, x, r) + r^{\frac{\alpha}{2}}\right)|y - z|. \tag{6.10}$$

Here $C = C(n, \kappa, \alpha, \lambda, \Lambda, r_0)$. *Finally, there is a constant* $C(\tau, r)$ *depending on* $n, \kappa, \alpha, r_0, \tau, r, \lambda, \Lambda$, *such that*

$$|\nabla u(y) - \nabla u(z)| \leq C(\tau, r)(\omega(u_x, x, r) + 1)|y - z|^{\beta}, \tag{6.11}$$

for any $y, z \in B(x, \tau r/10)$, *where as before* $\beta = \frac{\alpha}{n+2+\alpha}$.

The next lemma allows us to conclude that $(x, \rho) \in \mathcal{G}(\tau, C_0, 3, r_0)$, provided we have certain estimates on $b(u_x, x, r)$, and $\omega(u_x, x, r)$.

Lemma 6.4 *Let u be an almost minimizer for J in* Ω. *There exists* $K_2 = K_2(\lambda, \Lambda) \geq 2$ *such that for each choice of* $\gamma \in (0, 1)$, $\tau > 0$, *and* $C_0 \geq 1$, *we can find* r_0, η *small and* $K \geq 1$ *with the following property: if* $x \in \Omega$ *and* $r > 0$ *are such that* $0 < r \leq r_0$, $B(x, K_2 r) \subset \Omega$ *and*

$$|b(u_x, x, r)| \geq \gamma r(1 + \omega(u_x, x, r)), \tag{6.12}$$

and

$$\omega(u_x, x, r) \geq K, \tag{6.13}$$

then there exists $\rho \in \left(\frac{\eta r}{2}, \eta r\right)$ *such that* $(x, \rho) \in \mathcal{G}(\tau, C_0, 3, r_0)$.

7 Local Lipschitz Regularity for One-phase Almost Minimizers

The last intermediate result needed to obtain optimal regularity is the following:

Lemma 7.1 *Let* u *be an almost minimizer for* J^+ *in* Ω. *Let* $\theta \in (0, 1/2)$. *There exist* $\gamma > 0$, $K_1 > 1$, $\beta \in (0, 1)$, *and* $r_1 > 0$ *such that if* $x \in \Omega$ *and* $0 < r \leq r_1$ *are such that* $B(x, r) \subset \Omega_x$,

$$b(u_x, x, r) \leq \gamma r(1 + \omega(u_x, x, r)), \tag{7.1}$$

and

$$\omega(u_x, x, r) \geq K_1, \tag{7.2}$$

then

$$\omega(u_x, x, \theta r) \leq \beta\omega(u_x, x, r). \tag{7.3}$$

We are ready to combine all of the previous lemmas to obtain our main result for almost minimizers of the one-phase problem:

Theorem 7.1 *Let* u *be an almost minimizer for* J^+ *in* Ω. *Then* u *is locally Lipschitz in* Ω.

We want to show that there exist $r_2 > 0$ and $C_2 \geq 1$ (depending on $n, \kappa, \alpha, \lambda, \Lambda$) such that for each choice of $x_0 \in \Omega$ and $r_0 > 0$ such that $r_0 \leq r_2$ and $B(x_0, K_2 r_0) \subset \Omega$, where K_2 is as in Lemma 6.4,

$$|u(x) - u(y)| \leq C_2(\omega(u_{x_0}, x_0, 2r_0) + 1)|x - y| \quad \text{for } x, y \in B(x_0, r_0). \tag{7.4}$$

Sketch of the Proof Let (x, r) be such that $B(x, K_2 r) \subset \Omega$. We want to use the different lemmas above to find a pair (x, ρ) that allows us to control u. Pick $\theta = 1/3$, and let β, γ, K_1, r_1 be as in Lemma 7.1.

Let $\tau = \tau_2/2$, where $\tau_2 \in (0, \tau_1)$, and τ_1 is the constant that we get in Lemma 6.1 applied with $C_1 = 3$ and $r_0 = r_1$. Here τ_2 is the corresponding constant that appears in Lemma 6.3. Let now r_0, η, K be as in Lemma 6.4 applied to $C_0 = 10$, and to τ and γ as above. From Lemma 6.4, we get a small r_γ. Set

$$K_3 \geq \max(K_1, K), \quad \text{and} \quad r_2 \leq \min(r_1, r_\gamma). \tag{7.5}$$

Let $r \leq r_2$. The proof of the theorem then relies on the analysis of three cases:

Case 1:

$$\begin{cases} \omega(u_x, x, r) \geq K_3 \\ b(u_x, x, r) \geq \gamma r(1 + \omega(u_x, x, r)) \end{cases}. \tag{7.6}$$

Case 2:

$$\begin{cases} \omega(u_x, x, r) \geq K_3 \\ b(u_x, x, r) < \gamma r(1 + \omega(u_x, x, r)) \end{cases}. \tag{7.7}$$

Case 3:

$$\omega(u_x, x, r) < K_3. \tag{7.8}$$

Case 1 ends up yielding additional regularity, as by Lemma 6.3 we know that u is $C^{1,\beta}$ in a neighborhood of x.

In the two remaining cases, we set

$$r_k = \theta^k r = 3^{-k} r, \ k \geq 0.$$

Our task is to control $\omega(u_x, x, r_k)$. If the pair (x, r_k) ever satisfies (7.6), we denote k_{stop} the smallest integer such that (x, r_k) satisfies (7.6) (notice that $k \geq 1$, since we are not in Case 1). Otherwise, set $k_{\text{stop}} = \infty$.

Let $k < k_{\text{stop}}$ be given. If (x, r_k) satisfies (7.7), we can apply Lemma 7.1 to it. Therefore,

$$\omega(u_x, x, r_{k+1}) \leq \beta \omega(u_x, x, r_k). \tag{7.9}$$

Otherwise, (x, r_k) satisfies (7.8) (since $k < k_{\text{stop}}$). Then

$$\omega(u_x, x, r_{k+1}) = \left(\fint_{B(x, r_{k+1})} |\nabla u_x|^2 \right)^{1/2} \leq 3^{\frac{n}{2}} \omega(u_x, x, r_k) \leq 3^{\frac{n}{2}} K_3. \tag{7.10}$$

By (7.9) and (7.10), we obtain that for $0 \leq k \leq k_{\text{stop}}$,

$$\omega(u_x, x, r_k) \leq \max\left(\beta^k \omega(u_x, x, r), 3^{\frac{n}{2}} K_3 \right).$$

If $k_{\text{stop}} = \infty$, this implies that $\limsup_{k \to \infty} \omega(u_x, x, r_k) \leq 3^{\frac{n}{2}} K_3$. In particular, if x is a Lebesgue point of ∇u_x (hence a Lebesgue point for ∇u), $|\nabla u_x(x)| \leq 3^{n/2} K_3$.

This implies

$$|\nabla u(x)| \leq C3^{n/2}K_3. \tag{7.11}$$

If $k_{\text{stop}} < \infty$, we apply our argument from Case 1 to the pair $(x, r_{k_{\text{stop}}})$ and get that u is $C^{1,\beta}$ in a neighborhood of x. We end up concluding

$$|\nabla u(x)| \leq C'\omega(u_x, x, r) + C', \tag{7.12}$$

where C' depends on $n, \kappa, \alpha, \lambda, \Lambda$. We actually still have (7.12) in Case 1. Since (7.11) is better than (7.12), we proved that if $r \leq r_2$, (7.12) holds for almost every $x \in \Omega$ with $B(x, K_2 r) \subset \Omega$.

Now let $x_0 \in \Omega$ and $r_0 < r_2$ be such that $B(x_0, K_2 r_0) \subset \Omega$. For almost every $x \in B(x_0, r_0)$, (7.12) holds with $r = r_0/2$ (so that $B(x, K_2 r) \subset B(x_0, K_2 r_0)$) and so

$$|\nabla u(x)| \leq C'\omega(u_x, x, r) + C' \leq 2^{n/2}C'\omega(u_x, x_0, 2r_0) + C'. \tag{7.13}$$

Since we already know that u is in the Sobolev space $W^{1,2}_{\text{loc}}(B(x_0, r_0))$, we deduce from (7.13) that u is Lipschitz in $B(x_0, r_0)$ and (7.4) holds, proving Theorem 7.1.

8 Almost Monotonicity

In this section, we describe an analogue of the Alt–Caffarelli–Friedman [4] monotonicity formula for variable coefficient almost minimizers, which we obtained in [10]. Recall, for the remainder of this section, the notation $f^{\pm} = \max\{\pm f, 0\}$. In [4], it was shown that the quantity

$$\begin{aligned}\Phi(f, y, r) &\equiv \frac{1}{r^4}\left(\int_{B(y,r)} \frac{|\nabla f^+|^2}{|z-y|^{n-2}}dz\right)\left(\int_{B(y,r)} \frac{|\nabla f^-|^2}{|z-y|^{n-2}}dz\right) \\ &\equiv \frac{1}{r^4}\Phi_+(f, y, r)\Phi_-(f, y, r)\end{aligned} \tag{8.1}$$

is monotone increasing in r as long as $f(y) = 0$ and f is harmonic. While we cannot expect to get the same monotonicity, we proved an almost-monotonicity result in the style of [17].

We first proved the following estimate in [10].

Lemma 8.1 *Let u be an almost minimizer for J in Ω, and assume that $B(x, 2r) \subset \Omega$, where x is such that $A(x) = I$. Let $\varphi \in W^{1,2}(\Omega) \cap C(\Omega)$ be such that $\varphi(y) \geq 0$ everywhere, $\varphi(y) = 0$ on $\Omega \backslash B(x, r)$, and let $\lambda \in \mathbb{R}$ be such that*

$$|\lambda\varphi(y)| < 1, \text{ on } \Omega. \tag{8.2}$$

Then, for each choice of sign, ±,

$$0 \le Cr^{\alpha} J_{x,r}(u) + Cr^{\alpha+n} + 2\lambda \left[\int_{B(x,r)} \varphi |\nabla u^{\pm}|^2 + \int_{B(x,r)} u^{\pm} \left\langle \nabla u^{\pm}, \nabla \varphi \right\rangle \right]$$
$$+ \lambda^2 \left[\int_{B(x,r)} \varphi^2 |\nabla u^{\pm}|^2 + (u^{\pm})^2 |\nabla \varphi|^2 + 2\varphi u^{\pm} \left\langle \nabla u^{\pm}, \nabla \varphi \right\rangle \right], \tag{8.3}$$

where $C < \infty$ is a constant that depends only on $\kappa, n, \Lambda, \lambda$ and the $C^{0,\alpha}$ norm of A.

The following are the variable coefficient analogues of Lemmas 6.2, 6.3, and 6.4 in [17]. The proofs in [17] use Lemma 6.1, the continuity of almost minimizers, and the logarithmic growth of $\omega(x, r)$. In particular, the proofs go through virtually unchanged for almost minimizers with variable coefficients.

Lemma 8.2 *Still assume that $n \ge 3$. Let u be an almost minimizer for J in Ω and assume that $B(x_0, 4r_0) \subset \Omega$ and that $u(x_0) = 0$ and $A(x_0) = I$. Then, for $0 < r < \min(1, r_0)$ and for each choice of sign, ±,*

$$\left| \frac{c_n}{r^2} \Phi_{\pm}(u, x_0, r) - \frac{1}{n(n-2)} \fint_{B(x_0,r)} |\nabla u^{\pm}|^2 - \frac{1}{2} \fint_{\partial B(x_0,r)} \left(\frac{u^{\pm}}{r} \right)^2 \right|$$
$$\le Cr^{\frac{\alpha}{n+1}} \left(1 + \fint_{B(x_0, \tilde{C} r_0)} |\nabla u^2| + \log^2(r_0/r) + \log^2(1/r) \right). \tag{8.4}$$

Again, $c_n = (n(n-2)\omega_n)^{-1}$ and $C > 0$ depending only on $n, \Lambda, \lambda, \|A\|_{C^{0,\alpha}}$ and the almost-minimizing constants of u.

Lemma 8.3 *Let u be an almost minimizer for J in Ω, and assume that $B(x_0, 4r_0) \subset \Omega$ with $u(x_0) = 0$ and $A(x_0) = I$. For $0 < r < \frac{1}{2}\min(1, r_0)$, set $t \equiv t(r) \equiv \left(1 - \frac{r^{\alpha/4}}{10}\right) r$. Then for $0 < r < \min(1/2, r_0)$ and each choice of sign, ±,*

$$\left| \fint_{t(r)}^{r} \left(\int_{B(x_0,s)} |\nabla u^{\pm}(y)|^2 dy \right) ds - \fint_{t(r)}^{r} \left(\int_{\partial B(x_0,s)} u^{\pm} \frac{\partial u^{\pm}}{\partial n} \right) ds \right|$$
$$\le Cr^{n+\alpha/4} \left(1 + \fint_{B(x_0, \tilde{C} r_0)} |\nabla u|^2 + \log^2 \frac{r_0}{r} \right). \tag{8.5}$$

Here, $\partial u^{\pm}/\partial n$ denotes the radial derivative of $u^{\pm}$ and $C > 0$ depend only on $\|q_{\pm}\|_{\infty}, n, \Lambda, \lambda, \|A\|_{C^{0,\alpha}}$, and the almost-minimization constants.

Finally, we obtain an Alt–Caffarelli–Friedman almost-monotonicity type result.

Theorem 8.1 *Let u be an almost minimizer for J in Ω and let δ be such that $0 < \delta < \alpha/4(n+1)$. Let $B(x_0, 4r_0) \subset \Omega$ with $u(x_0) = 0$ and $A(x_0) = I$. Then there exists $C > 0$, depending on the usual parameters such that for $0 < s < r < \frac{1}{2}\min(1, r_0)$,*

$$\Phi(u, x_0, s) \leq \Phi(u, x_0, r) + C(x_0, r_0) r^{\delta}, \tag{8.6}$$

where

$$C(x_0, r_0) \equiv C + C\left(\fint_{B(x_0, 2r_0)} |\nabla u|^2\right)^2 + C((\log r_0)_+)^4. \tag{8.7}$$

9 Local Lipschitz Continuity for Two-phase Almost Minimizers

The proof of two-phase Lipschitz continuity follows the same blueprint as the one-phase case. We start with Lemma 9.1 that is an analogue of Lemma 7.1. However, the proof of Lemma 9.1 is a bit more involved as it requires the use of the two-phase almost-monotonicity formula, (8.6), to control oscillations. We state the appropriate version of that lemma here.

Lemma 9.1 *Let u be an almost minimizer for J in Ω and let $B_0 \equiv B(x_0, \lambda^{-1/2} r_0) \subset \Omega$ be given. Let $\theta \in (0, 1/3)$ and $\beta \in (0, 1)$. Then there exists $\gamma > 0$, $K_1 > 1$ and $r_1 > 0$ (which may depend on θ and β) such that if $x \in B(x_0, r_0)$ and $0 < r \leq r_1$ satisfy*

$$u_x(y) = 0 \textit{ for some } y \in B(x, 2r/3), \tag{9.1}$$

$$|b(u_x, x, r)| \leq \gamma r(1 + \omega(u_x, x, r)), \textit{ and} \tag{9.2}$$

$$\omega(u_x, x, r) \geq K_1. \tag{9.3}$$

Then,

$$\omega(u_x, x, \theta r) \leq \beta \omega(u_x, x, r). \tag{9.4}$$

We are now ready to prove our main result.

Theorem 9.1 *Let u be an almost minimizer for J in Ω. Then u is locally Lipschitz in Ω.*

The goal is actually to show a more precise estimate: that there exist $r_2 > 0$ and $C_2 \geq 1$ (depending on $n, \kappa, \alpha, \lambda, \Lambda$) such that for each choice of $x_0 \in \Omega$ and $r_0 > 0$ such that $r_0 \leq r_2$ and $B(x_0, K_2 r_0) \subset \Omega$ (with K_2 as in Lemma 6.4), then

$$|u(x) - u(y)| \leq C_2(\omega(u_{x_0}, x_0, 2r_0) + 1)|x - y| \text{ for } x, y \in B(x_0, r_0). \tag{9.5}$$

Idea of the Proof Let (x, r) be such that $B(x, K_2 r) \subset \Omega$. We want to use the different lemmas above to find a pair (x, ρ) that allows us to control u. Pick $\theta = 1/3, \beta = 1/2$ (smaller values would work as well), and let γ, K_1, r_1 be as in Lemma 9.1.

Pick $\tau = \tau_2/2$, where $\tau_2 \in (0, \tau_1)$, where τ_1 is the constant that we get in Lemma 6.1 applied with $C_1 = 3$ and $r_0 = r_1$. Here τ_2 is the corresponding constant that appears in Lemma 6.3. Let now r_0, η, K be as in Lemma 6.4 applied to $C_0 = 10$, and to τ and γ as above. From Lemma 6.4, we get a small r_γ. Set

$$K_3 \geq \max(K_1, K), \quad \text{and} \quad r_2 \leq \min(r_1, r_\gamma). \tag{9.6}$$

Let $r \leq r_2$. In the case of the two-phase problem, one has to consider four cases:

Case 0:

$$u_x(z) \neq 0, \qquad \forall z \in B(x, 2r/3). \tag{9.7}$$

Case 1: $u_x(z) = 0$ for some $z \in B(x, 2r/3)$ and

$$\begin{cases} \omega(u_x, x, r) \geq K_3 \\ b(u_x, x, r) \geq \gamma r(1 + \omega(u_x, x, r)). \end{cases} \tag{9.8}$$

Case 2: $u_x(z) = 0$ for some $z \in B(x, 2r/3)$ and

$$\begin{cases} \omega(u_x, x, r) \geq K_3 \\ b(u_x, x, r) < \gamma r(1 + \omega(u_x, x, r)). \end{cases} \tag{9.9}$$

Case 3: $u_x(z) = 0$ for some $z \in B(x, 2r/3)$ and

$$\omega(u_x, x, r) < K_3. \tag{9.10}$$

The proof now follows the same general strategy of that of Theorem 7.1.

Acknowledgments We thank an anonymous referee for providing valuable feedback on a preliminary version of the paper.

References

1. L. Ambrosio, Corso introduttivo alla teoria geometrica della misura ed alle superfici minime, Appunti dei Corsi Tenuti da Docenti della Scuola. [Notes of Courses Given by Teachers at the School], Scuola Normale Superiore, Pisa, 1997 (Italian).
2. G. Anzellotti, On the $C^{1,\alpha}$ regularity of ω-minima of quadratic functionals, *Boll. Unione Mat. Ital., VI. Ser., C, Anal. Funz. Appl.* **2** (1983), 195–212.
3. H. W. Alt & L. A. Caffarelli, Existence and regularity for a minimum problem with free boundary, *J. Reine Angew. Math.* **325** (1981), 105–144.
4. H. W. Alt, L. A. Caffarelli & A. Friedman, Variational problems with two phases and their free boundaries, *Trans. Amer. Math. Soc.* **282** (1984), 431–461.
5. F. J. Almgren Jr., Existence and regularity almost everywhere of solutions to elliptic variational problems with constraints, *Mem. Amer. Math. Soc.* 4 (1976), no. 165.
6. E. Bombieri, Regularity theory for almost minimal currents, *Arch. Rational Mech.* Anal. 78 (1982), no. 2, 99–130.
7. L.A. Caffarelli, D. Jerison & C. Kenig C, Global energy minimizers for free boundary problems and full regularity in three dimension, *Contemp. Math.*, 350, Amer. Math. Soc., Providence, RI (2004), 83–97.
8. L. A. Caffarelli, A Harnack inequality approach to the regularity of free boundaries. Part III: existence theory, compactness, and dependence on X. *Ann. del. Sc. Norm. Sup. di Pisa* **15** (1988) 583–602.
9. G. David, M. Engelstein & T. Toro, Free Boundary Regularity for almost-minimizers, *Adv. Math*, 350, (2019), 1109–1192.
10. G. David, M. Engelstein, M. Smit Vega Garcia & T. Toro, Regularity for almost-minimizers of variable coefficient Bernoulli-type functionals. *Math. Z.* 299, 2131–2169 (2021).
11. G. David, M. Engelstein, M. Smit Vega Garcia & T. Toro, Branch Points for (Almost-)Minimizers of Two-Phase Free Boundary Problems. arXiv:2204.05367 (2022).
12. D. De Silva & D. Jerison, A Singular Energy Minimizing Free Boundary, *J. Reine Angew. Math.,* 635 (2009), 1–22.
13. G. De Philippis, L. Spolaor & B. Velichkov, Regularity of the free boundary for the two-phase Bernoulli problem, *Invent. Math.* 225, 347–394 (2021).
14. D. De Silva & Ovidiu Savin, Thin one-phase almost minimizers. *Nonlinear Anal.* 193 (2020), 111507, 23 pp.
15. D. De Silva & O. Savin, Almost minimizers of the one-phase free boundary problem. *Comm. Partial Differential Equations* 45 (2020), no. 8, 913–930.
16. F. Duzaar, A. Gastel & J. F. Grotowski, Partial regularity for almost minimizers of quasi-convex integrals. *SIAM J. Math. Anal.* 32 (2000), no. 3, 665–687.
17. G. David & T. Toro, Regularity of almost minimizers with free boundary, *Calc. Var.* 54, no. 1, 455–524 (2015).
18. L. Esposito & G. Mingione, A regularity theorem for ω-minimizers of integral functionals, *Rend. Mat. Appl.* (7) 19 (1999), no. 1, 17–44.
19. S. Jeon & A. Petrosyan, Almost minimizers for the thin obstacle problem. *Calc. Var.* 60, 124 (2021).
20. S. Jeon & A. Petrosyan, Almost minimizers for certain fractional variational problems. *Algebra i Analiz*, 32 (2020), no. 4, 166–199.
21. S. Jeon, A. Petrosyan & M. Smit Vega Garcia, Almost minimizers for the thin obstacle problem with variable coefficients, arXiv:2007.07349.

22. D. Jerison & O. Savin, Some remarks on stability of cones for the one-phase free boundary problem. *Geom. Funct. Anal.* 25 (2015), no. 4, 1240–1257.
23. L. Esposito, F. Leonetti & G. Mingione, Sharp regularity for functionals with (p,q) growth. *J. Differential Equations* 204 (2004), no. 1, 5–55.
24. J. Goblet & W. Zhu, Regularity of Dirichlet nearly minimizing multiple-valued functions, J. Geom. Anal. 18 (2008), no. 3, 765–794.
25. O. S. de Queiroz & L. S. Tavares, Almost minimizers for semilinear free boundary problems with variable coefficients, *Mathematische Nachrichten.* **291** (2018), 1486–1501.
26. M. Giaquinta & E. Giusti, On the regularity of the minima of variational integrals, *Acta Math.* 148 (1982), 31–46.
27. M. Giaquinta & E. Giusti, Quasiminima, *Ann. Inst. H. Poincaré Anal. Non Linéaire* 1 (1984), no. 2, 79–107.
28. J. Lamboley & P. Sicbaldi, Existence and regularity of Faber Krahn minimizers in a Riemannian manifold, preprint, arXiv: 1907.08159.
29. G. Mingione, Regularity of minima: an invitation to the dark side of the calculus of variations, *Appl. Math.* 51 (2006), no. 4, 355–426.
30. L. Spolaor, B. Trey & B. Velichkov, Free boundary regularity for a multiphase shape optimization problem, preprint, arXiv: 1810.06963.
31. B. Trey, Lipschitz continuity of the eigenfunctions on optimal sets for functionals with variable coefficients, preprint, arXiv: 1909.12597.
32. G. Weiss, Partial regularity for a minimum problem with free boundary, *Journal Geom. Anal.* 9 (1999), 317–326.

Index

T. Mengesha, A. J. Salgado (eds.), *A³N²M: Approximation, Applications, and Analysis of Nonlocal, Nonlinear Models*, The IMA Volumes in Mathematics and its Applications 165, https://doi.org/10.1007/978-3-031-34089-5

The manufacturer's authorised representative in the EU is Springer Nature Customer Service Centre GmbH, Europaplatz 3, 69115 Heidelberg, Germany. If you have any concerns regarding our products, please contact ProductSafety@springernature.com

Printed and bound by CPI Group (UK) Ltd, Croydon, CR0 4YY
30/09/2026
02225318-0013